MOBILE AND PERSONAL SATELLITE COMMUNICATIONS
Proceedings of the First European Workshop on Mobile/Personal
Satcoms (EMPS'94)

F. Ananasso and F. Vatalaro

Mobile and Personal Satellite Communications

Proceedings of the First European
Workshop on Mobile/Personal Satcoms
(EMPS'94)

With 130 Figures

Springer-Verlag London Ltd.

Professor Fulvio Ananasso
Telecom Italia / Telespazio
Via Tiburtina, 965
00156 Roma
Italy

Professor Francesco Vatalaro
Universita' di Roma "Tor Vergata" (DIE)
Via Della Ricerca Scientifica
00133 Roma
Italy

ISBN 978-3-540-19933-5   ISBN 978-1-4471-3023-9 (eBook)
DOl 10.1007/978-1-4471-3023-9

British Library Cataloguing in Publication Data
A catalogue record for this book is available from the British Library

Library of Congress Cataloging-in-Publication Data
A catalog record for this book is available from the Library of Congress

© Springer-Verlag London 1995
   Originally published by Springer-Verlag London Limited 1995

Typesetting: Camera ready by authors
Printed by Athenæum Press, Gateshead, Tyne & Wear
69/3830-543210 Printed on acid-free paper

Table of Contents

Part IV: Channel and radiofrequency aspects

Part V: Panel papers

Foreword:
The Role of Satellites Within the Personal Communication Services Scenario

The evolution from *network-oriented* to *user-oriented* services has characterized the world of satellite communications in the 80's and early 90's, more and more attempting to reach the customer premises with Direct-To-Home (DTH) features in order to increase the penetration of satellite services into the user community. In parallel, during the last few years, a worldwide interest and unanimous consensus has arisen on Personal Communication Services (PCS), where satellites can play a crucial role in a *global* scenario for the provision of PCS all over the world.

Cellular systems (e.g., GSM, the well-known European digital standard), telepoint and cordless systems (e.g., CT2, DECT), as well as existing Mobile Satellite Systems (such as those of INMARSAT) using the geostationary orbit (GEO), and future Low/Medium Earth Orbit (LEO/MEO) systems for global personal communications: all of these are examples of PCS systems. Each of these systems is based on technologies designed and optimised for specific market segments and traffic scenarios. Telepoint/cordless systems match the requirement for wireless communications in *very high traffic density* environments (residential, in-building, public transportation, etc.). Cellular networks are the winning choice for *high-to-medium traffic density* areas (urban, suburban, and possibly rural environments), mainly in highly developed Countries. Satellite communications play their role whenever the terrestrial networks are either not competitive (*low traffic density* areas), not applicable (maritime and aeronautical services), or insufficiently deployed (less developed Countries).

For maritime and aeronautical communication services, where terminals are not strictly required to be "personal", the mature technologies of GEO satellite systems are likely to be the most suitable for present and future enhanced systems. GEO satellites can even be effectively utilized for setting up Satellite Personal Communication Networks (S-PCN). Even without considering portable solutions as the so-called "mini-M" standard (voice/data service over lap-top-size terminals), likely to be into service in 1996 with INMARSAT III satellites, there are several relevant examples of how to implement S-PCNs with large GEO spacecraft, although requiring at L-band heavily RF-powered multiple beam antennas (exceeding 6 m diameter) to service compact-size hand-held terminals. Examples are given by the architectures proposed by several companies to set up the *Standard P*, i.e. the personal communications system in the framework of INMARSAT Project 21. Other examples are provided by existing systems and systems to be deployed in the near term (such as the Australian AUSSAT and the American MSAT), and by systems for medium-long term, such as SPACEWAY and TRITIUM proposed by Hughes, and CELSAT.

However, for satellite PCNs other orbital configurations (i.e., LEO/MEO constellations) are being considered for the provision of personal communication services to hand-held terminals. Differently from GEO satellites, a LEO spacecraft only flies across an area for some tens of minutes a few times a day (depending on the orbital parameters), so that a real-time service is not feasible unless a complete constellation of LEO satellites is operational, in such a way to have at least one satellite in visibility at any moment. Conversely, the lower altitude brings the advantages of a lower transmission delay and a lower free space attenuation, making reasonable a more effective communication performance with lighter and less expensive user terminals. On these concepts is based the exciting challenge of systems as the 66-satellite IRIDIUM by Motorola, the 48-satellite GLOBALSTAR by Loral-Qualcomm.

From a commercial standpoint, S-PCNs are extremely attractive, since they may serve a significant sector of the cellular market, at least wherever the cellular coverage is poor, thus augmenting the cellular coverage. This market is growing up at an enormous rate. The cellular phones throughout the world were 7.8 million in April 1990, rising to more than 15.8 million in May 1992, and above 25 million in July 1993. Only across Europe 6 million analogue cellular users were present in February 1993 and about 8.5 million at the end of 1993: with GSM, started in 1992, more than 10 million subscribers are expected in Europe by 1996.

Conservative forecasts of worldwide cellular market indicate that America alone - mostly represented by USA - will pass from 17 million subscribers in 1993 to more than 60 million in the year 2005; Europe (8.5 million in 1993) will exceed 42 million, whereas the Asia-Pacific region is estimated to explode from the 6.5 million in 1993 to more than 67 million in 2005. Worldwide the 34 million subscribers of 1993 will likely become at least 125 million in 1999 and 188 million in the year 2005. Only 4–5 % of these users - which however means a "niche" market of several million users! - is expected to subscribe to S-PCNs like IRIDIUM, GLOBALSTAR, as well.

For these and other substantial reasons that you will discover going through these Proceedings, we think that the focus of the EMPS'94 workshop is extremely timely, also for the heavy influence that Europe can have on the market penetration of the S-PCNs proposed so far, and for the exploration and exploitation of new frontiers in satellite communications.

This volume is divided into five parts. The first part is devoted to an analysis of S-PCN services, markets and regulatory issues. The first paper by *F. Ananasso* considers both "big LEOs" and "little LEOs" potential markets, spectrum allocations and regulatory aspects. The second paper by *P. Porzio Giusto* and *G. Quaglione* dwells on main technical characteristics of some of the recently proposed LEO and MEO systems.

The second part addresses technical concepts at system and subsystem levels and key technologies for S-PCNs. The paper by *J. Ventura-Traveset et al.* deals with a transparent satellite payload concept based on extensive use of digital techniques. Then, the paper by *H. Kuhlen* treats the subject of multimedia integrated services and the possibilities of their provision via satellite systems. *De Gaudenzi et al.* address the Code Division Multiple Access (CDMA) technique and its application within mobile/personal satellite systems. Finally, in their paper, *M. Lisi et al.* analyse some of the most novel payload architectures proposable for GEO satellites.

In the third part processing and network aspects are covered. First, *E. Del Re et al.* consider technical aspects of the interworking between cellular networks and satellite systems. Then, *F. Delli Priscoli* concentrates on integration aspects in the framework of the future European Universal Mobile Telecommunications System (UMTS). In the following paper by *P. Capodieci et al.* focus is on the special architecture design of gateway earth stations suited to LEO satellite systems. *A. Böttcher et al.* dwell on several system aspects connected to the use of non-geostationary orbit systems, including in-space connectivity problems. Finally, *C. Cullen et al.* deal with networking and signalling aspects for S-PCNs.

The fourth part is devoted to channel and radiofrequency aspects. *G. E. Corazza et al.* consider alternatives for the statistical characterization of the mobile satellite channel. The paper by *G. Butt et al.* faces empirical and statistical channel modeling for mobile and personal satellite communications. Then, *E. Damosso et al.* analyse deterministic channel modelling via ray-tracing techniques and address the problem of channel measurements at Ka-band. Finally, in the paper by *E. Biglieri* advances in modulation, channel and source coding particularly applicable to the satellite mobile channel are reviewed.

The fifth and last part of these Proceedings collects some of the interesting speeches that were given during the two EMPS'94 panel sessions on market prospectives and on system alternatives. For completeness of information, the program of EMPS'94 is also provided as an appendix.

In conclusion, we invite you to read these Proceedings, confident that you will find them interesting and appropriate, and we look forward to seeing you at the next editions of the European Workshop on Mobile/Personal Satcoms!

Fulvio Ananasso   Francesco Vatalaro

Fulvio Ananasso received an Electronics and Electrical Communications Degree from the University of Rome in 1973, after which he joined Selenia (Rome) as a microwave designer in the Development Laboratory. He was involved in several military and civilian projects concerning microwave sub-systems for Radar, Avionics and Satellite Communications equipment. In 1981 he joined TELESPAZIO (Rome) - the Italian signatory to EUTELSAT, INMARSAT and INTELSAT - as Section Chief in the Space and Advanced Programs Division, with responsibilities related to satellite payload and digital transmission channel design. In 1987 he was appointed Associate Professor of Digital Signal Processing at the University of Rome Tor Vergata, Electronics Engineering Department. In 1990 he rejoined TELESPAZIO where he is now Director of the Advanced Studies and New Missions Division. He has performed a number of studies for National and International Organizations, including ESA, EUTELSAT, INMARSAT and INTELSAT. He is author of a Radio Systems book, a chapter ("Digital Transmission Channel") of the "Satellite Communication Systems Design" book by Plenum Publishing (1993) and over 100 technical papers on communication system technology.

Francesco Vatalaro received the Dr. Ing. degree in Electronics Engineering in 1977 from the University of Bologna, Italy. Then he was with Fondazione Ugo Bordoni at Pontecchio Marconi, involved in research activities on radio-systems for air navigation and control. In 1980 he joined the Central Laboratory of FACE Standard, Pomezia, researching microwave systems design and measurements. After, with Selenia Spazio, Roma, he was group leader of satellite ground segment radiosystems engineering, being mainly involved in the ITALSAT satellite programme. In 1987 he became Associate Professor of Radio Systems at Electronics Engineering Department of the University of Roma Tor Vergata. In the period 1987/1989 he was project manager of the ground segment of the European Data Relay System. F.Vatalaro was co-winner of the 1990 "Piero Fanti" INTELSAT / Telespazio international prize. His research interests include mobile and personal communication systems and spread spectrum systems. Professor Vatalaro is a senior member of the IEEE, and a member of the Italian AEI.

Part I

Services, markets and regulatory issues

System, Market and Regulatory Aspects for Satellite Personal Communications

Fulvio Ananasso
TELECOM ITALIA/TELESPAZIO
Via Tiburtina 965 - 00156 Roma (Italy)

Abstract

The paper touches upon a number of issues relevant to Personal Communications via Satellite. Rather than expanding on technological issues - deeply addressed in the literature -, an attempt is made to identify critical system/service aspects related to the deployment of Satellite Personal Communication Networks (S-PCN). Market and regulatory aspects are regarded as keeping pace with technological breakthroughs necessary to physically implement and deploy the systems. Presently proposed "little LEOs", "big LEOs" and "super LEOs" are briefly summarized, and the European Commission point of view on those systems is preliminarily outlined. Inmarsat plans to deploy the "Standard-P" system are also indicated.

1. INTRODUCTION

Satellite systems represent a vital technology for the provision of Personal Communication Services (PCS) on a global basis. Satellite communications play their role whenever terrestrial networks are either not competitive (low traffic density), not applicable (maritime and aeronautical services) or less/not developed at all.

For maritime and aeronautical communication services, where terminals are not strictly required to be "personal", the mature technologies of Geostationary Earth Orbit (GEO) satellite systems are probably the most suitable for present and future enhanced systems. GEO satellites can also be effectively utilized for setting up S-PCNs. Even without considering portable solutions as the "mini-M" standard (voice/data service over lap-top-size terminals), likely to be into service in 1996 with INMARSAT III satellites, there are several relevant examples of how to implement S-PCNs with (large) GEO spacecraft, although requiring (at L-band) heavily RF powered multiple beam antennas (exceeding 6-meter diameter) to service compact-size hand-held terminals. Examples are given by the architectures proposed by several Companies to set up "Standard P", to provide personal communications at the turn of the century in the framework of INMARSAT Project 21. Hughes-proposed SPACEWAY and TRITIUM systems, CELSAT,... exploit similar features.

However, for Satellite PCNs even other orbital configuration (Low/Medium Earth Orbit (LEO/MEO) constellations) are being considered for the provision of Personal Communication Sevices to hand-held terminals. Differently from GEO satellites, LEO spacecraft only fly across the service area for some tens of minutes a few times a day (depending upon the orbital parameters), so that a real-time service is not allowed unless a complete constellation of LEOs is operational [1]. Whenever a real-time operation (e.g. voice) is requested, several spacecraft (from about ten to as much as several hundred) have to be put into orbit, in such a way to have at least one satellite in visibility for 100 % of the time. Conversely, the lower altitude (typically from 700 km upwards as opposed to 36,000 km GEO) permits more effective communication performances with lighter/less complex user terminals, due to substantially lower link attenuation (15-to-20 dB depending upon the orbit).

2. EVOLUTION OF MOBILE SATELLITE COMMUNICATIONS

The first generation of Mobile Satellite Systems (MSS) was dominated, in the commercial arena, by L-band (1.5-1.6 GHz) INternational MARitime telecommunication SATellite organization (INMARSAT) network, featuring roughly 33,000 terminals worldwide at the end of 1993 [2]. Main users are maritime (INMARSAT-A), and are supported by GEO satellites.

Starting from the late 80's, U.S. S.Diego-based Qualcomm Inc. launched the OmniTRACS service both in North America (via GSTAR spacecrafts) and Europe (via EUTELSAT satellites, exploiting the *EutelTRACS* brand name). They provide two-way messaging and Automatic Position Reporting (APR) features by exploiting measurements carried out by a pair of Ku-band (14/12 GHz) GEO satellites [3]. Target market is the long-distance road haulage industry, and roughly 70,000 trucks were equipped in mid 1994 with Qualcomm terminals, more than 65,000 being sold in USA (to 315 trucking companies[1]) and only 5,000 in Europe. The growth has been substantially restricted, at least in Europe, by the high cost of equipment, user charges and expectation of 2nd generation, pan-european GSM cellular network. This is somewhat applicable also to INMARSAT services.

From the ESA side, the PRODAT system provides a high quality electronic messaging service through satellite between, on one hand, mobile stations installed on trucks or other land mobiles and, on the other hand, fixed stations connected to the system through terrestrial networks [4][5].

For the following generation of MSS, it was realized that the key to improving satellite penetration is reducing size/cost of user terminal and tariffs, in addition to easing internetworking with terrestrial systems. Several regional initiatives have been undertaken to provide L-band capacity (in addition to INMARSAT network) over North America (AMSC/TMI), Australia (OPTUS) and Japan (DoCoMo and MPT, where S-band, Ka-band and millimetre waves are going to be experimented).

---

1 In April 1994 the U.S.Federal Communications Commission (FCC) granted Qualcomm a permission for construction and operation of 156,600 terminals (increased from the original 80,600 units).

In Europe, the European Mobile Satellite (EMS) payload will be delivered into orbit at the beginning of 1996 on-board the second flight unit of the Italian Ka-band satellite ITALSAT [6]. It will be complemented in 1997 by L-band Land Mobile (LLM) payload embarked on Advanced Relay and TEchnology MISsion (ARTEMIS) spacecraft. INMARSAT wise, the *Mini-M* initiative will soon (1996) permit to utilize worldwide compact lap-top-size terminals in conjunction with INMARSAT III multiple beam spacecrafts.

3. POTENTIAL MARKET FOR SATELLITES IN PERSONAL COMMUNICATIONS

From a commercial standpoint, Mobile Satellite Services (MSS) providing Personal Communications are extremely attractive since they may serve a significant segment of the (cellular) market - at least wherever cellular coverage is poor, thus *augmenting* the cellular coverage -. This market is growing up at an enormous rate [7][8][9]. The cellular phones throughout the world were 7.8 million in April 1990, raising more than 15.8 million in May 1992 and more than 25 million in July 1993. Only across Europe 6 million cellular users were present in February 1993 and about 8.5 million at the end of 1993, although every nation adopts individual (analogue) standards which generally do not permit roaming at European level. This will be overcome by the widespreading usage of (digital) Global System for Mobile communications (GSM), started in 1992, having 1.1 million subscribers at the end of 1993 and expecting in Europe more than 10 million subscribers by 1996.

Conservative forecasts of Worldwide Cellular Market (Fig.1), indicate that America alone - mostly represented by USA - will pass from the 1993 17 M subscribers to more than 60 millions in the year 2005; Europe (8.5 millions in 1993) will exceed 42 millions, whereas Asia-Pacific Region is estimated to explode from the 1993 6.5 millions to more than 67 millions in 2005. Worldwide, the 34 million subscribers of 1993 will likely become at least 125 millions in 1999 and 188 millions in the year 2005. Only a few percent of those subscribers - which means however a "niche" market of several million users! - is espected to be (even) subscribers of Satellite Personal Communication Networks (S-PCN) like IRIDIUM, GLOB-ALSTAR,...

4. FREQUENCY SPECTRUM/REGULATORY ASPECTS

4.1. Spectrum Allocation

The problem of radio frequency management for non-GEO satellite applications has been addressed by the February 1992 World Administrative Radio Conference (WARC-92) in Malaga, Spain, and partly re-considered in the framework of WRC-93, a preparatory meeting held in Geneva in November 1993 to anticipate WARC-95 activities (Geneva, October 1995).

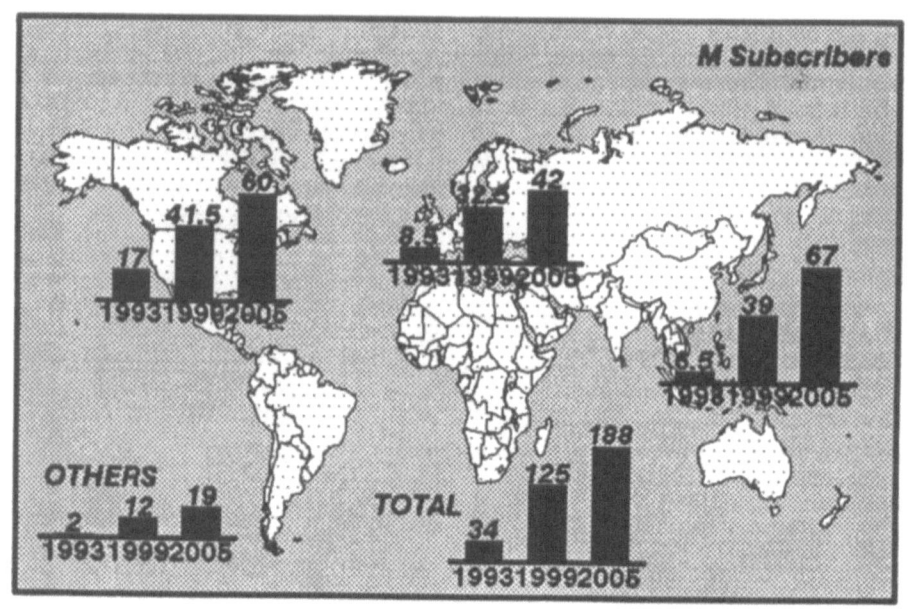

FIG.1. CELLULAR PHONES IN THE WORLD.

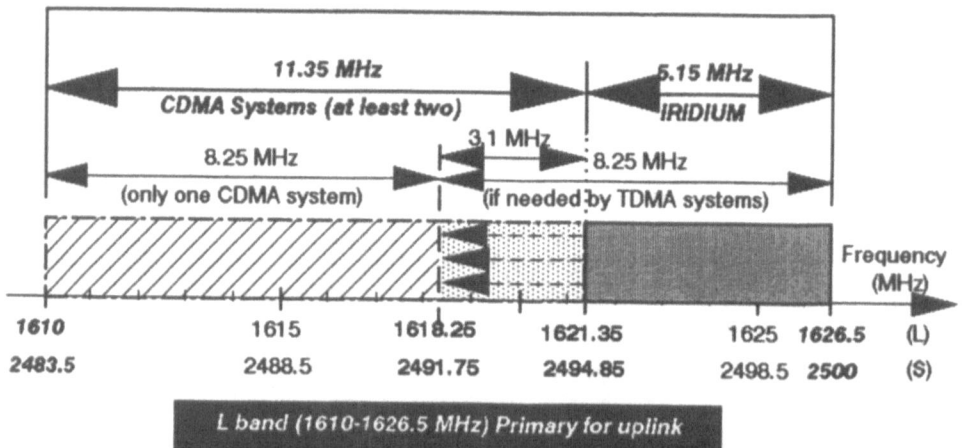

FCC PROPOSED SPECTRUM SEGMENTATION FOR "Big LEOs"
(NPRM, Notice of Proposed Rule Making, January 19, 1994)
FIG.2.

timely implement the system. This waiver would have to be granted by the first half of 1994 in order not to delay the start of service. A similar waiver was requested in April 1994 by Loral Qualcomm Satellite Services (LQSS), that in the meantime had achieved a $ 275M first round equity financing for its system GLOBALSTAR, thus facing similar start of service limitations as in the IRIDIUM case. However, as of October 1994, such waivers have not been granted yet, but rather deferred to the end of licensing process decision.

Whatever the reason was for not granting the 319(d) waiver, at least two applicants, the ones having their systems financed (Motorola and LQSS), strongly lobbied the FCC/Congress to quickly arrive to Report and Order (R&O), due October 14, 1994, and including final spectrum sharing plan. This would enable granting of operational licenses, including interim construction waivers, by 105 days (January 31, 1995).

The R&O is expected to substantially confirm the tentative "service rules" the FCC issued on January 19, 1994 via a Notice for Proposed Rule Making (NPRM), substantially featuring a "band segmentation" approach [10]. All of the companies who filed to provided MSS service in the 1610-1626.5 MHz band before June 3, 1991 were given the opportunity to adjust their applications to accommodate the specific new rules. In particular, according to the proposed plan, the 1610-1626.5/2483.5-2500 MHz band will be divided into two parts: the lower 11.35 MHz, to be shared by CDMA systems, and the remaining 5.15 MHz for IRIDIUM (Fig.2). If only one CDMA applicant is awarded a license, the (CDMA) spectrum would be reduced to 8.25 MHz. The remaining 3.10 MHz would then be left unassigned until another CDMA applicant is given a licence, or assigned to IRIDIUM - thus summing up to 8.25 MHz - if it can prove to need this extra capacity to optimize the system. In the event applicants agree with this plan - comments from the open public were accepted by June 20, 1994 -, licenses could be granted by the first quarter of 1995. Otherwise, the spectrum would be divided by auctions/competitive biddings, lotteries or comparative hearings.

The service requirements for "threshold design qualification standards" include:

*   continuous voice coverage over the entire globe (*except the poles*) at least 75% of the time;
*   continuous voice coverage over the USA, 100% of time;
*   "strict" financial qualification, i.e. each applicant must have the financial ability to construct and launch the system;
*   ability to operate on co-primary basis with radio-astronomy (1610-1613.8 MHz);
*   use of Ka feeder link spectrum; co-ordinate among other Ka band applicants for Fixed Satellite Systems (FSS) and 28 GHz "cellular" television (LMDS, Local exchange carrier Message Distribution System).

Licensing procedures, as for little LEOs, would encompass the following conditions:

*   "blanket" licensing for transceiver terminals (mobile services operators);

One of the most relevant decisions taken at WARC-92 was to allocate the Radio Determination Satellite System (RDSS) 1610-1626.5 MHz (L-band) and 2483.5-2500 MHz (S-bands) slots to LEO satellite services on a *worldwide, primary basis*. This would enable LEO systems to have a reasonable amount of spectrum (i.e. capacity) to (potentially) serve a substantial number of subscribers and reach/exceed the break-even point to remunerate the service provision. LEO systems proposed so far plan to utilize either VHF/UHF bands below 1 GHz (from 137 MHz to 450 MHz) - "Little LEOs" -, L/S Band - "Big LEOs"- or Ka band - "Super LEOs" - (see later Section 5).

Concerning S-band mobile telecommunications, the radio frequencies allocated to Future Public Land Mobile Telecommunication Systems (FPLMTS) occupy 230 MHz at S-band from 1885-to-2025 MHz (140 MHz) and 2110-to-2200 MHz (90 MHz); 30 MHz (1980-to-2010 MHz) + 30 MHz (2170-to-2200 MHz) are provided to MSS for use globally on a secondary basis after 2005 (1996 in USA) - 100 MHz more are provided in Region 2 -.

According to preliminary agreements reached at WRC-93, many delegations will ask at WARC-95 to anticipate to 1996 - as for USA - the release of the previously mentioned 30+30 MHz paired core mobile satellite band (1980-to-2010 MHz and 2170-to-2200 MHz). This might bring INMARSAT to use those frequencies for its Project 21, instead of either the usual 1.5-1.6 GHz band used by the Organization GEO satellites or the RDSS band used by "Big LEOs".

4.2. USA Regulatory Situation

The problem of U.S. Federal Communication Commission (FCC) licensing has become extremely crucial for the development of LEO systems, despite the fact that their "global" nature would rather need international consensus than go-no go issues related to a single country, although significant as USA. Indeed, with present international regulations, **every** country is playing an important role in granting licenses across its territory, to make it possible that the mentioned "globality" be exploited.

However, all the Big-LEO applicants consider of vital importance to be able to access even the US market, which requires FCC approval, i.e. license for constructing, launching and operating the system. At the moment, all the proponents (to the exclusion of Loral GLOBALSTAR) have applied for and been granted experimental licenses. Motorola obtained in December 1992 an experimental license to construct and launch five satellites, necessary to in-orbit testing and validation of the intersatellite link concept (every spacecraft communicates with the preceeding and following ones in the same orbit, plus one at either adjacent orbits to the left and right).

Since virtually all Big LEO applicants plan to enter into commercial service in the 1997-2000 timeframe, there is the need to start building the complete system as soon as possible; as already pointed out, this would only be possible in the presence of an operational license. In the absence of which, Motorola, the sponsor of the first financed system (IRIDIUM, $ 1.6 bn in two rounds equity financing), asked FCC in 1993 to be granted a $ 30M waiver - 319(d) - for procuring and developing long-lead hardware necessary to

- first satellite under construction by 1 year from license;
- launches completed by 4 years from license;
- start of services by 6 years from license;
- licenses will expire after 10 years;
- modifications of satellites/services because of new technologies require a request for modification of licensing rules.

It is worth mentioning that, as of June 1993, ORBCOMM, STARSYS and VITA ("Little" LEOs below 1 GHz) have obtained from the FCC experimental licences (VITA also got a "pioneer's preference" distinction[2]); Leosat Corporation, who submitted in September 1990 an application to provide intelligent mobile services, has been negotiating with FCC to remove some bureaucratic obstacles (the filing with FCC was not accompanied by FCC fee form 155). Concerning the "Big" LEOs above 1 GHz, four out of five applicants have been granted experimental licenses, but no pioneer's preference distinction has been awarded.

4.3. Standardization/Regulatory Activities In Europe

The "Big LEO" systems (mostly IRIDIUM and GLOBALSTAR) have attracted much attention in Europe within the Regulatory and Standards Organizations. Both IRIDIUM and GLOBALSTAR, although of U.S. origin, have a strong European presence; the former is backed by Italy's STET, Financial Holding for Telecommunications (6th TLC carrier in the world) and Germany's VEBACOM (the 3rd DCS-1800 cellular licensee); the latter has a strong industrial presence from Europe, including French Alcatel, Italian Alenia and German Deutsche Aerospace, part of the European Alliance of Loral Space System which own the majority of LQSS, the backer of GLOBAL-STAR. On the other hand, the IRIDIUM European Consortium also include industrial activities, performed by STET's TELESPAZIO in the field of Constellation Control Facilities, System Engineering, Telecommunication Network Management (TMN), Gateway Engineering, Installation and Operation, Business Support System (Billing and Administration),.....

Europe is carefully considering the S-PCN issue. As already pointed out, at least two systems with significant European industrial interest (IRIDIUM and GLOBALSTAR) are expected to be operational by the turn of the Century. The general approach is in favour of a fair competition between alternative systems, all of them should be permitted to coexist, no ban for any system should be tolerated. In few words, a true competitive environment is envisaged for S-PCN in Europe. Local monopolies and protective actions shall be no longer allowed.

It must be stressed, however, that the European Commission raised substantial criticism about the FCC band segmentation approach, identified some unfair conditions against European operators and threatened reciprocity measure to boycott that plan

---

2 This gives the "pioneer" firm a position of preference in starting the service, in view of the novelty of the proposed system.

pushing European administrations not to grant licenses in their territory. This is another issue under negotiation, adding uncertainty to the practical implementation of global personal communications.

5. LITTLE LEOs, BIG LEOs, SUPER LEOs

According to what previously outlined in Section 3, recent projections indicate 4-to-6 million MSS subscribers in the world by the year 2000, and 12-to-15 million by 2008, which represents 4-5 % of the total PCN market. Even this small "niche" market is a very lucrative one, mostly addressed by the large capacity, worldwide "big LEO" systems[3] operating at L/S band and offering a full range of telephony-based services (including voice, data and facsimile), but also by "little LEOs" below 1 GHz[4], substantially used for store-and-forward/messaging purposes (e.g. automobiles, truck fleets,...). It has also to be mentioned the INMARSAT Project 21 for Personal Communications ("Standard P"), addressed later in this Section.

Furthermore, in February 1992 - that is more than eight months after the June 3, 1991 FCC deadline for MSS applications - Celsat Inc. (San Diego, California) filed with FCC to provide hybrid satellite/ground-based cellular services (including, voice, data, image and position determination) to mobile and fixed users. Celsat intends to use two (+ one) GEO satellites to cover with more than 100 beams United States, utilizing 32-to-37 MHz bandwidth in either S-band (2.1-2.129 GHz and 2.428 GHz) or L and S-band combined (1.611-1.625 GHz and 2.4835-2.5 GHz). The system can offer 55,000 circuits, which sum up to more than 1 million circuits including the ground capacity. The system cost reaches about $ 600 M, and envisages less than 25 cents a minute subscriber charge.

More recently, Kirkland (WA)-based Teledesic Corp. (formerly Calling Communications Corp.) has filed before the U.S. Federal Communication Commission (FCC) a permission to construct, launch and operate an 840-spacecraft LEO constellation (434 miles or 700 km altitude, U.S.$ 9bn estimated cost) to provide (fixed and mobile) voice and data services to countries/regions lacking telecommunication infrastructures [11][12]. The targeted market is the 40 million people worldwide who are on waiting list to get telephone services (16.5 years in Pakistan, .8 years in China, ...), basically in countries with less than 5 telephone lines per 100 inhabitants.

"Big LEOs" (apart from elliptic orbits of Ellipso and "Super LEOs" as Teledesic) envisage circular orbits from 780 km (Iridium) to 10373 km (Odyssey) altitude, 6-to-66 spacecraft constellations. "Little

---

3 Motorola IRIDIUM, conservatively targeting about 1 % of the total cellular market and 0.5 % of the paging market, Loral GLOBAL-STAR, TRW ODYSSEY, Constellation Communication Inc.(CCI) ARIES and Mobile Communications Holding Inc. (MCHI) ELLIPSO.
4 North American ORBCOMM, STARSYS and LEOSAT, the Russian GONETS, the French TAOS, the Italian TEMISAT and German SAFIR.

LEOs" all use circular orbits from 785 km (Orbcomm) to 1390 km (Gonets) altitude, 18-to-36 spacecraft constellations (Tables 1 and 2)[8][9].

Lastly, it has to be mentioned the Hughes SPACEWAY S-PCN system, a GEO constellation filed twice before FCC, the first time in December 1993 (2 satellites, $ 660M cost), then in July 1994 (9-to-17 satellites, $ 3.2bn cost for the 9 spacecraft configuration) [13].

Concerning the INMARSAT "Project 21" (the system for Personal Communications of the 21st Century), after a couple of year deadlock within the Council, delaying the implementation of the system , the 74-nation Organization decided last February 1994 to go ahead with the $ 2.4bn INMARSAT-P Personal Communication System. The system will utilize an Intermediate Circular Orbit (ICO, or MEO) configuration, envisaging a 10,000-12,000 km altitude constellation made up of 12-15 spacecraft.

The important decision taken at that Council, in addition to the mentioned selection of an ICO constellation, was to spin-off a new, limited liability company ("INMARSAT-P") to provide satellite-delivered PCS: the new company would be owned 15-to-20% by INMARSAT directly and 50-to-55% by the INMARSAT Signatories, for a total of 70% (or more) ownership by the Organization. The remaining 30% would be funded by private investors buying stakes on the market.

Detailed time schedule for setting up the company envisaged from May to July 1994 strategic work with investment banker, in order to have draft prospectus on the affiliate company ready to discuss at July 1994 INMARSAT Council Meeting. A special Council Meeting has been held in September 1994, enabling the establishing of the affiliate by December 1994. This would enable to raise about $ 500M initial capital investment, to be followed shortly by an equal amount of money, necessary to access with equitable conditions the debt market. In parallel, $ 10M have been spent during 1994 to refine the ICO system concepts developed by several companies - including Matra Marconi, GE Astro and TRW - under INMARSAT Contracts in 1993 and 1994. The company is to be established by the end of 1994, with services to be offered by the year 2000.

Concerning frequency spectrum, INMARSAT goal is to utilize FPLMTS bands for mobile user link (S-band), provided that WARC-95 releases their usage all over the world much earlier than the presently scheduled 2006 timeframe. Feeder links would preferably be using frequencies below 16 GHz (C-band, for instance, would enable the Organization to exploit preexisting Shore (Coast-Earth) Stations (CES) for "Gateway" purposes.

6. CONCLUSIONS

The paper has expanded upon some issues relevant to Personal Communications via Satellite. An attempt has been made to identify critical system/service aspects related to the deployment of Satellite Personal Communication Networks (S-PCN). Those aspects are being addressed in the framework of several Projects sponsored by ITU, ETSI, EEC,..., substantially dealing with equipment standardization

SYSTEM	ORBIT ALTITUDE	INCLINATION	PERIOD	ORBITAL PLANES	SATELLITES PER PLANE	TOTAL # OF SATELLITES
GONETS	Circular 1390 km	83°	113.56'	6	6	36
LEOSAT	Circular 970 km	40°	104.47'	3	6	18
ORBCOMM	Circular 785 km	45° 70°	100.13'	3 1	8 2	24 2
SAFIR	Circular 690 km	98.04°	100'	1	6	6
STARSYS	Circular 1300 km	60°	111.59'	4	6	24
TAOS	Circular 1208 km	57°	109.59'	5	1	5
TEMISAT	Circular 950 km	82.5°	110'	1	2	2
VITASAT	Circular 800 km	98.7427°	101.07'	1	2	2
ARIES	Circular 1018 km	90°	105.5'	4	12	48
TELEDESIC	Circular 700 km	98.2°	98.77'	21	40	840
ELLIPSO BOREALIS	Elliptic(520/ /7800 km	116.5°	180'	3	5	15
ELLIPSO CONCORDIA	Circular 7800 km	0°	280'	1	9	9
GLOBALSTAR	Circular 1389 km	47° 52°	113.53'	8 8	3 6	24 48
IRIDIUM	Circular 780 km	86.4°	100.13'	6	11	66
ODYSSEY	Circular 10373 km	55°	359.53'	3	4	12

TABLE 1. ORBITAL PARAMETERS FOR "LITTLE LEOs" (TOP) AND "BIG LEOs" (BOTTOM).

SYSTEM	FREQUENCY		SERVICE		ESTIMATED COST
	User Link	Feeder Link	Voice (kbps)	Data (kbps)	(U.S.$)
GONETS	VHF/UHF	?	NO	4.8	300 M
LEOSAT	VHF	UHF	NO	4.8	100 M
ORBCOMM	VHF/UHF	VHF	NO	2.4/4.8(up/d) 56 feeder link	< 200 M
SAFIR	VHF/UHF	VHF/UHF	NO	0.3/9.6	20 M
STARSYS	VHF	VHF	NO	4.8 up/9.6 down	< 200 M
TAOS	VHF	VHF	NO	1.2 inbound 14 outbound	< 200 M
TEMISAT	VHF/UHF	VHF/UHF	NO	2.4/9.6	10 M
VITASAT	VHF/UHF	VHF/UHF	NO	9.6	10 M
ARIES	L/S band	C band	YES (4.8)	2.4	< 500 M
TELEDESIC	Ka band	Ka band	YES	16-2048	9,000 M
ELLIPSO	L/S/C band	L/S/C band	YES (4.8)	0.3-9.6	600 M
GLOBALSTAR	L band uplink S band down	C band	YES (2.4/4.8/9.6)	9.6	1,700 M (48 sats)
IRIDIUM	L band	Ka band	YES (2.4/4.8)	2.4	3,370 M
ODYSSEY	L/S band	Ka band	YES (4.8)	9.6	1,300 M (9 sats) 1,800 M (12 sats)

TABLE 2. SOME COMMUNICATION PARAMETERS AND COST.

and type approval, frequency spectrum allocation and assignment (i.e. licensing), mutual recognition of licences, multiple-entry provision and so forth [14][15]. In other words, system, market and regulatory aspects have been addressed as keeping pace with technological breakthroughs necessary to physically implement and deploy the systems. Presently proposed "little LEOs", "big LEOs" and "super LEOs" have been briefly summarized, and the European point of view on those systems preliminarily outlined. Inmarsat plans to deploy the "Standard-P" system have also been indicated.

REFERENCES

[1] G.Maral, J.J.De Ridder, B.G.Evans, M.Richharia: "Low Earth Orbit Satellite Systems for Communications", International Journal of Satellite Communications, Vol.9, No.4, July-August 1991

[2] INMARSAT Product Portfolio, 1 June 1993

[3] A.Salmasi: "An Overview of the OmniTRACS - The First Operational Mobile Ku-band Satellite Communications System", 1st International Mobile Satellite Conference, Pasadena (CA, USA), June 1988

[4] E. Kristiansen, A. Jongejans: "Accessing the Mobile Communication System "Prodat" from Very Small Satellite Terminals", ESA Bulletin, No.66, May 1991, p.95

[5] PRODAT-2 System Presentation, ESA/ESTEC Document, September 1, 1993

[6] European Space Agency (ESA): "Mobile Satellite Business Network (MSBN): System Requirement Specification", ESA-ES-TEC, 1992.

[7] F.Ananasso, G.Rondinelli, P.Palmucci, B.Pavesi: "Small Satellite Applications: A New Perspective in Satellite Communications", 14th AIAA International Communication Satellite Systems Conference, Washington DC (USA), March 1992

[8] F.Ananasso, M.Carosi: "Architecture and Networking Issues in Satellite Systems for Personal Communications", International Journal of Satellite Communications, Special Issue on *Personal Communications Via Satellite*, January 1994

[9] F.Ananasso: "From Big GEOs to Small Satellites: a Step Forward in User-Friendly Satellite Services", AIAA 15th International Communications Satellite Systems Conference, S.Diego (CA, USA), February 1994

[10] Mobile Satellite News, Vol.6, No.3, February 2, 1994

[11] E.D. Tuck, D. P. Patterson, J. R. Stuart, M. H. Lawrence: "The Calling[SM] Network: a Global Telephony Utility", 1993 International Mobile Satellite Conference (IMSC-93), Pasadena (CA, USA), June 1993

[12] Application of Teledesic Corporation for a Low Earth Orbit Satellite System in the Fixed Satellite Service (FCC Filing), March 21, 1994

[13] Applications of Hughes Communications Galaxy, Inc. for Authority to Construct, Launch and Operate SPACEWAY, a Global Interconnected Network of Geostationary Ka Band Fixed Communications Satellites (FCC Filing), December 3, 1993 and July 26, 1994

[14] **F.Ananasso, F.Delli Priscoli:** "Technology and Networking Issues in 3rd Generation Satellite Personal Communication Networks", 3rd International Conference on Universal Personal Communications (ICUPC-94), S.Diego (CA, USA), September 1994

[15] **F.Ananasso, F.Delli Priscoli:** "Technology Challenges in TDMA Approach to 3rd Generation Personal Communication Services", IEEE 1994 Global Telecommunications Conference (GLOBE-COM-94), S.Francisco (CA, USA), December 1994

14

Technical Alternatives for Satellite Mobile Networks

Pietro Porzio Giusto[1], Giuseppe Quaglione[2]

[1]CSELT - Via G. Reiss Romoli, 274, 10148 Torino (Italy)
[2]TELESPAZIO - Via Tiburtina, 965, 00156 Roma (Italy)

Abstract

Market surveys show that satellite networks for mobile personal communications can account for a significant demand and offer healthy returns on investments. Most architectures proposed for them are based on low-earth orbits or medium-earth orbits, as from low altitudes it is possible to obtain small beam footprints, allowing large frequency reuse factors and transmit power requirements suitable for hand-held telephone sets. The choice of the values for the many parameters involved in the system design depends primarily on the area to be served and the capacity to be provided.

1. Introduction

Mobile communications by satellite were first used for maritime service. In 1979 the international organisation INMARSAT (INternational MARitime SATellite organisation) was established to provide communications to ships world-wide and in 1982 the INMARSAT network began operation officially with international commercial service [1]. Although it required large, expensive terminals its market grew enough to offer healthy returns on capital. Roughly 30,000 ship earth stations are currently in operation.

Subsequently aeronautic satellite communications were introduced. The first system was launched in 1989 by Skyphone; it allows air travellers to make calls over the eastern part of the USA, the Atlantic Ocean, Europe, Africa and part of the Middle East [2]. The calls are relayed by an Inmarsat satellite. In 1990 Inmarsat itself started offering in-flight telephone services and nowadays a number of companies offer telephony on air liners to passengers and crew [3].

Until now satellites have been little used for mobile communications on land. The primary use on land has been by aid organisations and media reporters where conventional communications means were not accessible. But recent surveys show that there is most likely a significant market for a global satellite mobile communication network. In fact, a number of companies and travelling business people need access to telecommunications services from places where terrestrial links are not feasible, or involve high costs, or cannot be opened in due time. They include construction and exploration sites, branch offices of multinational corporations, as well as isolated and less developed areas. For those cases, satellites can offer suitable solutions.

Satellite networks for global telecommunications are expected to account for 1% of the cellular market by the year 2002 [4]. Investments in this area can be profitable, as they are aiming at services that in many cases cannot be delivered by other means. Meanwhile the technology progress has made it possible to conceive satellite networks for data and voice transmission with small, low-cost, hand-portable terminals, in line with the evolution towards universal mobile personal communications. A number of companies got involved with proposals to provide and operate networks of that kind, and WARC 92 (World Administrative Radio Conference) allocated spectrum to such services.

However, the political and technical issues related to them are wide and complex. The proponents have to seek agreements with many operators and choose the best trade-off about network architecture, access technique, transmission method, etc. One of the most important and controversial point is the type of orbit to be adopted, as NGSOs (Non-Geostationary Orbits) seem to be more suitable than GSO (Geostationary Orbit) and among NGEOs a number of alternatives have to be compared, namely LEOs (Low-Earth Orbits), MEOs (Medium-Earth Orbits), and HEOs (Highly Elliptical Orbits). This is discussed in section 2, showing that LEO and MEO systems appear to be most attractive for hand-held services on a global scale.

Among the various proposals based on LEOs an MEOs, section 3 describes the three ones that currently can best represent the major architecture alternatives for a global, non-geostationary satellite mobile network: *Iridium, Globalstar,* and *Odyssey.* Section 4 mentions other system with some interesting characteristic. Finally section 5 offers some comparison to help identify the technical and operational characteristics that are peculiar of the most significant solutions.

2. Geostationary orbit vs. non-geostationary orbits

A geostationary satellite is placed on a circular orbit, lying in the equatorial plane, at an altitude of 35,786 km, so that it makes one complete revolution in 24 hours and appears to remain stationary to an observer placed on the surface of the Earth. It can relay communications between distant earth stations using fixed antennas 24 hours a day, offering the advantage of fixed earth-to-satellite geometry and obviating the need for satellite tracking. But geostationary satellites cannot provide service above latitudes of approximately 70°, and involve a long propagation delay (about 270 ms, one way). Moreover, from the geostationary orbit it is difficult to provide small "coverage cells" (very large on-board antennas would be required), in order to be able to reuse the available radio channels many times and, what is more important, to limit the requirements for the mobile terminals transmit power at levels suitable for hand-held telephone sets. In addition, launch to geostationary orbit requires more launch vehicle performance, power and, as a corollary, cost.

Elliptical orbits are more tailored for regional or continental services (e.g., Europe), as they can exhibit quasi-stationary properties and can be designed to cover the target area with large elevation propagation paths (satellites are used for service on a small arc of the orbit, around the apogee), leading to potentially better link availability in urban areas. The disadvantages of elliptical orbits include "zoom" effects (changing the size of the beam footprints as the satellite travels along its service arc), frequent passages through the Van Allen belts, and long propagation paths (involving long propagation delays and difficulties in realising small coverage cells).

For global mobile personal communications, circular low-earth orbits or medium-earth orbits are more suitable. These involve some problems due to the satellite motion relative to the Earth, such as Doppler shifts (changing the apparent frequency of the radio waves to and from the satellite) and the need for switching to another satellite as the serving one disappears below the horizon; on the other hand they feature low propagation delays, actual global coverage, if desired, and allow small coverage cells to be obtained with relatively small on-board antennas. In fact most satellite networks recently proposed for land mobile services are based on LEO or MEO constellations.

The design parameters of LEO and MEO systems for a given service area include:

- orbital height;
- number, inclination, and spacing of the orbital planes;
- number and spacing of the satellites on each orbital plane;
- relative phasing of the satellites on adjacent orbital planes.

In addition, the service availability has to be considered for different environments (rural areas, urban areas, etc.).

The above parameters influence the constellation characteristics and costs; for example, the closer the satellite to the Earth, the faster their speed (in particular the speed relative to the Earth surface); the lower the orbit, the larger the number of satellites to provide global coverage. An extensive discussion of those relationships and economic trade-offs is out of the scope of this work; the next section highlights the major characteristics of three typical systems to show the variety of the possible combinations.

3. Major architecture examples

The first announcement of plans to launch and operate a satellite network for cellular communications was given by Motorola Satellite Corp. in June 1990. In November 1990 Ellipsat Corp. filed an FCC (Federal Communications Commission) application for a system of six satellites into elliptical orbits. In June 1991 applications for *Globalstar*, *Odyssey*, and *Aries* were filed by Loral Qualcomm Satellite Services, TRW Inc., and Constellation Communications, respectively. Other proposals followed, but we deal only with *Iridium*, *Globalstar*, and *Odyssey*, as they are representative of the most promising alternatives.

3.1 Iridium

Iridium is planned to provide both cellular communications and position location services to literally any point on the Earth.

It derived its name from that of the metallic element having an atomic number of 77, as originally the network was to have a constellation of 77 satellites [5, 6, 7]. Improvements in on-board technology and deployment of satellites led to a 66 satellite constellation, but the original name was retained.

The *Iridium* satellites are placed on 6 orbital planes, which are inclined by 86° with respect to the equatorial plane (it is a quasi polar constellation). Each orbital plane contains 11 equally spaced satellites, travelling a circular orbit at an altitude of 780 km. Satellites on adjacent planes are co-rotating and interleaved (see fig. 1), with the exception of those on the last orbital plane (plane 6) with respect to those on the first orbital plane (plane 1), which happen to be counter-rotating. The spacing between the adjacent planes in the range 1-6 is 31.6°, so the spacing between plane 6 and plane 1 is 22°. This arrangement optimises the visibility of the satellites from the Earth surface.

Each *Iridium* satellite can communicate and route network traffic to the satellites that are fore and aft of it in the same orbital plane and the satellites in the adjacent two orbital planes (fig. 1). Crosslinks operate at 23 GHz, using four antennas; therefore a maximum of four crosslinks can operate at the same time.

Iridium is designed to accommodate up to 250 gateways and two network operation control stations (one in Italy), which will provide sufficient redundancy to guarantee continuous operation in the event of failures. At present *Iridium* is considering to operate about 15 gateways. Each gateway will have two tracking antennas to communicate, using radio links at 30/20 GHz, with the "active" satellite as well as with the satellite coming into view. Each gateway will also

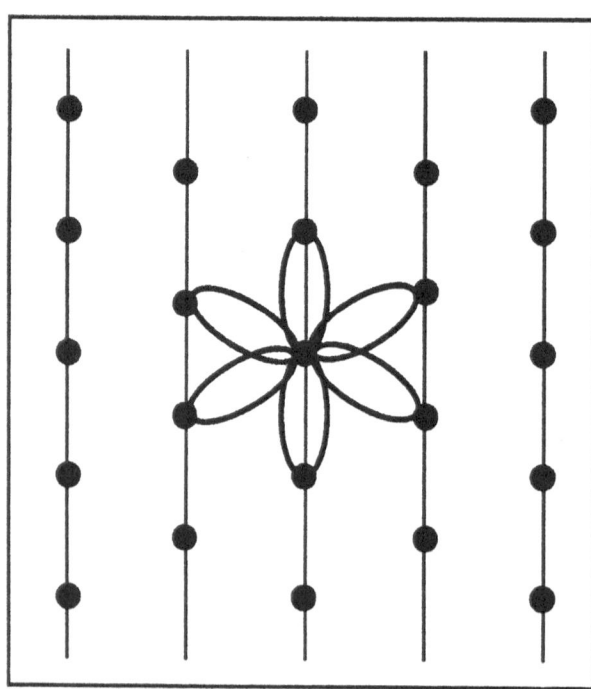

Fig. 1 - The six possible links connecting each satellite in the Iridium *constellation; only four crosslinks can be active at the same time.*

connect the mobile user with the public switched network.

The intersatellite links form a network allowing calls between *Iridium* users to be routed without using any part of terrestrial networks; in the case of calls with a terminal of a fixed network the intersatellite links can be used to route the call through the *Iridium* gateway minimising the terrestrial part of the circuit.

Since the satellites are at a low altitude, the propagation delays are roughly the same as those experienced in a terrestrial network. However the actual transmission delays are longer, because of the time required for processing signals and arranging and re-arranging transmission frames (90 ms in the mobile-to-satellite link, 9 ms in each crosslink or satellite-to-gateway link).

Each satellite generates 48 spot beams to form a continuous overlapping pattern of coverage cells; the cell diameter is around 700 km. Only 2150 beams, out of the total of 3168, are active, as some of them are switched off around the Earth poles, where overlappings among patterns generated by different satellites increase.

The mobile terminal satellite links operate at 1,6 GHz on both the up-link and the down-link, using a TDMA (Time Division Multiple Access) multiple access techniques mixed with an FDMA (Frequency Division Multiple Access) technique and TDD (Time Division Duplex).

Voice is coded at 4.8 kbit/s, but voice coders at 2.4 kbit/s can be accommodated as well, thus doubling the systems capacity. Discontinuous transmission, inhibiting signal emissions on voice inactivity, is used to reduce interferences and increase capacity. Data services will be transmitted at 2.4 kbit/s.

A mobile terminal can remain in a coverage cell for a time period of the order of 1 minute (the coverage cells travel at about 400 km/minute), therefore handovers are needed to maintain active calls. On the other hand, handover procedures can be simplified, as the motion of the mobile terminal can be neglected with respect to that of the cell pattern, while the motion of the coverage cells is predetermined and the position of the mobile terminal is known.

Iridium needs advanced technologies for on-board signal processing, intersatellite communications, satellite station keeping, and network operation and maintenance. It is a highly sophisticated network of interacting satellites, whose architecture details shall continue to be investigated by Motorola and its associated Partners, including the STET Companies like Telespazio and CSELT. On the other hand, it fully exploits the peculiarities of satellite communications systems, with minimum need for co-operation with terrestrial networks and minimum cost involved in the terrestrial links.

3.2 Globalstar

Unlike *Iridium*, *Globalstar* is not designed to provide full global service and will not operate a complex network of intersatellite links. In fact *Globalstar* is designed to provide service in the latitude range of about -70° to +70° and exploit to the maximum possible extent the existing terrestrial infrastructures. The complete space segment will consist of a total of 48 satellites. They are placed on 8 orbital planes

(6 satellites per plane), inclined by 52° on the equator [8, 9, 10, 11]. Each satellite will carry sixteen spot beams. The orbit altitude is about 1400 km. The service area is covered with a minimum elevation angle greater than 15°, with two satellites over the horizon at almost any time.

Mobile terminals will access the satellite links by a synchronous CDMA (Code Division Multiple Access) technique in order to facilitate the handover procedures. The satellite transponders are transparent, no intersatellite link will be provided, and the mobile terminals can communicate directly only with a gateway that is visible from the serving satellite. Therefore mobile to mobile communications always involve a double hop. Propagation delays however are short: for a mobile-to-mobile communication it is less than 72 ms (one way, excluding the time required for signal processing) and propagation on terrestrial links.

Voice will be coded at a variable bit rate (1 to 8 kbit/s), depending on the speech activity and background noise.

Globalstar could be deployed in a reduced version, covering only continental USA. In tab. 1 values in brackets show the parameters for that case.

3.3 Odyssey

The *Odyssey* system [12-13] is based on a MEO constellation with twelve satellites at an altitude of 10,354 km. The satellites are placed on three orbital planes (four satellites per plane) with inclination of 55°.

Each satellite generates 19 spot beams, which are continuously steered to best cover the populated regions; the network is not designed to provide service over the oceanic or less populated areas. In the target service area the minimum elevation angle is 30° and at least two satellites are visible at any time.

As the velocity of the satellites is low (the orbital period is about 360 minutes as compared to 100 minutes of the *Iridium* system) and the beams are relatively large and continually re-oriented to remain pointed towards the same portion of the area visible from the serving satellite, in most cases a user will stay within one cell for the entire duration of a call, thus precluding the need for "handover" procedures.

Odyssey will integrate and cooperate with terrestrial networks; mobile-to-mobile calls will be routed to the appropriate gateway through dedicated interconnection lines.

Continental USA coverage will require two gateway stations, one on the East coast and one on the West coast. Each gateway station will be equipped with four antennas, separates by 30 km to provide site diversity; they will be able to track three serving satellites and acquire a rising new one.

Odyssey will provide telephony, data services (real time, store and forward), positioning, position reporting, and message service.

It should be noted that the *Inmarsat Project 21*, if implemented, will use a constellation of MEO satellites similar to the *Odyssey* proposal.

Other aspects are discussed in section 5, comparing some characteristics of the considered systems.

4. Other systems

4.1 Jocos (Juggler Orbit COnStellation)

A new constellation type, called Jocos (Juggler Orbit COnStellation) has been recently proposed [14], with the aim of minimising the number of satellites needed to cover the major land masses. It is based on geosynchronous orbits with a nodal period of eight sidereal hours and a ground track intersecting the equator orthogonally; this implies an altitude of 13900 km and an inclination of the orbital planes of 70°.

Ref. [14] shows that six satellites on suitable Jocos orbits can provide coverage of the major land masses 24 hours a day.

4.2 Teledesic

In March 1994 a plan to launch an ambitious global satellite system, capable of delivering voice and high-rate data services through a LEO satellite constellation, was unveiled and the relevant application with the FCC was filed [15] by the "Teledesic Corporation". The project was proposed by the Calling Communications Corporation [16], so it is known as either "Teledesic" or "Calling" network.

That network is based on 840 satellites (21 orbital planes inclined by 98.2°, 40 satellites per plane) operating at an altitude of about 700 km. It is designed primarily to offer fixed, high-quality services through user terminals that, depending on their class, can have antenna diameters from 16 cm to 1.8 m (the smallest ones will thus benefit by some mobility). Their average transmit power varies from about 0.01 W to 4.7 W.

The channels operate at multiples of the 16+2 kbit/s basic rate (16 kbit/s for traffic, 2 kbit/s for associated signalling and control) up to 2048+256 kbit/s. For instance, a set of eight basic channels supports the equivalent of an ISDN "2B+D" link. The Teledesic network also supports some fixed-site GigaLink Terminals, that operate at multiples of 155.52 Mbit/s up to 1.24 Gbit/s. The quality of service will be comparable to that of today's terrestrial communications systems in terms of transmission delays (it will be of the same order of magnitude of fiber-optic links), bit error probability (10^{-9}), and link availability (99.9%). Assuming to use all the basic channels for high-quality telephony service, the network capacity is around 0.5 E/km^{-2}.

The Teledesic network uses a fast packet switching technology (with packets containing 512 bits), based on ATM (Asynchronous Transfer Mode), with a distributed adaptive routing algorithm. Each satellite, which is a switching node of the network, is normally linked with eight neighbouring satellites to form a mesh, and can independently route each packet along the paths that currently offers the least expected delay to its destination. The packets to a given destination can thus experience different propagation delays; they are buffered, and if necessary resequenced, at the destination terminal to equalise timing variations.

The Teledesic constellation is designed to ensure that at least one satellite is above a 40° elevation angle, 24 hours a day, over the service area, which includes the entire Earth surface between 72° north and south latitude and part of higher latitudes (95% of the Earth's surface). The coverage is made by mapping the service area into a fixed grid of 20,000 supercells, each consisting of nine square cells of 53 by 53 km. Although the satellites pass over at high speed, the cell boundaries are fixed with respect the Earth's surface, as the satellite antenna beams are steered to compensate for the satellite motion.

The service up-link operates at 30 GHz, using a combination of FDMA (Frequency Division Multiple Access), TDMA (Time Division Multiple Access) and SDMA (Space Division Multiple Access): the cells of each supercell are spatially scanned in a regular cycle by the satellite transmit and receive beams, apportioning a time slot to each cell; within each cell time slot, each terminal is assigned one or more frequency slots, according to the requested capacity. In addition a checkboard pattern of left and right circular polarisation is used to reduce interference between adjacent supercells.

The service down-link operates at 20 GHz, using ATDMA (Asynchronous Time Division Multiple Access): the satellites transmit to each cell the packets addressed to the terminals that are currently in it; each individual terminal selects those addressed to itself by examining the packet address field.

Satellite components and subsystems are designed for high-volume production, thus introducing new features in the space market.

5. Comparisons

Tab. 1 shows some parameters of the three systems that will be compared as representatives of the major alternatives for universal mobile communications. The differences among them are relevant to geographical coverage, co-operation with terrestrial networks, overall propagation delay, frequency bands and multiple access technique, elevation angle and fade margin.

5.1 Geographical coverage

Iridium provides a truly global coverage, while the other systems are tailored for areas offering significant communications traffic. However, both *Globalstar* and *Odyssey* can offer an incremental start-up to region in the mid-latitudes with a limited number of satellites (24 for *Globalstar* and 6 or 9 for *Odyssey*).

5.2 Co-operation with terrestrial networks

All the three systems are designed to use a number of gateway stations to interfac the public switched networks and envisage dual-mode handsets capable to operate on either the terrestrial cellular channels or satellite channels.

Globalstar and *Odyssey* require maximum co-operation with terrestrial networks, because the user circuits established through a particular satellite enter the PSTN (Public Switched Telephone Network) at a gateway station located within the region served by the satellite and then use the terrestrial networks, even for intercontinental calls. On the contrary, *Iridium* is designed on the basis of the OBP-ISL (On-Board Processing - Inter Satellite Link) technologies, which allow the individual calls to be routed through the space segment to the gateway station closest to the fixed party, making minimum use of terrestrial infrastructures. This has an obvious consequence on the overall cost of the calls and on the overall time delay experienced by the information signals, as explained in the next paragraph.

In the case of *Iridium* the minimum number of gateways is dictated by the traffic volume during initial service offering, considering reasonable assumptions on the local operators' needs and terrestrial tails. However, in principle, *Iridium* could provide a true global service with only one gateway in operation, since it uses gateways only for call setup functions and as the interface into the public switched networks. On the other hand, in *Globalstar* and *Odyssey* systems the gateways provide connectivity with the system satellites and the subscribers in the local regions, therefore a minimum number of gateways is required to provide coverage of the service area and service cannot be offered in areas where no gateway is in operation.

The need for having a gateway in the visibility area of the serving satellite involves constraints on the maximum distance between the gateway stations, depending on the orbit altitude and the handover procedure. The requirements on the gateway locations are particularly strict for the *Globalstar* system, as the diameter or the visibility area of the *Globalstar* satellites is lower than 8000 km. Moreover, if a simple handover procedure is desired or no dedicated inter-gateway lines are available, the allowed maximum distance between gateways is much lower than the diameter of the visibility area, because handovers have to be performed when both the serving satellite and the handover satellite have view of both the mobile user and the serving gateway station. Therefore the maximum distance between gateways depends on the extension of the common visibility area of the satellites, which is a function of the latitude, as the *Globalstar* system is designed to cover the most populated regions better than the others. It turns out that the maximum distance between *Globalstar* stations cannot exceed values ranging from less than 1000 km up to about 3500 km. In fact, to provide a reasonable coverage, the *Globalstar* system is designed to have approximately 125 ground stations [10].

5.3 Overall propagation delay

The overall propagation delay for telephone conversation is due to four main components:

- the propagation delay due to the distance from the satellite to Earth (5-10 ms for LEO, 70-80 ms for MEO);

- the delay due to terrestrial networks (5-20 ms over land, up to 100 ms for transoceanic cables);

- the processing delay due to transmission systems (e.g., frame length of TDMA systems) and OBP;

- the processing time for voice coding and decoding (typically 60-80 ms for a 4.8 kbit/s system).

The last component is independent of the type of satellite constellation and transmission system.

Iridium has the shortest contribution from the satellite path and from the terrestrial tails, however it has a significant contribution due to the TDMA frame (90 ms in L-band links and 9 ms in the Ka band links).

In general it can be concluded that *Globalstar*, being based on transparent transponders on LEO, should have some advantage over the other systems. However, for a mobile-to-mobile connection, *Iridium* does not require a double hop through a gateway station.

5.4 Frequency bands and multiple access technique

Iridium uses the band 1616.0-1626.5 MHz for both the mobile-to-satellite and satellite-to-mobile links, as it adopts Time Division Duplex (TDD).

Globalstar and *Odyssey* use the 1610-1626.5 MHz band for the up-link and the 2483.5-2500.0 MHz band for the down-link. According to the access technique, modulation, frequency reuse pattern, and number of active coverage cells, the total frequency reuse factor shown in tab. 1 is obtained. Considering cell size, number of voice channels per cell, and cell overlappings, shoud have about the same capacity, in the order of few tenths of millierlangs per square kilometre, that is much below that of the terrestrial cellular networks.

Globalstar and *Odyssey* are planning to use a CDMA access technique, therefore they can share the same frequency bands, while *Iridium* must rely on a band segmentation approach to coexist with other systems, because it uses a TDMA technique and the L-band has the secondary status in the down-link. However, the S-band envisaged by *Globalstar* and *Odyssey* in the down-link is also allocated to the ISM (Industrial, Scientific and Medical) applications. Therefore mobile users in populated urban areas may experience levels of cumulative interference exceeding the thermal noise of the receiver with some degradation of the service.

The traffic capacity available from each satellite is dependent on many factors, like the primary power from the solar cells and the batteries, the modulation and the multiple access system, the frequency reuse arrangement and the channels needed to support overhead functions such as handover and registration. The figures reported in tab. 1 are the best estimated values on the basis of the scarce data available. To obtain the overall capacity of the system, the number of voice circuits per satellite should be multiplied by the number of operational satellites of the constellation. In general terms it can be stated that the *Iridium* system is more power limited rather than band-limited, due to the use of FDMA/TDMA techniques, while the *Globalstar* primary limitation should be the self-interference noise due to the reuse of the same

Parameter \ System	IRIDIUM	GLOBALSTAR (*)	ODYSSEY
No. satellites	66+7	48 (24)+8	12+3
Nominal altitude (km)	780	1389	10354
Orbit inclination	86°	52° (47°)	55°
Coverage	Global	within ±70° latit.	Major land masses
Satellite mass (kg)	≈ 700	≈ 400	1134
Satellite Power (W)	1200	1000	1800
Serv. uplink band (MHz)	1616.0-1626.5	1610.0-1626.5	1610.0-1626.5
Serv. downlink b. (MHz)	1616.0-1626.5	2483.5-2500.0	2483.5-2500.0
Feeder link bands	Ka	C	Ka
No. spot per satellite	48	16	19
No. coverage cells	3168 (2150	768 (384) (&)	228
Frequency reuse pattern	12	1	3
Frequency reuse factor	180	not available (&)	76
Intersatellite links/OBP	YES (†)	NO	NO
Access technique	FDMA/TDMA	CDMA	CDMA
Modulation	QPSK	QPSK	BPSK
No. voice circuit per sat.	1100 (+)	1354 (#)	2300
Min. elevation angle (^)	15°	15°	30°
Fade margin (dB)	16	7	not available
Satellite lifetime (years)	5	7.5	10

(*) In brackets the values relevant to a reduced constellation, covering Continental U.S., Central Europe and Japan.
(&) Depending on the actual overlap between beams.
(†) For call routing through the satellite network.
(+) Peak value, power limited.
(#) Self-interference limited.
(^) Minimum elevation angle of the satellites for the 90% of the time, as seen from a latitude of 48°.

Tab.1 - Some parameters of three proposed satellite mobile networks frequency bands in adjacent beams. Also the *Odyssey* appears to be more power limited.

5.5 Elevation angle and fade margin

A very important characteristic for a satellite personal communications system is the service availability, as the shadowing effects due to natural or artificial obstacles, like terrain, trees and buildings, cause significant signal fades.

Two parameters are essential to ensure a good service availability and a reasonable building penetration: the minimum elevation angle from the user to the satellite and the propagation fading margin designed into the link budget.

From the point of view of the elevation angle, the ODYSSEY constellation has some advantage due to the geometry of the MEO orbits: the minimum elevation angle for a user at a medium latitude (48°) is around 30° for more than 90% of the time, as compared to about 15° elevation for the LEO orbits.

However the IRIDIUM design is aiming to achieve about 16 dB of average fading margin for voice and about 35 dB for paging services, as compared to less than 10 dB for voice communications of the other two systems; on the other hand, *Globalstar* is designed with substantial overlappings of the coverage areas, allowing the selection of a good link with a high probability.

It is difficult to make a direct comparison of the overall link availability, due to the lack of an internationally agreed model for the link margin required at L-band for urban, suburban or rural environments.

6. Conclusions

The market perspective is spurring a number of initiatives to provide mobile communications services through satellite networks.

The most suitable architecture is based on low-earth orbits (LEOs) or medium-earth orbits (MEOs), as LEOs and MEOs allow the generation of small coverage cells, yielding large frequency reuse factors and transmit power levels suitable for hand-held telephone sets.

Among the LEO and MEO networks proposed for mobile communications, *Iridium*, *Globalstar* and *Odyssey* are representatives of the major alternatives.

The choice of the values for the many parameters involved in the system design depends primarily on the area to be served. *Iridium*, aiming at a real global coverage and by-passing the terrestrial networks, is the most complex, but it offers the best performance.

Even though the market for satellite mobile communication is relatively small, if compared with that of land mobile cellular networks, investments in this area can be profitable, as the offered services cannot be delivered by other means in most of their service area. In fact, low-earth orbit satellites can herald the introduction of truly ubiquitous, wireless communications.

References

[1] P. Wood: "Mobile Satellite Services for Travellers", IEEE Communications Magazine, Nov. 1991, pp. 32-35.

[2] J.C. Schoenenberger: "Telephones in the Sky" - Electronics & Communications Engineering Journal, March/April 1989.

[3] Space News: "Satellite Phone System To Be Tested on Airplanes" - March 23-29, 1992.

[4] W. Hager, P. Bartholomé, D. Gillick, S. White: "Hearings on Non-Geostationary Mobile Satellite Systems (Low Eath Orbiting Satellite Systems: LEOs)", Rapporteur's Report, Commission of the European Communities, Brussels, 9-10 Nov. 1992.

[5] "Low earth orbit global cellular communications network", IRIDIUM System Presentation (Motorola), 8th European Satellite Communications Conference, London, Dec. 1991.

[6] J.G. Aiken, P.A. Swan, R.J. Leopold: "LEO satellite - based telecommunication network concepts", Workshop on Advanced Network and Technology Concepts for Mobile, Micro and Personal Communications, JPL Laboratory Pasadena CA, May 30 - 31, 1991.

[7] J.E. Hatleid, L. Casey: "The Iridium^{TM} System Personal Communications Anytime, Anyplace", IMSC '93, June 16-18, 1993, Pasadena, CA, Proc. pp. 285-290.

[8] GLOBALSTAR SYSTEM APPLICATION, Application of Loral Cellular System Corp. before the Federal Communication Commission, Washington D.C., June 3, 1991.

[9] R.A. Wiedeman, A.J. Viterbi: "The Globalstar Mobile Satellite System for Worldwide Personal Communications", IMSC '93, June 16-18, 1993, Pasadena, CA, Proc. pp. 291-296.

[10] A.J. Navarra: "Globalstar", Mobile Communications International, Issue 15, Autumn 1993, pp. 46-49.

[11] A.J. Navarra: "Crisscrossing the World", Global Telephony, Feb. 1994, pp. 24-28.

[12] R. Rusch, P. Cress, M. Horstein, R. Huang, E. Wiswell: "ODYSSEY, a constellation for personal communications", A.I.A.A. 14th Conference, March 1992, Washington D.C.

[13] C.J. Spitzer: "Odyssey Personal Communications Satellite System", IMSC '93, June 16-18, 1993, Pasadena, CA, Proc. pp. 297-302.

[14] G. Pennoni: "Jocos-Satellites For Global Mobile Communications", Globecom '94, San Francisco, CA, Nov. 27-Dec. 1, 1994, to be presented.

[15] "Application of Teledesic Corporation for a Low Earth Orbit Satellite System in the Fixed Satellite Service", Application to the US Federal Communications Commission, Washington, D.C., March 21, 1994.

[16] E.F. Tuck, D.P. Patterson, J.R. Stuart, M.H. Lawrence: "The Calling^{SM} Network: a Global Wireless Communication System", International Journal of Satellite Communications, Vol. 12, No 1, Jan.-Feb., 1994, pp. 45-61.

Part II

Concepts, systems and key technologies

A Digital Transparent Satellite Payload Concept for Personal Mobile Satellite Communications: A VLSI Technology Review

J.Ventura-Traveset, M.Hollreiser
I.Stojkovic and F.Pets

European Space Agency, ESTEC, RF System Division
P.O. Box 299, 2200 AG Noordwijk, The Netherlands

Abstract. *This paper deals with a transparent digital satellite payload concept based on the implementation of digital demultiplexing, narrowband Digital Beamforming and digital on-board switching. Such a payload architecture has been identified as a valid candidate for the provision of Universal Mobile Telecommunication Services (UMTS) by Medium Altitude Earth and Geostationary Orbit (MEO and GEO) satellites. The paper includes a description and technological issues of the different subunits that compose the mobile payload and the European Space Agency (ESA) related activities. Emphasis is put on the digital part of the payload (the Payload Processing Unit), and in particular on the digital technology and VLSI considerations for its hardware implementation.*

1 Introduction

Land-mobile satellite communications are evolving towards providing compatibility with the services offered by terrestrial digital cellular systems and complementing them in low population density areas and developing countries, where terrestrial coverage can not be provided economically. Satellite opportunities arise when considering wide roaming requirements, and in particular to provide world-wide user roaming capability, since only about 15% of the Earth's geographical area is expected to be covered by cellular terrestrial services by the year 2000 (see Fig.1).

Key factors for improving the satellite penetration include the cost/size reduction of the user terminals, reasonable user tariffs and the interoperability with the terrestrial systems. These key issues have a direct impact on the performance and flexibility required by future satellite payloads for the provision of Universal Mobile Telecommunication Services (UMTS).

Within this revolutionary scenario, the European Space Agency (ESA) is actively pursuing system studies and technology developments for all different orbital systems proposed (GEO, MEO, LEO and HEO). Some of the results have been published recently ([1] to [5]). In particular, in [1], an advanced transparent repeater concept was introduced with on-board Digital Signal Processing operations that include narrowband Digital Beamforming (DBF) and Digital Multiplexing and Demultiplexing operations. Such a payload architecture has been identified as a valid candidate for the provision of Universal Mobile T-elecommunication systems with GEO and MEO orbital systems, and will be the center of our discussion in this paper. While emphasis in [1] was put on the detailed description of all different sub-units of the payload (antenna, RF and digital parts) we will focus here on the digital part, and in particular on the VLSI aspects of the digital processor unit implementation. Thus, a major section is devoted to the VLSI technology considerations for Digital Signal Processing in Space, including the problems related to ASIC design, manufacturing and testing and Multichip Module (MCM) techniques for compact and low mass architectures.

Fig. 1. Forecast on terrestrial cellular coverage by the year 2000.

The organization of the paper is as follows: Section 2 presents the general payload architecture concept and identifies the different subsystems that will be further discussed in the subsequent sections. Sections 3 to 6 will summarize the fundamental points on technology and functionality of the different payload sub-units. Section 7, is the core of the paper, with a major discussion on the different aspects of VLSI space technology for the hardware implementation of the DSP sub-units that the payload architecture encompasses. Finally, section 8 summarizes the main points of the paper.

2 Repeater Architecture

A number of different architectures have been recently studied and compared on characteristics such as payload mass, DC power consumption, flexibility in matching traffic to beams, power-bandwidth flexibility, frequency reuse potential, compactness of feeder link, etc. The generic architectures studied include: 1) the well established architecture with SAW filter channelisation, crossbar MMIC switching and RF beamforming (like in Inmarsat-III, EMS, LLM); 2) transparent architecture with all-digital or hybrid (SAW-CFT/digital) channelising, routing and beamforming; and 3) architectures with on-board regeneration.

System parameters involved in personal satellite communications drive towards solutions with many beams in the coverage area, primarily because of the low antenna gain and limited transmission power at the user terminal. This is true for all orbital altitudes, but becomes more pronounced for high orbits. High number of beams, high number of antenna feed elements and the need for high granularity to retain feeder link compactness, all drive towards solutions based on digital signal processing: The transparent payload architecture with narrowband digital beamforming, routing and channelisation was found to be a well matched solution. If there are no bandwidth constraints on the feeder link spectrum, the requirements on the granularity of the demultiplexers can be relaxed and thus architectures based on wideband beamforming can be more efficient than the narrowband ones.

This paper limits its scope to transparent architectures with narrowband digital beamforming, routing and channelisation. The main advantages of this approach can be summarized as follows: very high number of overlapped beams and use of near-peak antenna gain leads to best RF power efficiency and satellite G/T; fine demultiplexing granularity and digital switching lead to high flexibility to match traffic needs; improved frequency reuse through use of peak antenna gain and interference cancellation techniques; phase and delay adjustments of the feed chains to calibrate or correct the antenna system can be implemented in the digital BFN; one hop mobile-to-mobile connections. Although the payload processor in this case is adapted to operate with narrowband access schemes (like FDMA or narrowband TDMA), the inherent transparency enables the introduction of services based on CDMA.

The concept of this transparent digital payload is illustrated in Fig.2. In addition to the antenna sub-systems, the payload is basically composed of three more blocks: the feeder RF-front-end, the Payload Processor Unit (PPU)– with the Forward (FWD) and Return (RTN) processors – and the RF-front-end at the mobile side. Mobile-to-mobile single hop connection can be achieved by routing traffic from the return to the forward processors. In the following sections, a discussion on key design and implementation aspects of those units is included, with a major section devoted to the VLSI space technology considerations for the DSP sub-units hardware implementation.

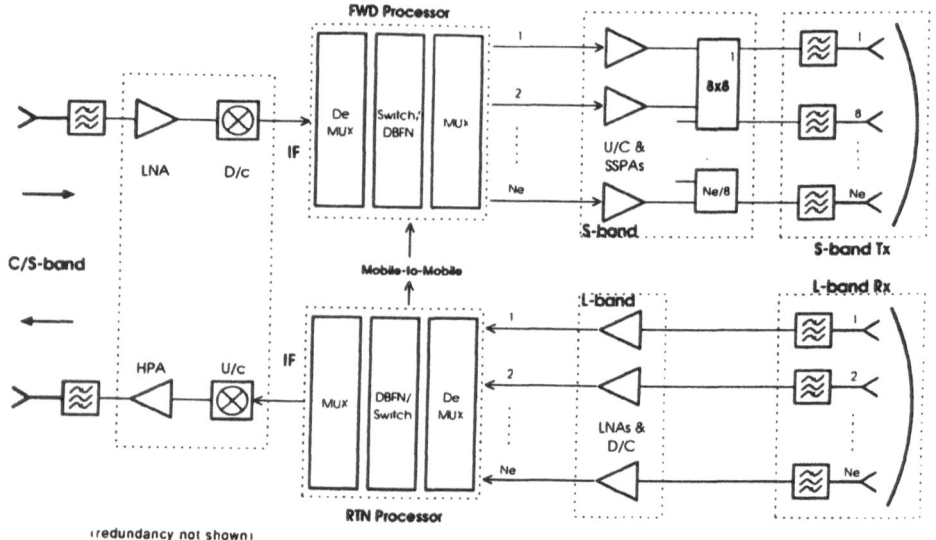

Fig. 2. Generalised architecture for a satellite payload in GEO/MEO orbits providing Personal Communication services.

3 S/C Antenna Designs

Different spacecraft antenna designs have to be applied to GEO and MEO constellations, due mainly to the large difference in their respective scanning angles. While from a geostationary orbit the scanning angle is confined to approximately $\pm 9°$, from the MEO orbits this angle could be up to $\pm 30°$

An optimised solution for GEO has been found to be a defocused reflector antenna. One of the design drivers is that the number of antenna elements should be minimised, as it has a direct effect on the power requirement of the PPU (while the number of beams does not affect significantly the power consumption of the payload). On the other hand, it is convenient to use many elements per beam so as to be able to spread the power over the numerous amplifiers (SSPAs) and thus retain power/beam flexibility with small Butler(-like) matrices, or avoid such matrices altogether. A defocused reflector antenna combines the relatively small number of elements with a simultaneous spreading of power over a number of feed elements.

For MEO constellations, planar array solutions are well adapted to the coverage requirements. The large scan requirements for a MEO payload antenna make use of reflector antennas difficult. Minimizing the number of radiating elements, therefore, has to be compromised and solutions are sought with direct radiating arrays.

Internal ESA/ESTEC studies resulted in baseline designs for GEO [2] and MEO [3] payloads based on transparent digital architectures.

For MEO constellations separate Tx/Rx arrays currently provide the simplest solution to dual frequency operation, albeit with the need to deploy one of the arrays. Of the two dual frequency array options studied, the interleaved array is preferred at this time. However, this is a new area and the array performance is difficult to predict due to unknowns regarding the radiating elements design and performance in the dual element environment with digital processing.

For GEO constellations the optimal solution is a focused array-fed reflector antenna, presenting a compromise between the lowest number of feed elements and the spreading of power among them.

Key technologies, required to be developed to space standard, are large deployable reflector antennas and planar phased array technology for either dual frequency or interleaved antennas.

4 TX and RX Subsystems

As we have noted in section 2, MEO/GEO payload architectures for personal communications are driven by the large number of beams to be generated. As a consequence, these architectures are characterized by high demultiplexing granularity and antennas with many radiating elements. The number of radiating elements to be considered ranges from at least 50-60 (for MEO payloads) to well over 100 (for the GEO payloads) in each direction. Consequently, all RF elements in the antenna chains are replicated many times for both the forward and the return link repeaters. The mass and power consumption of these elements, within a transparent digital payload architecture, is quite significant, typically forming up to one quarter of total payload mass and at least three quarters of the overall payload power consumption. Therefore, any savings on mass and improvements in power efficiency achieved in these blocks will have a pronounced effect.

Main contributing elements in the transmit and receive chains on the mobile side are SSPAs, upconverters and output filters at L or S-band (for transmit), and input filters, LNAs and downconverters at the return link side at L-band (for receive). Because of the high incidence of these blocks, huge savings can be achieved in both mass and power by paying appropriate attention to their design. The possible approach for further development would encompass some or all of the following:

- Integrating the three transmit blocks (upconverter/SSPA/filter) as a single miniaturized package, thus reducing the mass.
- Improving the SSPA efficiency, particularly under multicarrier or pulsed conditions of operation.
- Reducing the power consumption of the upconverter block.
- Integrating the three receive blocks (filter/LNA/downconverter) into a common package, or even a certain number of these into a common package.
- Sharing the power supply circuitry among a number of units

The European Space Agency (ESA) is exerting a significant effort aimed at developing equipments which satisfy the requirements of personal communications payloads. As a step in this direction, an L-band transmit/receive module based on European technology is developed under an ESA contract [6]. Table 1 shows the overall system parameters associated to this development.

Transmit subsystem:	Gain	40 dB
	Output power	12 W
	Secondary Efficiency	> 30%
	Mass	< 0.15 Kg
Receive Subsystem:	Gain	40 dB
	Noise Figure target	1.5 dB
	DC Power	1.1 W
	Mass	< 0.11 Kg

Table 1. Basic parameters of L-band transmit/receive module [6].

As a results of efforts through the ESPRIT program, significant progress has been achieved in the solid state arena, where several GaAs manufacturers are very active in Europe and MMICs can be delivered at competitive price and quality levels. Likewise, the availability of low noise HEMTs and medium power MESFETs is good. Making use of these technologies, miniaturized low noise amplifiers (LNAs), operating at L and Ku bands, have recently been developed under ESA contract [7]. They incorporate discrete HEMT front ends and MMIC gain blocks, simultaneously achieving very low noise performance and a high degree of miniaturization.

New solid state technologies, particularly those targeted at increased efficiency and power capability of SSPAs, have become available. Good examples are Heterojunction Bipolar Transistors (HBT) and Pseudomorphic Multi-junction HEMTs (PM-HEMT), both of which ESA is presently evaluating for their applicability to space.

Finally, improvements in filter design and manufacture are leading towards tuning-less approach, at least for wide-band applications, and new technologies will be applied to reduce their mass and size.

5 Multiplexing and Demultiplexing Technology

Demultiplexing and multiplexing operations respectively precede and follow the digital beamformers in the forward and return processors. The function of the demultiplexer, in both forward and return links, is to extract each individual channel from the input FDM multiplexed stream for its subsequent processing which obviously includes the frequency mapping and beamforming operations. Alternatively, the multiplexer will generate the complementary operation, i.e. the FDM multiplexed stream generation, after the beamforming and mapping processing have been performed.

Depending on the system specifications and the mission technology constraints, solutions for the mux/demux routing operations range from fully analog ones (by means of SAW/CFT devices) through hybrid analog/digital architectures to fully-digital ones.

In the following two sub-sections digital and analog multiplexing and demultiplexing operations will be reviewed at the individual block level.

5.1 Digital Demultiplexing

Introduction: The operations of multiplexing and demultiplexing basically require the same considerations when dealing with a digital implementation solution. Based on that, we will consider only the demultiplexing operation, the same conclusions being applicable to the multiplexer counterpart.

The use of digital solutions for a demultiplexing operation is unavoidable for such situations where a fine channel granularity (around $\leq 100KHz$) is required. In those situations, SAW/CFT demultiplexing solutions are not suitable due to the size limitations of the quartz-crystal forming the SAW chirp delay line (see section 5.2). Digital solutions, on the other hand, seriously compete with hybrid analog/digital solutions for such situations in which the total band to be processed is of the order of few MHz (around $\leq 10MHz$). For larger processing bandwidth an analog pre-demux processing is generally more convenient.

In our discussion, we shall concentrate on the last demultiplexing units (or stages) from the several demultiplexing stages that conform the channelization function of a payload, since they usually are the most complex devices and since their principles are also applicable to their previous demultiplexing stages.

Algorithms: In a first general classification, two possible algorithm approaches must be considered:

1. Per-channel Processing approach.
2. Block Processing approach.

In the per-channel approach each channel is processed independently from all the others. In a block processing approach, though, all channels are processed in common, sharing particular elements such as filtering or a FFT operation.

Block-processing solutions are very much superior (in terms of complexity) than the per-channel ones for such situations requiring the demultiplexing of uniform slots (all slots with equal bandwidth) or where the individual channel-slot bandwidths are related by an integer factor (e.g., related by factor of two's). We are assuming here, irrespectively of the access scheme used, that the channels to be processed by our satellite are grouped in uniform slots of equal bandwidth, and that each slot is individually beamformed in the narrowband beamforming operation. With those two assumptions, then, block processing solutions are the only ones to be considered[1].

For block-processing approach, two families of algorithms shall be considered:

1. FFT-oriented algorithms.
2. Tree-oriented algorithms.

FFT-oriented algorithms are basically the DSP implementation of the SAW-CFT processors described in section 5.2. The generalized architecture (see Fig.3) consists of a polyphase bank of filters, followed by a FFT-block of a number of points equal to the number of channels to be demultiplexed (or double if a complex representation is used) and a second filter stage (channel-definition filter) separate for each channel.

Conceptually, the demultiplexer structure could be performed with a single filter stage preceding the FFT block, so avoiding the channel-definition stage. However, for our mobile satellite concept, a single-filter-stage approach would result in impractical long filters. A reported variation [8] on this general two-stages polyphase-FFT algorithm consists of replacing the second bank of filters (the filter-definition stage) with a single front-end network based on an interpolated Hilbert transform FIR filter. A similar solution, recently proposed in an ESA contract [9], consist on a first-stage tight filter separation between even and odd channels in the FDM multiplex (via a pair of imaged lowpass and highpass halfband filters) such that the filter requirements of the second stage polyphase-FFT (consisting now in 2 blocks, one for the processing of even and one for the processing of odd channels) are very relaxed.

For *Tree-oriented algorithms*, on the other hand, the demultiplexing operation is based on a successive division of the input signal into smaller frequency bands at each stage of the tree using two complementary halfband filters and a 4-point DFT. The common processing cell is a Polyphase-FFT structure for two channels, which is repeated as many times as needed (Fig. 4 illustrates for the case of N=16 channels [10]).

---

[1] Note that other payload solutions, e.g with wideband beamforming, may require flexible demultiplexers to allocate several slots to a given beam, and thus other solutions, including fixed analog filters, may be more suited.

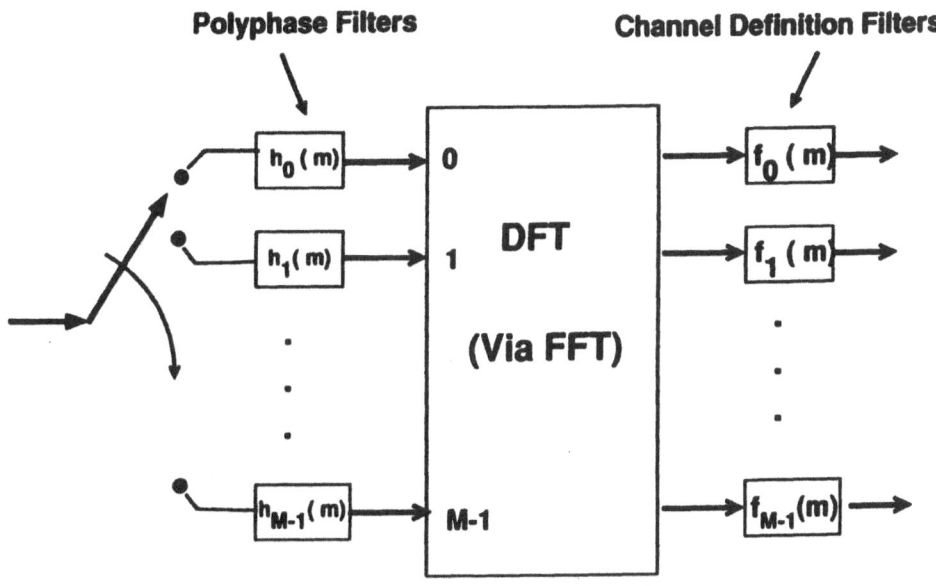

Fig. 3. Illustration of the Polyphase FFT demultiplexing approach.

Comparison of the algorithms: The selection of the most adequate algorithm from the two above mentioned families is very much depending on the system requirements.

As a minimum, the following system parameters must be considered and properly trade-off:

1. *The number of channels* to demultiplex.
2. The *antialiasing filter shape factor* which determines the portion of wasted slots in the input stream.
3. The required *flexibility.*
4. *Redundancy* considerations.
5. The final ASIC *hardware design.*

For the trade-off analysis of the above parameters and the two algorithms, the reader is referred to [1].

ESA activities in Digital Multiplexing/demultiplexing: First pioneering ESA hardware activity on this field was the development of a 16-channel demultiplexer

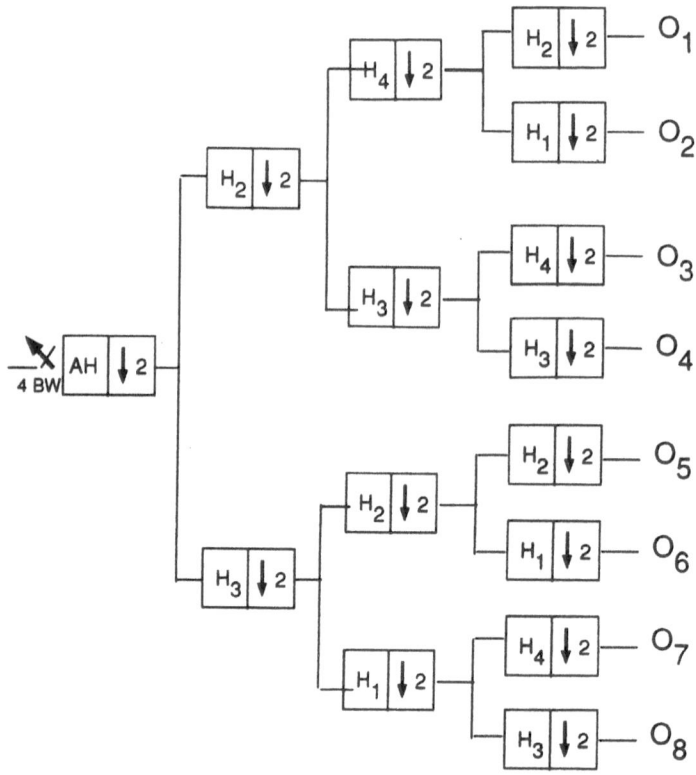

Fig. 4. Illustration of the tree-based demultiplexing approach ($N_c=8$).

ASIC for on-board applications [10] based on the tree-algorithms. For completeness, the design of the dual 16-channel multiplexer is presently under development under another ESA contract [11]. Furthermore, ESA is now pursuing [9] the ASIC implementation of a dual multiplexing/demultiplexing architecture based on the Polyphase-FFT algorithm, to be integrated in a demonstration Payload Processor Unit (PPU) based on narrowband beamforming. Finally, a similar payload concept has been studied for ESA under [12] and the associated demultiplexer technology and ASIC implementation is under development in [13].

5.2 SAW-CFT Demultiplexing

General Considerations: The hybrid analogue /digital demultiplexer described here relies on an analogue transmultiplexing process based on the Chirp Fourier

Transform. It is basically an analogue technique, which can be used to perform relatively narrowband frequency demultiplexing (and multiplexing) to a large number of slots in a single hardware unit. The coarse demultiplexing achieved in this way is further enhanced in terms of narrower slots and very small guardbands, by digital means.

The application of this method for a mobile and personal communications satellite repeater is novel and the idea has been pursued by ESA through a number of development contracts.

The principle of the Chirp Fourier Transform is based on the processing of a linear frequency modulated signal by a dispersive delay line used in its inverse mode to compress such a signal into a short pulse. The input signal is mixed with a linearly swept local oscillator (chirp) and downconverted to the IF band (the operating frequency band of the dispersive filter). Each signal frequency component present in the input band during the LO sweep moves across the IF band as a sweeping tone that appears at the IF as a linear frequency modulated signal, i.e. a chirp. The sweep rate of the local oscillator is matched to the delay slope of the SAW dispersive delay line, so that a time compressed pulse appears at the output of the processor for every frequency component of the input. The overall result is a domain conversion of frequency to time. The output signal, in practice, has a considerably higher bandwidth than the input signal, but is compressed in time.

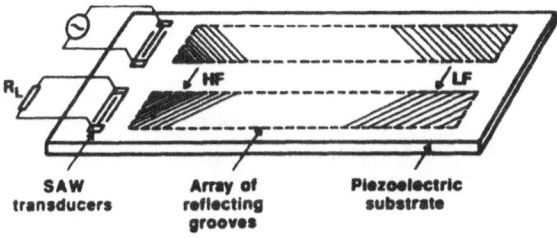

Fig. 5. SAW reflecting Array Compressor (RAC) [15].

The most significant component of a CFT processor, from a functional, as well as a design aspect, is the SAW dispersive delay line (Reflecting Array Compressor, RAC), whose basic structure is illustrated in Fig.5. It's primary characteristic, the linearly variable group delay with frequency, has to be maintained in a relatively large frequency band, within the specified environmental conditions (temperature range, mechanical environment etc). Research projects conducted by ESA/ESTEC indicate that an operating input bandwidth in excess of 30MHz, i.e. the full mobile allocation in L-band, is achievable.

An Example of a CFT for Mobile Application: The technical complexity of the CFT is summed up by the time-bandwidth product. In general, the wider the bandwidth of the SAW device (dispersive delay line),

- the shorter the SAW device,
- the shorter the impulse response,
- less degree of selectivity, i.e. wider minimum frequency slots and guard-bands.

In an on-going ESA project [14], it has been demonstrated that temperature stable quartz SAW substrates allow time bandwidth products up to 3000 to be realised for SAW chirp lines with typical weight functions.

Table 2 shows typical parameters of a CFT unit suitable for the processing of the full L-band mobile frequency allocation (34MHz) and optimized to perform, in conjunction with a fine digital demultiplexer, the breakdown of the input spectrum to slots of 30KHz [15].

The fine digital demultiplexing needed for further demultiplexing of each CFT slot is based on the same principles as described in section 5.1. The order of the demultiplexing will be low for the CFT case which will favor the use of the tree-oriented algorithm. To make an efficient overall processor, the DSP chips must be capable of handling a large number of CFT slots in time multiplex. Compared to the chips for the all-digital case, the chips will need more memory to be capable for time multiplexing between a large number of CFT slots, but will be considerably simpler due to the low order demultiplexing.

Current State of CFT Technology in Europe and ESA Involvement: The technology has been successfully operationally demonstrated and there are a number of units currently used for experimentation. However, there are still problems which need to be overcome before submission for space qualification. The main problem encountered up to now is related with packaging and the change of parameters when exposed to mechanical shocks or random vibration. To this end, novel developments for the correction of the phase and amplitude responses are considered involving real time measurements.

The present research activity supported by ESA, aimed at improving the packaging is due to be finalized soon. If the results are encouraging and activities on CFT technology continue, space qualification could be expected in 1996, according to ESA estimates. Finally, in a separate contract [9] a SAW CFT device is being developed with the aim of eventually demonstrating its operation in a full payload processor environment.

Conclusion: Based on earlier ESA studies, the CFT-based (de)multiplexing principles and implementation parameters required to handle the situations which are expected to be encountered for MEO/GEO mobile and personal communications payloads are well known and the hardware has been demonstrated. Our systems analyses have shown that CFT multiplexing and demultiplexing, although capable of coping with a very wide range of input bandwidths and slot sizes, yields best results, compared to competitive all-digital techniques, when

Bpass = 34MHz	processed bandwidth
fc ≃ 200MHz	centre frequency of SAW chirp line
k ≃ 0.85	guard factor for CFT output frames
sampling	complex
Tw ≃ 5µs	duration of dispersive response
Ts ≃ 2µs	CFT frame repetition rate
B_{SAW}	≃100MHz
CFT slot bandwidth	240KHz
ripple in pass band	needs to be compensated by simple DSP filter
Suppression of aliasing	40 dB
Processing delay	< 30µs (for full CFT/ICFT chain including cover filters)
Order of Digital Demux	8
Maximum phase drift	≤ 4.2° for ±30K temperature
Power consumption	1.35W (including all amplification and digital demux to 30KHz slots)
Total mass	275g

Table 2. Typical parameters of a CFT unit suitable for a full L-band mobile processing.

the overall processed bandwidth is relatively large. The breakpoint depends on a number of system parameters and ranges between several megahertz to 15 MHz. Processing of bandwidths in excess of 15MHz will almost certainly be more efficient, especially in terms of DC power requirement, employing CFT techniques, then using an all-digital approach. Demultiplexing to slots narrower than approximately 250KHz, although perfectly achievable with CFT only, will be better performed if the CFT process is augmented by digital demultiplexing, because smaller SAW devices can be used and much tighter guardbands can be achieved.

6 Digital Beamforming

General description: The basic idea of digital beamforming is the same as for a normal RF or IF beamforming: A signal impinging on the array antenna will arrive at different time instants to the different radiating elements, depending on its direction of arrival. Taking the signal present at one of the elements as a reference, and if the signal can be considered narrowband, the different times of arrival to each element are equivalent to a difference of phase[2].

The function of the beamformer is to first phase-shift each radiating element signal for the exact amount corresponding to the expected direction of arrival of the wanted signal and then add all the signals. In this way, only the signals arriving from the expected direction of arrival will be added in phase, whilst any other signal will be subject to less enhancement (or even cancellation), depending on their direction of arrival. The thermal noise generated at each radiating element branch, being uncorrelated between different radiating elements, will be added incoherently, resulting in an ideal improvement of 10*log(Nr) in signal to noise ratio, Nr being the number of radiating elements.

In order to control the sidelobes of the radiation pattern, a certain amount of amplitude control can be introduced in the beamformer branches. The same ideas apply in both receive and transmit directions.

Digital beamforming performs this function in the digital domain, after sampling the complex envelope of the radiating element signals at the proper sampling rate determined by the Nyquist criterion. The basic block diagram of a receive digital beamformer is shown in Fig. 6. The signals are first amplified, downconverted and sampled (either at baseband or at a suitable IF frequency) and then multiplied by complex coefficients and added. The modulus and argument of the complex coefficients are equivalent to the amplitude and phase shift tapering on RF beamformers.

Advantages of digital narrowband beamforming applied to Mobile Communications Payloads:

1. Power consumption and mass for the digital beamforming function depends linearly mainly on the total bandwidth processed and on the number of radiating elements and does not depend on the number of beams. This is the major difference compared with classical RF beamforming; where the number of beams to be implemented is very high, RF beamforming would not lead to practical implementations.

2. In the limiting case, a single channel per beam could be implemented without a major cost increase in beamforming. From the beamforming point of

---

[2] This is the assumption made for narrowband beamforming which is considered throughout this paper.

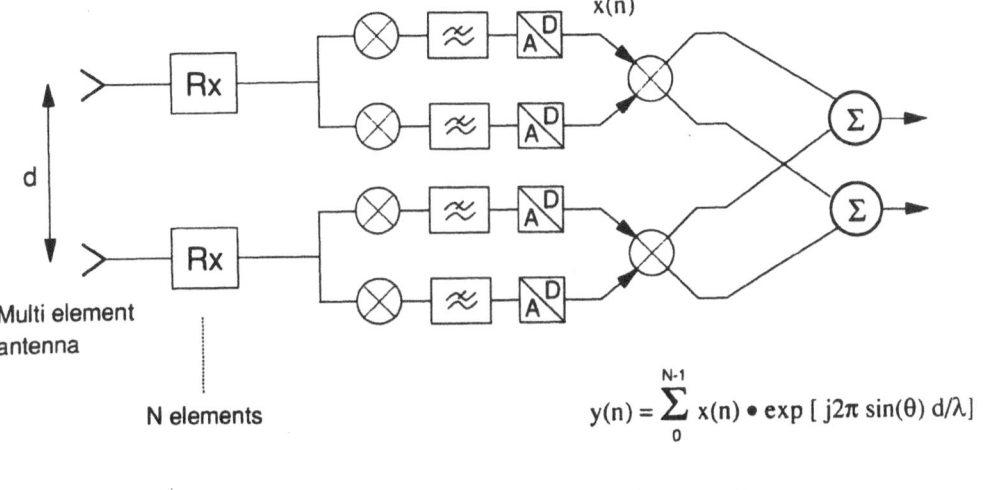

$$y(n) = \sum_{0}^{N-1} x(n) \bullet \exp[\, j2\pi \sin(\theta)\, d/\lambda\,]$$

Θ = Beam angle

Fig. 6. Basic block diagram of a digital beamformer.

view, the advantages of the implementation of a single channel per beam are twofold:

(a) Assuming a fixed antenna size, to achieve a certain value of EIRP, less RF power per channel is necessary on board, with the associated reduction in total power consumption.

(b) The frequency reuse of the available spectrum can be increased (see Fig. 7) by a factor of approximately 2.3.

3. The calibration and compensation of phase and amplitude errors originating in the analog part of the antenna and beamformer (LNAs, up/down converters, SSPAs, diplexers, antenna deformations, ...) becomes much easier.

4. Active interference suppression techniques can be easily implemented on board to reduce the cochannel interference due to signals using the same frequency arriving from different beams on different directions.

5. Digital beamforming related techniques allow the possibility to find the direction of arrival of the different signals to the antenna aperture. This function permits the correct pointing of the beams to the mobile users and thus the optimization of the antenna gain. A description on the different algorithm techniques to perform the function of detection of arrival can be found in [1].

ESA Activities in DBF: ESA hardware activities on DBF were started with the development of a first DBF ASIC under ESA contract [16]. This ASIC accepts a time multiplex of 8 bit sampled complex signals with a maximum total rate of 32MHz and performs the beamforming operation (complex multiplication and accumulation) with 11 bit complex coefficients, providing a 16 bit beam output.

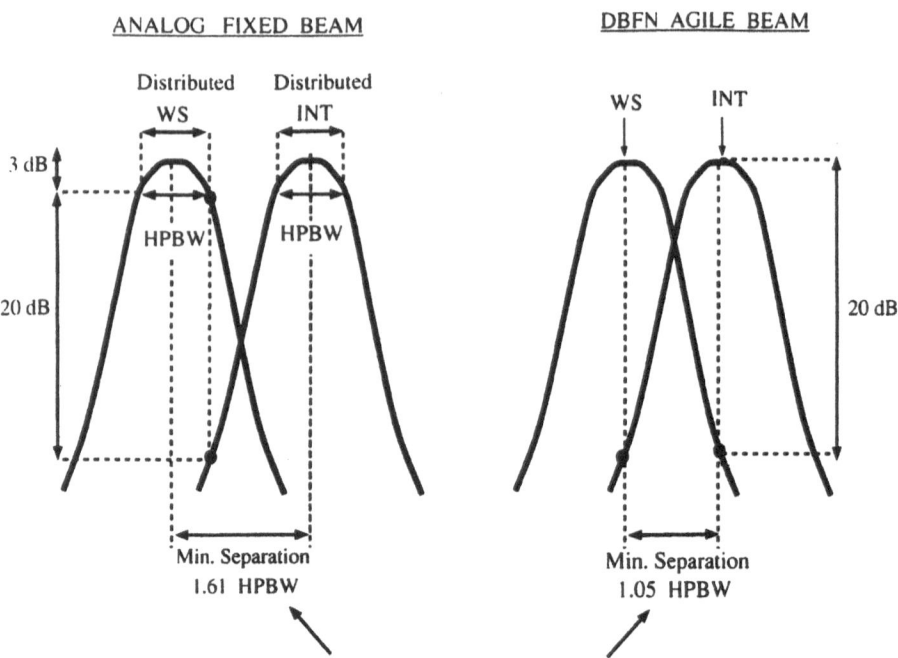

Fig. 7. Illustration on the frequency reuse factor increase with DBF.

This ASIC has been manufactured using a 1μ standard cell CMOS technology, and its power consumption at maximum 32MHz multiplexed rate is 0.75W. In addition, ESA is presently developing an on-board oriented dual Tx/Rx DBFN ASIC [9] to be integrated in a demonstrative Payload Processor Unit (PPU) (mentioned in section 5.1) which should prove the concept of the digital transparent payload we are considering here. Finally, a similar activity, related to the ASIC implementation of a narrowband DBF, has recently been initiated under[13], which includes the integration of several of those basic DBF ASICs into a Multichip Module (MCM) package (see section 7).

7   Digital Technology

General Considerations: Historically signal processing systems and equipments for space and ground applications were purely analogue. To meet the combined requirements for increased functionality and improved performance the analogue signal processing technology is increasingly being replaced by digital technology which is implemented via the use of ASICs. DSP is at the centre of today's rapid advances in communication and earth observation, and as we have seen in previous sections, is also de core of our mobile digital payload concept.

Without the tremendous advances in VLSI technology in terms of shrinkage, complexity and performance improvements as well as in related areas, such as CAD tools, packaging and assembly technology, these developments would not have been possible and DSP would play a much less important role. Only these high performance technologies allow integration of complex systems, as our mobile digital payload concept, with acceptable mass, power consumption and volume, while maintaining high speed and reliability.

High complexity communication and remote sensing systems place extreme demands on the technology in terms of architecture, circuit design and process. Major design challenges include CAD tools for simulation, synthesis and verification, packaging and power control.

Technologies: During the last 25 years progress in microelectronics was achieved through reduction of the minimum feature size. ASICs, led by DRAMs and microprocessors, have tagged along, a generation behind, but reaping the same benefits in cost, speed, size and quality. Space qualified technologies are roughly 4-5 years behind the commercially available ones. Actual flight hardware is being developed roughly 4 years in advance to the flight. Therefore, by the time the ASIC will actually fly, the technology used will be roughly 8-9 years old (Compared to commercial technology).

ASICs come in all sizes and combinations of features, but can typically be categorised into five different groups (see Fig.8):

1. Filed Programmable Gate Arrays (FPGA)
2. Gate Array or Sea of Gates
3. Composite Arrays
4. Standard Cell and Compiled Cell Chips
5. Full Custom Chips.

From 1) to 5) the integration capability, performance, NRE cost, development time and manufacturing time increase and cost per part decreases due to reduced silicon area. The structural choice depends on time to market, design expertise, desired performance level, planned production volume, power consumption, area and cost trade off.

Gate Arrays and Sea of Gates have by far the biggest market share [17], followed by Standard Cell Designs. FPGAs recently are receiving increasing interest. FPGAs are currently the largest growing segment in semiconductor industry [17]. The main advantage is the user configurability in the field. This results in lower NRE costs, shorter development and manufacturing (programming) cycles for prototyping, ease of redesign and reprogrammability, and low start-up cost. On the other side only lower operating speed and lower complexity compared to mask programmable Gate Arrays can be achieved. Actel FPGAs are under

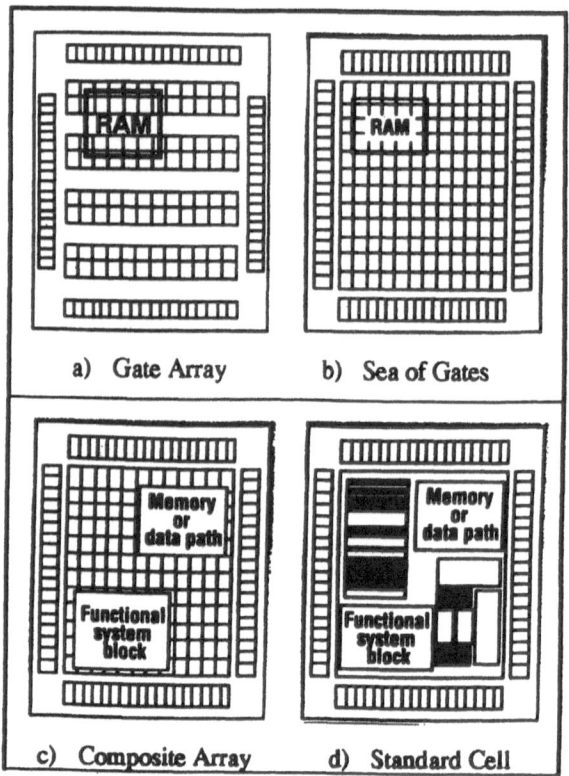

Fig. 8. ASIC Categories.

ESA evaluation at the moment, but also the Xilinx architecture, for which Harris plans an implementation on their SOS process, might become interesting.

After having discussed the main ASIC categories we come to the actual technologies. Important selection criteria for choosing a technology during an ASIC design are:

- integration capabilities
- Power dissipation
- Maturity of supported design tools
- Maturity of technology
- Maturity and complexity of libraries
- Embedded blocks and macrocells availability.
- I/O availability.

For space application, *radiation tolerance* and general qualification status are of additional importance.

Space environment imposes special requirements on components. In particular for long term missions, a tolerance to a total radiation dose of 50-100kRads and a high resistance to single event upset (SEU) are desirable. To this respect, technologies such as SOS or SOI usually have some advantage, however by changing process parameters and optimising layout techniques also for standard CMOS processes high radiation tolerances can be achieved. Due to it's semi-insulating substrate and relative wide energy gap, GaAs has higher radiation tolerances and can work at higher operating temperatures.

There is a large range of technologies available in Europe. It is not the intention to provide a complete overview, but rather, we will concentrate on the most important ones suitable for the DSP function implementation of digital mobile payloads. At present two technologies have obtained the so called "capability approval" [17]. These are the SOS4 2.0μm Standard Cell based technology from ABB Hafo with 10k gates and the 2.5μm CMOS SOS technology from GEC Plessey Semiconductor offering Standard Cell as well as Gate Array design with 8k gates. Both technologies are radiation hard with minimum levels of 100kRad total dose. Programmes have been started for the 1.25μm SOS5 Standard Cell based technology from ABB Hafo with 40k gates, the 1.5μm SOS Sea of Gates technology from GEC Plessey Semiconductor with 26k gates and the 1μm SCMOS1-RT radiation tolerant (min. Level 30kRad) technology from Matra MHS with 35k gates. Matra MHS has completed the evaluation testing phase and is supposed to finish the capability approval testing phase end of 1994.

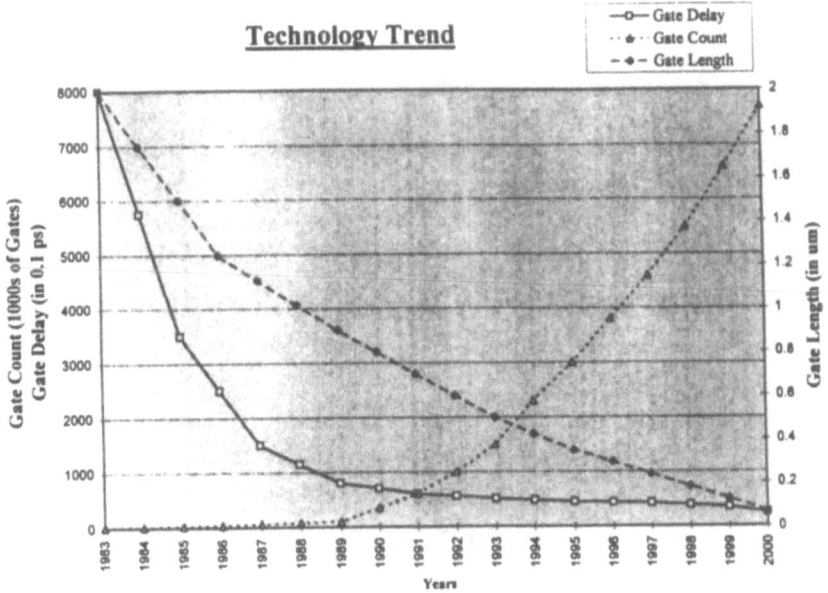

Fig. 9. Technology trend evolution.

Scaled CMOS seems to be the ideal technology for the low power revolution. Improved power consumption, performance and density can be achieved simultaneously [17]. An order of magnitude and more is achievable for the power-delay product. Fig.9 shows some trends starting from 1983 and extra-polating to the year 2000 for channel length, gate delays and integration complexity. As supply voltages are reduced new I/O buffers schemes have to be introduced in order to cope with the lower voltage swings and to avoid the power hungry and noisy TTL compatible signal levels [17]. It is foreseen that we might use $0.6\mu m$ very soon here in Europe for DSP research contracts and, on a later stage, for real on-board hardware.

MCM: It can be anticipated that for the payload concept we are considering here, the number of ASICs that will be required will be of the order of several hundreds with high number of interconnections, and power dissipation of the order of few hundred Watts. In order to reduce the mass of the digital part of the payload it becomes clear that the use of packaging methods based on the use of thin film multi-chip module (MCM) will become an essential issue. MCMs and 3D packaging have recently become very popular in the context of ever higher function density, reduced size, weight and power consumption and improved performance and reliability. As the cost per transistor in the future might increase [17], instead of decreasing as it has been in the past, MCMs and 3D packaging might provide an alternative to advanced processes in order to increase the active chip area.

Fig. 10. MCM basic structure.

MCMs are components, generally in bare die form, mounted on a multi-layer substrate including interconnections and dielectric (see Fig.10. Four different categories can be identified: MCM-L based on technologies such as high density laminated PCBs, MCM-C based on thick film hybrids. MCM-D based on multi layer thin film technology, and MCM-S based on Silicon substrates and silicon foundry processes. Mounting for all of these technologies is possible in three different ways: flip chip die with solder bumps, TAB or wire bond. Standard packages are primarily in Kovar. MCMs can be assembled on to a PCB. Several substrates can also be included into a single very large MCM package. 3D packaging allows very dense assembling [17]. MCM for space application is treated for what it is: a passive part and an assembly technology [17]. There are several ESA studies ongoing at the moment for the different techniques. There is not a single best method. Trade off has to be performed case by case. Aspects to consider are TCE of the substrate and package, thermal conductivity and density, testability, reliability, availability, mass and size. Testing is a very important issue for MCMs mainly due to the usage of bare dies. The new dimension for MCMs is that it is much more difficult to test bare dies than packaged ones as done for standard PCB design. Boundary Scan technique will be one of the working horses for testing as well as Built in self Test (BIST) techniques.

Architecture Design Techniques: In developing VLSI architectures for DSP the designer is faced with a number of key choices. These include: full custom circuits and layout vs. cell based design, programmable vs. dedicated hardware architectures, bit serial vs. bit parallel, selection of numbering system, regular arrays architectures, synchronous vs. asynchronous communication, central vs. distributed storage, analogue vs. digital and monolithic vs. module based systems. These are great topics for heated debate but there is no one universally correct choice. The correct architecture will depend heavily on the application and related algorithm and its constraints on performance, power, size, mass, flexibility, compatibility, market expectations and, of course, cost.

The roughly 50 ASICs designed in the Radio Frequency System Division at ESA/ESTEC in the past 5 years work almost without exception according to a more or less parallel, pipelined and multiplexed architecture with little control and high performance. In cases were designers faced more complex algorithms, with high control, but low speed requirements, commercial DSPs were used. Internally we also looked into systolic array architectures. [17].

Circuit Design Techniques: There are plenty of circuit design techniques only in CMOS [17]. We can distinguish two main classes: the static and the dynamic [17]. The main representative of the static group is the standard design technique based on complementary pairs for each input. This is the working horse of all gate arrays, sea of gates and standard cell designs. It is a robust technique with high noise margins. Dynamic techniques allow great reduction of area. Recently, extremely high speed combined with reasonable power consumption have been reported. Dynamic techniques usually require full custom design approach as they are very load sensitive and have to be designed and optimised carefully. However for DSP, which usually employs regular structures, full custom design techniques might be justifiable, especially if very high performance combined with low power consumption is required. The power-delay product usually can be improved by a factor of 5-10 [17]. BiCMOS attempts to combine the advantages of CMOS such as low power consumption and high noise immunity and the advantage of Bipolar, which is high drive-capability, however at the expense of high power consumption [17]. In Si-Bipolar technology the logic families can be categorized into saturated and non-saturated logic families. The best known saturated technique is TTL and the non-saturated technique is ECL. It allows very high speed in combination with very high power consumption. On GaAs DCFL, BDCFL and SCFL are the most common ones [17] although there is a long list of others. Fig.11 illustrates for each one of the above mentioned techniques.

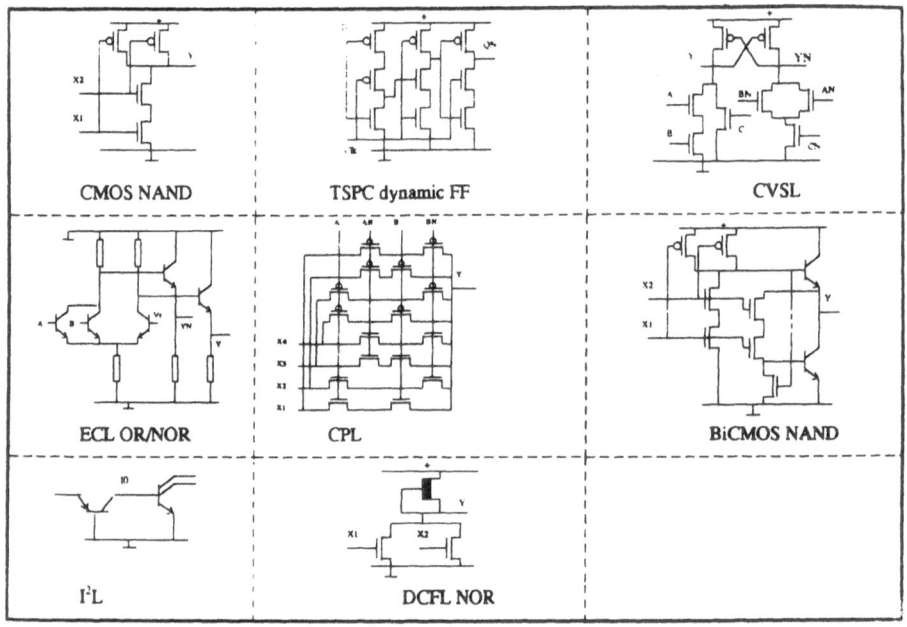

Fig. 11. Circuit design techniques.

All these different techniques mainly trade off speed, power consumption, voltage swing, noise margin and complexity against each other.

There are several critical issues during circuit design as complexity and performance increases. One of the major problems is clock scheme, clock distribution and I/O interfaces. These problems are discussed in detail in [17] and different solutions, among them PLLs, are provided. Quality clock distribution is critical to high performance. It also provides a challenge for layout CAD. PLLs allow to generate higher on-chip clocks than are fed into the ASIC. Due to the low speed, input clock termination is avoided and packaging parasitics in the on-chip path can be neglected. At the same time RFI emissions are reduced. PLLs allow internal phase alignment, delay calibration, clock multiplication and generation of multiple clock phases.

Testing: The problems raised by production faults are still a highly studied topic. The larger the chips the more there is a need for including on-chip testability features and access during board level testing.

Most foundries have already implemented some form of the Join Test Action Group (JTAG) test-port and test-controller. a standard which has been introduced in 1990. This 1149.1 IEEE standard boundary scan, which is a board level scan path, is a solution to the problems posed by growing circuit complexity and card miniaturisation, particularly via SMT and MCM. It was designed to allow three different kinds of testing [17]: internal testing, external testing and sampling test

Boundary scan technique goes perfectly well with BIST test at ASIC level. The most well known BIST technique is based on a deterministic LFSR pattern generator and on MISRs for the signature analysis. Regular blocks such as RAMs, ROMs, etc. usually have their own built in test.

On ASIC level a quite promising technique is the Circular Self-Test Path (CSTP). It offers rather low complexity [17]. The standard test-strategies on ASIC level can be divided into ad hoc and structured techniques. Among the ad hoc techniques there is multiplexed isolation to increase controllability and observability of internal nodes. The most well known structural technique is Scan Path and it's diversions. Recently two other techniques are receiving attention: IDDQ and cross check testing [17].

All the above techniques serve to detect production faults as early as possible. Once the devices are on the PCB or MCM it will take much more effort to localize a defect device.

Power Consumption: Power consumption for Space applications is an essential factor, especially as the number of ASICs for future payloads which are considered at the moment as part of different studies might be in the order of several hundred parts per single payload. In a CMOS circuit there are three main contribution to power consumption: Dynamic Power, Short Circuit Current and Leakage Power. By far the main contributor is the dynamic power consumption.

Power estimates, without calculating the actual Ceff , which done by hand is a rather complex task for a large circuit, are always very inaccurate and can be wrong by 100% and more [17]. In fact, area or gate count and power consumption are not necessarily related [17].

In order to minimise the power consumption, the supply voltage, the effective switching capacitance Ceff as well as the operating frequency should be minimised. Clearly the reduction of the supply voltage has the largest impact. On the different levels of the design (system-, algorithm-, architecture-, circuit/logic- and technology-level) the reduction of power can be achieved by different means [17]. I/O buffers play an important role in terms of power consumption especially as frequency increases [17].

Design Tools: Design tools support the development on all the different levels. The main tools are introduced in [17], evolution are shown and outlook into the future is given. VHDL is becoming a key technology. Improved behavioural VHDL simulation, together with possibly architectural synthesis and VHDL gate level simulation might provide an integrated and homogeneous method of describing, simulating and synthesizing from behavioural down to gate level. This homogeneous multi-level simulation environments are essential for future DSP design tools [17].

The driver for VLSI was computer industry. This is reflected in the type of tools available nowadays. Up to architecture level, tools are meanwhile nicely integrated. However system electronics such as in telecommunication or in earth-observation is defined at a much higher level and is characterized by a high degree of heterogeneity: ASICs are mixed with standard components, analogue is mixed with digital circuitry, hardware is mixed with software, DSP is mixed with more protocol oriented processing, etc.. Tools are needed to do estimations concerning complexity, power consumption, mass and size. This area is hardly addressed at all in today s CAD tools. The higher one moves the more specific design tools and methodologies will be needed. There will not be the one and only tool good for telecommunication engineers, computer systems, remote sensing, etc..

8 Summary

This paper has presented a technology overview for the different sub-systems of a transparent repeater concept with on-board Digital Signal Processing operations, including narrowband Digital Beamforming and Digital Multiplexing and Demultiplexing blocks.

Emphasis in the discussion have been put on the DSP sub-systems that conform the core of the payload, and in particular, on the VLSI technology considerations for its hardware implementation.

This repeater concept is adequate for the provision of Universal Mobile T-elecommunication Services by Medium Altitude Earth and Geostationary Orbit satellites.

The development of such a payload architecture, is in line with present and planned activities of ESA for MEO and GEO orbit systems, which have also been implicitly discussed in the paper.

Glossary of Acronyms

AWGN: ADDITIVE WHITE GAUSSIAN NOISE CHANNEL
ASIC: APPLICATION SPECIFIC INTEGRATED CIRCUIT
AWGN: ADDITIVE WHITE GAUSSIAN NOISE
BDCFL: BUFFERED DIRECTED COUPLED FET LOGIC
BFN: BEAM FORMING NETWORK
BICMOS: BIPOLAR CMOS
BIST: BUILT IN SELF TEST
BW: BAND WIDTH
CAD: COMPUTER AIDED DESIGN
C/I: CARRIER LEVEL RELATIVE TO INTERFERENCE LEVEL
C/N: CARRIER LEVEL RELATIVE TO NOISE LEVEL
CDMA: CODE DIVISION MULTIPLE ACCESS
CFT: CHIRP FOURIER TRANSFORM
CMOS: COMPLEMENTARY METAL OXIDE SEMICONDUCTOR
CSTP: CIRCULAR SELF-TEST PATH
DBF: DIGITAL BEAM FORMING
DBFN: DIGITAL BEAM FORMING NETWORK
DC: DIRECT CURRENT
DCFL: DIRECT COUPLED FET LOGIC
DEBS: DYNAMICALLY EFFICIENT BIASING SCHEME
DSP: DIGITAL SIGNAL PROCESSING
ECL: EMMITTER-COUPLED LOGIC
EMS: EUROPEAN MOBILE SATELLITE
EoC: EDGE OF COVERAGE
ESA: EUROPEAN SPACE AGENCY
ESTEC: EUROPEAN SPACE RESEARCH AND TECHNOLOGY CENTRE
F/TDMA: FREQUENCY / TIME DIVISION MULTIPLE ACCESS
FAFR: FOCUSED ARRAY-FED REFLECTOR (ANTENNA)
FDMA: FREQUENCY DIVISION MULTIPLE ACCESS
FFT: FAST FOURIER TRANSFORM
FPGA: FIELD PROGRAMMABLE GATE ARRAYS
GaAs: GALLIUM ARSENIC
GEO: GEOSTATIONARY EARTH ORBIT
HBT: HETEROJUNCTION BIPOLAR TRANSISTOR
HEMT: HIGH ELECTRON-MOBILITY TRANSISTOR
HEO: HIGHLY INCLINED ELLIPTICAL ORBIT
HPA: HIGH POWER AMPLIFIER
ICFT: INVERSE CHIRP FOURIER TRANSFORM
I/O: INPUT/OUTPUT
JTAG: JOIN TEST ACTION GROUP
LEO: LOW EARTH ORBIT
LFSR: LINEAR FEEDBACK SHIFT REGISTER
LLM: L-BAND LAND MOBILE (PAYLOAD)
LNA: LOW NOISE AMPLIFIER

MAGSS:	MEDIUM ALTITUDE GLOBAL SATELLITE SYSTEM
MEO:	MEDIUM ALTITUDE EARTH ORBIT
MESFET:	METAL SEMICONDUCTOR FIELD EFFECT TRANSISTOR
MIC:	MICROWAVE INTEGRATED CIRCUIT
MISR:	MULTIPLE INPUT SHIFT REGISTER
MMIC:	MONOLITHIC MICROWAVE INTEGRATED CIRCUIT
NB-DBF:	NARROW-BAND DIGITAL BEAMFORMING
NB-TDMA:	NARROW-BAND TIME DIVISION MULTIPLE ACCESS
NPR:	NOISE POWER RATIO
NRE:	NON RECURRING ENGINEERING
PCB:	PRINTED CIRCUIT BOARD
PLL:	PHASE LOCKED LOOP
PM-HEMT:	PSEUDOMORPHIC MULTIJUNCTION HIGH ELECTRON-MOBILITY TRANSISTOR
PPU:	PAYLOAD PROCESSING UNIT
RF:	RADIO FREQUENCY
Rx:	RECEIVE(R)
SAW:	SURFACE ACOUSTIC WAVE
SCFL:	SOURCE COUPLED FET LOGIC
SDR:	SIGNAL TO DISTORTION RATIO
SEU:	SINGLE EVENT UPSET
SMT:	SURFACE MOUNTED TECHNOLOGY
SNR:	SIGNAL TO NOISE RATIO
SOI:	SILICON ON INSULATOR
SOS:	SILICON ON SAPPHIRE
SSPA:	SOLID STATE POWER AMPLIFIER
TAB:	TAPE AUTOMATED BONDING
TDM:	TIME DIVISION MULTIPLEX
TDMA:	TIME DIVISION MULTIPLE ACCESS
TTL:	TRANSISTOR TRANSISTOR LOGIC
Tx:	TRANSMIT(TER)
UMTS:	UNIVERSAL MOBILE TELECOMMUNICATION SYSTEM
VHDL:	VHSIC HARDWARE DESCRIPTION LANGUAGE
VHSIC:	VERY HIGH SPEED INTEGRATED CIRCUIT
VLSI:	VERY LARGE SCALE INTEGRATION

References

1. J.Ventura-Traveset et al.' "Key Payload Technologies for Future Satellite Personal Communications: A European Perspective", *Submitted for publication to the International Journal in satellite Communications*, Sept. 1994.

2. J.Benedicto et al., "Geostationary Payload Concepts for Personal Satellite Communications", *IMSC'93*, Pasadena, June 1993.

3. P.Rastrilla et al., "Medium Altitude Payload Concepts for Personal Satellite Communications", *AIAA 15th International Communications Satellite Systems Conference*, San Diego, USA, Feb/March 1994.

4. J.Ventura-Traveset et al., "A Technology Review for Future Satellite Personal Communication Payloads", *AIAA 15th International Communications Satellite Systems Conference*, San Diego, USA, Feb/March 1994.

5. J.Benedicto et al., "Regional and Global Personal Communication Satellite Systems", *IEEE Symposium on PIMRC'93*, September 1993, Yokohama, Japan.

6. Matra Marconi Space UK limited, "Transmit/Receive Module" *ESA Contract 10418/93/NL/DS*, presently on-going ESA contract.

7. DASA, "Final Report on Technologies for LNA" *ESA Contract 9290/90/NL/US*.

8. F.J. Lake et al., "An Efficient Architecture for Digital Multiplexing and Demultiplexing using a Hilbert Transform Network", *Proc. Second Int. Workshop on DSP Techniques applied to Space Communications*, 1990, ESA WPP-019

9. British Aerospace, Alcatel Espace, Frobe, AME, "Study and Development of Digital Beamforming Techniques", *ESA Contract 10381/93/NL/JV*, presently on-going ESA contract.

10. ANT Bosch Telecom: "Final Report of VLSI Development of a Digital Demultiplexer", *ESA Contract: 9178/90/NL/US(SC)*, 1994.

11. ANT & British Aerospace, "Design of a 16-Channel frequency Multiplexer (FMUX) ASIC", *Rider to ESA Contract 10381/93/NL/JV*.

12. Matra Marconi Space UK, "ASIC Development for Use in Digital Transmultiplexers" *Future ESA Contract*

13. Matra Marconi Space UK, "ASIC Development for Use in Digital Beamforming Networks" *Future ESA Contract*

14. Delab and Frobe Radio, "Developments Related to SAW Applications for satellites, Work-Order No.2: Mounting on SAW Chirp Lines", *ESA Contract 9107/NL/DS*.

15. P.M. Bakken et al., "SAW-based Chirp Fourier Transform and its Application to Analogue On Board Signal Processing", *International Journal on Satellite Communications*, Vol.7, pp 283-293, 1989.

16. ERA Technology, "Breadboarding of a Digital Beamforming Network", *ESA Contract 8714/90/NL*. Final report.

17. M. Hollreiser, "VLSI for DSP in Space", *Proc. of Fourth International Workshop on Digital Signal Processing Techniques Applied to Space Communications*, London, Sept. 26-28, 1994

Multi-media Satellite Mobile Services and Systems

Hanspeter Kuhlen
Deutsche Aerospace AG (DASA)
Munich, Federal Republic of Germany

"My interest is in the future because I am going to spend the rest of my life there"
(Charles F. Kettering)

1. The brave new Multi-Media world

"Multi-media" is one of the fashionable catchwords which is used in many ways, either as an indication of the dawning of a great time where we reach the land of milk and honey or on the contrary where it seems to be the end of the world, actually a late "Orwell 1984".

Multi-media is the ultimate state of communication. It is the interaction of a human being with either one or a group of other human beings or a machine, a computer, anywhere on the globe, close or far, exchanging or retrieving information by means of a dialogue. In a multi-media communication, information is transferred in an entirety of unprecedented completeness by combining visual, audio, oral, and manual elements such as still or moving pictures, graphics, movies, sound, music, data and any combination of them.

This paper will emphasise on services of a potential future mobile satellite communications market, discuss the associated user terminals as well as the corresponding systems which will provide the transmission infrastructure for a multi-media environment in a moving vehicle. Compared to the throughput capacity of a glassfibre link a satellite can hardly compete, but in contributing the genuine advantages of satellites "mobility, broadcast and wide area coverage" it becomes obvious that satellites will play an important role by complementing the terrestrial broadband networks.

The provision of multi-media services in a mobile environment merges several formerly independent markets: computing, (mobile) communications, photography, video and audio processing and presentation, music production. Mobile communications will soon become a multi-media environment which will be no longer limited to two-way voice and low rate data services. The multi-media environment will extend the provision of several high-rate audio, video and data services which are presently only available through fixed networks.

The term "multi-media", originating from a personal computer environment, normally requires a two-way communication channel. However, this definition may be extended so that it also includes a pseudo or virtual return channel. For instance, rather than retrieving information from a mailbox or database by individual dialogues, it can be more efficient to broadcast such information as long as it is of potential interest to many other subscribers.

The operation is comparable to the Videotext services but due to much higher bitrates of up to a few hundred kbit/s it results in much higher update rates and information transfer. The virtual dialogue occurs only between the subscriber and its local terminal, which selectively stores and displays only those parts of the bitstream which are presently of relevance to him. If an additional on-line dialogue becomes necessary, it may flow through a regular (mobile) telephone.

2. Transmission requirements for multi-media services

The necessary infrastructure to cope with the multi-media services will comprise several systems, ground and space based, each optimised for a special group of services but mutually complementary. It includes:

- High bitrate services: 10's to 100's of Mbit/s for TV, trunk telephony, hi-speed LAN/WAN/MAN services. Asynchronous Transfer Modes (ATM) on glassfibre links.

- Medium bitrate services: 64 kbit/s to 10Mbit/s for services in the field of mass file transfer, audio- and data broadcast, but also for mesh and star net configured system topologies (VSATs).

- Low bitrate services: 100's of bit/s to 64kbit/s for voice, still picture, graphics (e.g. meteo information, road maps with current updates on traffic load, detours etc.), navigation, position determination, provision of universal/ local time and date reference.

Within telecommunications the mobile communications has received a major attention with incredible growth rates. For instance, the German GSM networks D1 (DeTeMobil) and D2 (Mannesmann Mobilfunk) alone attracted a subscriber community of nearly 2 million paying customers after a start-up period of less than two years. While the terrestrial mobile networks mainly serve the areas of high subscriber density, satellite based systems such as Globalstar and Iridium extend the coverage and service availability globally.

Although the main category of services to be provided in the initial phase of commercial operation will be voice services, the data services including fax exchange are expected to grow significantly in quantity but also in quality (direct computer retrieval/storage with attached printer/scanner). Electronic mail services will very soon reach the same level of utilisation as the fax services today since it will be possible to communicate directly multi-media material, which has been produced on a PC: documents, data base information, video, graphic and sound objects etc..

In this paper we will concentrate more on the mobile rather than the fixed environment. At a first glance one might assume that only services in the lower bitrate categories seem to be reasonable for mobile applications. This is not necessarily so. Current investigations are heading for a provision of high, medium and low bitrate transmission channel for digital audio and data broadcast to mobile subscriber from terrestrial and space based transmitters. A multiplex with a total uncoded bitrate of more than 2 Mbit/s can be dynamically reconfigured to cope with a mix of services. Each data stream can be individually encoded to provide different grades of service dependent bit error performance.

The catchword here is Digital Audio Broadcasting (DAB$^{®}$). Today, most parameters of the DAB$^{®}$ system are defined (Eureca 147) and a draft version has entered the standardisation process under ETSI [1]. The main parameters defined in the Eureka 147 programme are shown in Table 1. A terrestrial roll-out of the services will commence in 1995 in Denmark, France, Germany, Sweden, Switzerland and the UK. Other European countries will follow.

Although digital broadcasting systems do exist (e.g. Digital Satellite Radio-DSR), new system concepts are under investigations using the subcarrier band of the direct-to-home TV satellites. SES Astra will shortly commence transmissions in ADR (Astra Digital Radio) and Eutelsat will use SCPC techniques (Allsat) to provide digital radio to European households. Due to significant improvements in the area of data reduction strategies (MPEG) the source encoded bitrates for video and audio signals have been reduced to 10Mbit/s and below. Latest reports from Intelsat and other operators anticipate bitrates as low as 6Mbit/s sufficient for the transmission of TV signal for home quality. Similar reductions are reported from the audio world. Here, 128 kbit/s for a full stereo audio signal in near CD-quality can be considered state-of-the-art.

However, all present transmissions from geostationary platforms are provided in the Ku-Band under elevation angles of less than 30° in the central and northern European latitudes. None of these systems have been designed or optimised for the fading characteristics of mobile propagation channel.

Several options exist to overcome the special deficiencies of a typical mobile channel, namely the dynamic frequency selective fading due to multipath propagation. One solution is to spread the transmitted signals over a larger bandwidth (spread spectrum). A second possibility is to increase the coding effort and use sophistication in the user terminal. For instance, in the case of DAB, the encoded audio and data signals are modulated unto over hundreds of subcarriers to achieve a frequency spreading and interleaving to provide a time spreading (e.g. 1536 carriers in Mode I and 192 in Mode III) at a price of higher complexity. Each carrier is modulated with a differential 4-PSK modulation. The coding and modulation scheme of the entire multiplex is called Coded Orthogonal Frequency Division Multiplex (COFDM).

A third option to increase the mobile channel performance is to increase the probability of direct line of sight conditions. This can be achieved by placing the transmitter almost perpendicular above the service area. This conclusion has led to a concept named Archimedes which has been investigated under ESA sponsorship in the last couple of years. Archimedes is a constellation based on elliptical orbit planes 63.4° inclined against the equator (HEO). These orbit constellations can provide apogees at high latitudes over northern latitudes.

Due to the non-synchronous movement of the satellite with respect to the earth, more than one satellite is necessary to provide a continuous service. Continuity is achieved by placing fellow satellites on appropriate orbits making sure by smooth hand-over that always one active satellite maintains in view of the service area. After one orbital period, each satellite will illuminate different areas of the earth from different apogee positions. Appropriate design of orbital parameters leads to three service areas in Europe, East Asia and North America. Therefore, a system like Archimedes would not only provide a platform for innovative mobile services but would in addition offer a capacity sharing among the three main economic regions of the world.

The satellite remains for several hours around the apogee in a quasi-geostationary position relative to the earth. During this period the satellite shows a characteristic similar to a geostationary satellite but, different to this, illuminates the northern service areas under very high elevation angles of between 50° and 90°. Different to circular orbit constellations (LEO/MEO) such as for instance Globalstar, Odyssey and Iridium, HEO constellations are far better of for serving dedicated service areas. A brief description of the actual Archimedes concept, as it is under investigation in a DASA led study group, will follow in section 5.

The European standard for digital audio broadcasting (Eureka 147- DAB[®]) has been developed in three modes to cope with different propagation conditions for terrestrial and space systems applications. While Modes I and II are considered for terrestrial transmissions, Mode III considers the satellite propagation environment. However, user terminals shall be capable of decoding all modes. Therefore, a future radio subscriber can select from a variety of audio programmes and data services offered by local, regional and European programmers. Further details on the parameter variation for the Modes are shown in Table 1. DAB[®] actually spreads a single broadcast or data channel over a bandwidth of 1.536 MHz.

Bandwidth is an extremely precious, because limited, resource. Therefore, several broadcast   and data channels are grouped into a so-called multiplex (block, ensemble) in order to maintain a bandwidth efficiency at least equal to that of VHF-FM transmissions. Each multiplex can comprise between six and for instance 18 sound programmes or the equivalent in data services, depending on the required sound quality is highest, i.e. 192 kbit/s, for a broadcast of six stereo programmes in full CD quality or lowest, with 64 kbit/s but for 18 programmes.

It is one of the most prominent advantages of the DAB® standard that the instantaneous composition of the multiplex remains flexible, in other words its actual channel assignment can be changed by shortest notice. This allows very innovative services and programmes. For instance, programmes, where the voice contributions are transmitted separately, say in three different languages at a bitrate of 64kbit/s each, will join the channel capacities for (CD-) music parts by using the combined bitrate capacity of 192kbit/s. Programmes of this type are already under evaluation for pan-European broadcast services (European Digital Radio - EDR).

Parameter	Mode I	Mode II	Mode III	Remark
T_F	96 msec	24msec	24msec	Frame duration
T_{Null}	1296,9 μsec	324,2 μsec	168,4 μsec	Null symbol duration
T_S	1246,1 μsec	311,52 μsec	155,76 μsec	Overall symbol duration
t_S	1000 μsec	250 μsec	125 μsec	useful symbol duration
$\Delta = T_S - t_S$	246,1 μsec	61,52 μsec	30,76 μsec	guard intervall duration
J	76 (+1)	76 (+1)	153 (+1)	no. of symbols per frame
J_{data}	72	72	144	no. of data symbols per frame
N_{max}	2048	512	256	max. no. of carrier
N_{data}	1536	384	192	used no. of carriers
$R_b T_F = 2 J_{data} N_{data}$	221184	55296	55296	no. of data bits per frame
R_b	2,304 Mbit/s	2,304 Mbit/s	2,304 Mbit/s	bit rate (uncoded)
$B = N_{data} / t_S$	1,536 MHz	1,536 MHz	1,536 MHz	bandwidth

Table 1: Digital Audio Broadcasting (DAB®) Modes (Eureka 147)

In any case, independent of the internal constellation, one multiplex will not exceed a raw bitrate of 2.304 Mbit/s. The resulting multiplex signal occupies in any case a bandwidth of 1.536 MHz. This results in an equivalent bandwidth of 256 kHz per programme if e.g. six programmes are transmitted, which leads to the same order of bandwidth required as a present VHF/FM stereo composite signal (300kHz).

Thus, with the appropriate link design the mobile radio terminals will receive this bitrate and will have access to all services in a multiplex. Thus, the multiplex concept provides sufficient flexibility to transmit either several audio programmes of lower (e.g. 64kbit/s) or fewer programmes of higher (192kbit/s) e.g. CD sound quality [1].

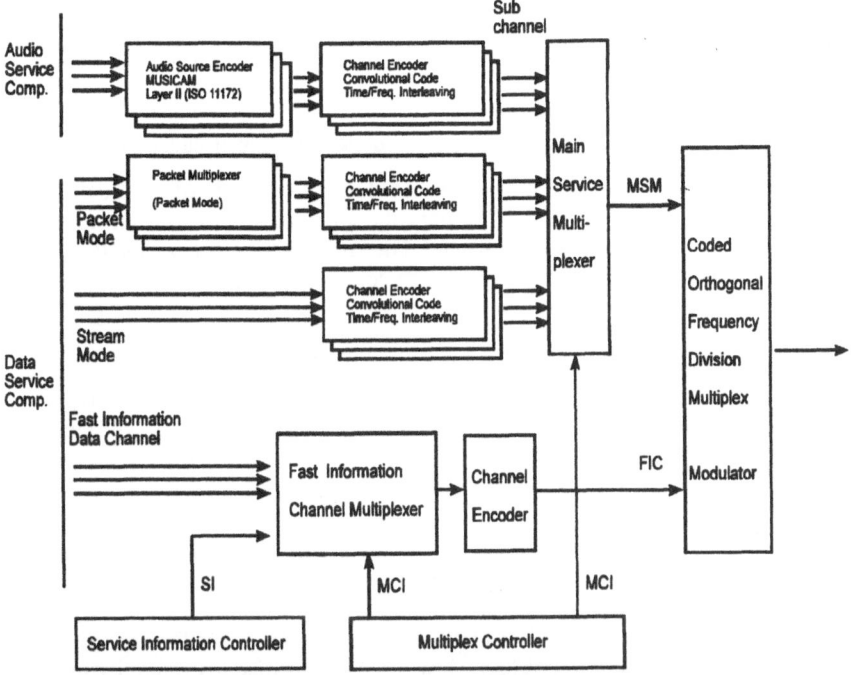

Figure 1: DAB® Multiplex as defined by Eureka 147

Figure 1 shows the various options to construct a multiplex [2]. The option of broadcasting data either permanently in a dedicated channel or during pauses of the main programme or multiplexed into some of the sound programmes opens such a wide field of applications that it will change the traditional way to broadcast radio significantly.

The multiplex controller allows transitions from one service or data flow to another within a six second time intervall. Within this short interval the multiplex can be reconfigured,   either   ad-hoc   or programmable. It is assumed, that this flexibility will be one of the great driver for the acceptance of DAB$^{®}$, which differentiates DAB$^{®}$ from all predecessors including Digital Satellite Radio (DSR).

DAB$^{®}$ also provides robustness in the mobile channel due to time and frequency interleaving to overcome signal interruptions. Unfortunately time interleaving, although very attractive in compensating burst errors, has its limits when it takes too long to change from one programme to another. It would be annoying to a user, when a programme change lasts tens of seconds. Table 2, published by Phillips, shows the main parameter of DAB$^{®}$ in comparison to the traditional FM and DSR systems.

After several years of intensive lab research and field testing it can be concluded that the digital DAB$^{®}$ standard as well as the corresponding hard- and software components have now reached a state of maturity which allows the introduction on a broader scale, the real market place. The introduction of a totally new transmission infrastructure without any upward compatibility from the existing user terminals is a critical phase.

Presently, several pilot networks in the VHF range (220MHz/ TV-Channel 12) and in the L-Band (1.48GHz) are in the phase of implementation and regular transmissions of audio and data services are scheduled to commence in 1995. In the long run, all services will be digital. This trend of all telecommunications services will without doubt also affect the broadcasting services.

Several factors will determine the grade of acceptance as well as the rate of penetration into the existing analogue world: the availability (coverage), the attractiveness of services (no repetition of anyhow existing services), the attractiveness of equipment (handling) and, most of all, the prices for services and equipment. All elements are interrelated: low price means quantity, quantity means attractive services, attractive services means subscriber, subscriber want availability and low price. It is impossible to fulfil all criteria at once. It is therefore expected, that the final transition from the analogue to the fully digital world will take at least ten years or more - but it will come. The contribution from the satellite segment shall at least be the provision of wide coverage.

Table 2 shows the main advantages but also the critical issues, which still require improvement. For instance, the bandwidth efficiency. Major improvements are expected particularly in the area of data reduction. The present state-of-the-art for a transmission of a stereo signal in almost CD quality is reported to be 128kbit/s.

Considering that 256kbit/s were required about three years for the almost the same quality, provides an impression on the speed of development. In this point the standard shows its great strengths in that it remains open for any changes and improvements without requiring a redesign of equipment or components.

Parameter	System		
	VHF/FM + RDS	DSR	DAB
Transmission	terrestrial	satellite/cable network	home/ mobile
Reception	home + mobile	home only	home + mobile
Frequency range	88 - 108 MHz	12 GHz (Ku)	225 MHz/ 1.48GHz
RF sensitivity	$\geq$ 40 dBμV	$\geq$ 48 dBμV	$\geq$ 24 dBμV
IF Frequency	10.7 MHz	118/40 MHz	36MHz (t.b.d.)
IF Bandwidth	300 kHz/ 1 Prg	14MHz/ 16 Prg	1.536MHz/ n Prg
Modulation method	FM + DPSK (RDS)	Diff. 4-PSK	COFDM
Multipath resistance	no	no	yes (single fre. netw.)
Noise protection	no	yes	yes
Audio bandwidth	15 kHz	15 kHz	20 kHz
Audio S/N (1k sine)	60 dB	95 dB	> 100 dB
Audio delay	< 1 msec	4 msec	< 500 msec
Audio processing	analogue	digital	digital
Data processing	digital	digital	digital
Data capacity	1.2 kbit/s	11 kbit/s	$\geq$ 32 kbit/s
Error protection	CRC	Block code	convolutional code
Decoder dissipation	$\leq$ 200 mW	$\leq$ 500 mW	$\leq$ 2500 mW
Decoder complexity	Low	Medium	High

Table 2: Comparison of analogue FM and digital audio broadcast DAB[®] [3]

3. Mobility in a changing world

When trying to estimate the potential needs for future broadcast and data applications as well as services it is important to identify first potential profiles of future customers. So let us briefly look at the changes which are already in progress and try to predict where they will lead. Technology can change the way of living and the way of living inspires technology.

John Naisbitt, a famous American trend analyst, in his book "Megatrends" [4] identified already in 1984 ten major trends which were about to change the economic, social and cultural life in the United States. Today, ten years after, we must admit that most of these predictions did prove to be correct not only for the US but also for Europe, yet more important, they are still valid. Six of the trend predictions can be related to telecommunications or are closely interdependent with the developments in telecommunication characterised by digitalisation and global networking. Those are

- The change from an industrial to an information society => information has become a product.

- The change from forced technology to high tech with a high human touch => people don't buy technology, they buy solutions to problems. Technical products must show an increased user friendliness. Personal computer have almost achieved an equal "home acceptance" as video and audio entertainment equipment.

- The change from national economy to global economy => GATT and the three main economic zones are settled: Nafta, EU, Asean. Communications within and among the economies are becoming more crucial success factors.

- The change from centralisation to decentralisation => "lean" is the new key word in production and management as well as in stronger moves toward federalism. Regions and ethnic groups require more self determination. Global availability and access of computerised Telecommunications Networks (e.g. Internet, CompuServe$^{®}$) supports this trend.

- The change from hierarchy to networking => the change from master-slave to peer-to-peer structures.

- The change from either/or to multiple options => competition is the new buzzword, after termination of more and more monopolies not only in the area of telecommunications.

Decentralisation and global business engagements automatically create mobility. Mobility has become one of the key trends in the late 80s and early 90s. It has created a new way of living in most countries of the former "western" world, but also in Japan and other East Asian countries. Today we see this way of living rapidly floating into eastern European countries and the CIS as well. People accept to move, they spend a fortune to buy brand clothing, fancy vehicles (off-road cars, mountain bikes) and spend their holidays all over the world.

They buy and consume corresponding media with supporting information and entertainment such as specialised magazines, periodicals, and TV programmes.

The walkman was an early indicator for the trend towards mobility and it was sold to the millions. Walkman is not only for entertainment but can also be used for education. It suddenly became possible to combine outdoor activities with business e.g. learning a language, a professional skill or other mental exercises.

After the end of the east-west confrontation and the dawning of new and large potential markets accompanied by technological innovations and improvements the rate of change is accelerating exponentially, shifting so fast that it is sometimes impossible to even make short-term predictions accurately. Probably the best example is the annual progress of performance of personal computers. If I only compare the performance of my present home computer (486/50MHz) to the one I had less than ten years ago, it is difficult to explain to my youngest son (aged 16) what was considered then to be "fast" or "mass memory".

Considering what happened only in the last five years in the computer and telecommunications industry, it is almost impossible to extrapolate this to the next five years. Simply because the process of diversification and innovation of services has just begun, according to the announcements of the main players in this business. Liberalised markets, together with the termination of voice monopolies in many European countries in 1998 will create business opportunities which are even unknown today.

In this light it is no surprise that statistics found out that in another ten years from now, at least 25% of all current "knowledge" and accepted "practices" will be obsolete. This may be even less in the arena of software and digital processing for data reduction, protocol development, and so on. The overall performance of the hard- and software allows to combine and process in real-time data, voice and video signals in multi tasking environments at lowest costs with a global applicability.

4. Multi-media mobile services

As a matter of fact, many of the multi-media features mentioned above, are already available in present versions of workstations and high-end personal computer. Most of the required technologies, hardware and software, have already reached a surprisingly high state of maturity. With appropriate PC-cards very comfortable audio, video and data processing can be performed. With presently available hard- and software, a Personal Computer can be turned into a TV-, radio-, or game terminal, a data communications terminal connecting the user to the world (e.g. via Internet, CompuServe etc.), a sound processor for a musical instrument (or perform like a musical instrument), a video processor creating animated slide shows or home-made movies (including morphing and other effects) or a designer place for technical, architectural and artist's layouts.

It is therefore not difficult to predict, that future "Media Terminals" will be such a computer, where only the loaded software and the connected periphery determines its actual function.

The prices for multi-media sound and video components almost collapsed due to the size of the global market driven by fierce competition. This includes elements such as high speed CPU's, CD-ROMs, true colour high resolution screens, mass storage devices, sound and video processing components and sophisticated interface cards ranging from serial/ parallel data interfaces which can be connected via modem to a telephone network or directly via ISDN to a LAN- or to an air interface to receive terrestrial, satellite or cable TV/audio/data broadcast services.

There is still room for improvement: lower power consumption, smaller boxes and user friendly keyboards, while at the same time increasing processing power. It seems that the technology no longer is the bottleneck. What really is required to receive more attention is the development of services and applications as well as increased network capacity to cater for a very fast growing user community.

With the advent of digital services in the mobile arena by the introduction of GSM, new networks became available which allowed to offer almost all services of the fixed networks also to the mobile environment. Thus, differences in "mobile" or "fixed" services become less and less important. Only the limitations in the provision of bandwidth make adaptations necessary, e.g. taking into account on-going developments in data reduction. Converting an ISDN voice signal at 64kbit/s into a 13kbit/s or less to become more bandwidth efficient in the air interface.

So, let us briefly look at the categories of bitrates required for given types of services. For instance the voice telephony services in the ISDN regime are transmitted in 64kbit/s due to the 8kHz sampling rate at an 8bit per sample resolution. In the GSM channel, source encoded voice is transmitted at 13kbit/s (6.5kbit/s half rate) and, for instance, in Globalstar with a dynamic voice bitrate of between 1.2 and 9.6kbit/s, voice activity controlled.

In the case of video transmission, bitrates may vary between tens and hundreds of Mbit/s. However, video telephones are already available at bitrates of as low as 64kbit/s and less than 20kbit/s for special applications. The present figure for planning a next generation TV broadcast is 6 Mbit/s for a good TV signal. Here, data reduction (MPEG) is the key for an effective multiplication of TV broadcast channel without increasing the transmission channel capacity.

What are the kind of services expected in a mobile multi-media environment ?

Strictly speaking, multi-media services do require a (real-time) response channel. However, forward and return link may need not to follow the same paths in both directions. So, they can have different bitrates, different information transferred (e.g. data one way, audio return) or even a different propagation medium. While listening to the radio programme one could also respond via the (mobile) telephone to announce a selection in case of menus, a decision in case of polling, a response in case of raised questions.

Once the broadcast services are offered to the mobile subscriber, it goes without saying that inherently also fixed subscriber are potential customer.

Since the mobile telephone networks are anyhow optimised to provide mobile services at lowest costs, it would make little sense commercially when trying to implement a return channel into a system which is optimised for broadcasting. Therefore, it is assumed that a typical infrastructure in a mobile terminal (car, travel bus, train, etc.)

The range of services will comprise single traditional services as well as combinations:

- **Voice, digitally encoded**

  real-time telephony, non-real time voice mail, vocoded voice, artificial voice, medium quality voice (news) including sound processing and editing

- **Sound**

  Entertaining music in CD quality, accompanying sound to a picture or TV signal, emergency warning or information messages, creation of different warning tones, etc..

- **Picture**

  Graphics, animated graphics showing a process, illustrating statements by animated graphics, slide show for entertainment and advertising, side information to the on-going programme, simulation of a scenario, dynamically updated road maps with traffic information, cyber space games and applications, lotteries, transmitting information files, drawings, tables, etc. to many remote plants/ construction sites etc. True colour TV pictures including video processing and editing for video telephone, video conferencing, video-on-demand showing e.g. evening events upon arrival in a city.

- **Position Determination, Time, Date and Synch Services**

  Navigation information, time, date, reference signal of high precision for system which run more efficient in a synchronised mode (monitoring, in-situ measurements, global TV synchronisation)

- **Emergency Data Broadcast**

  Searching people, specialists, equipment, blood and organ donators, help, advise

- **Mix of Services**

  Combinations of these services really result in a multi-media session. Jointly working on a single paper, table or drawing but from two or more locations simultaneously. Everybody can see each other and discuss/comment the on-going process. All may have access to a common data base where further reference material is retrieved, while two of the communicators go on completing the paper, the third party can search for supporting material. This will lead to totally new experiences offering unlimited opportunities.

But not only multi-directional, multi-party services will be developed. Also innovations in the broadcast services are expected. Future broadcasters will utilise and combine more media components than today. To give a few examples: Could you imagine these situations where for instance data is associated to audio programmes ?

- Broadcast programme for product information (what's new ?): while listening to the presentation of the products you receive data information about technical details, availability, accessories, places where to buy etc. In a mobile environment, where to find the nearest repair shop, gas station, restaurant, hotel etc.

- Broadcast of special music with techno effects (light, smell effects): while listening to the presenter or cyber jockey (moderator) you receive side information about the music, messages, further data, photos, graphics, simulations, animations, etc.. downloaded into a cyber helmet over your head.. Everybody can have the feeling of diving, flying, space walking, sky diving, mountain climbing, car racing, etc. in his virtual reality world.

- Broadcast for contemplation and relaxation: you listen to the music accompanied by a slide show of appropriate photographs. This can also include broadcast programmes for various cultural interests (hobbies, religious programmes, cooking advise, gardening, etc.): while listening to the presentations you receive additional graphic/ data information.

- Broadcast of sport events (soccer, gymnastics, tennis, skiing..) with simultaneous provision of current or final results of the other activities, particularly during big sport events (Olympics, World/ European/ National Championships, ..).

- Multi-lingual programme broadcast on three simultaneous low bitrate channel, which are combined the next hour for music transmission in CD-quality.

- Broadcast programme for computer software (entertainment, business, information, social, cultural, artists, designer software): While or shortly before the presentation of new software products, the new (demo) version is transmitted under shareware conditions, similar to sampler CD's. With the software available in the resident computer, one can follow the explanations and recommendations of the presenter by testing everything on your own computer. With the addition of life music performance, winning games, competitions or other entertainment attractions the software shows will probably have a good score in audience appreciation. If you like the received software programme, just call   the producer or dealer, pay by credit card and receive the authorisation number to use the programme right away. This   service alone can change the distribution of software significantly. In case of broadcast through a continental satellite system, e.g. Archimedes, this programme could potentially be received by 35 million people Europe wide.

- Broadcasting for education: programmes and courses, software supported. A fireworks of new options.

As said before, "Return channel" are possible through the computer networks (Internet®, CompuServe® and other commercial or amateur packet radio networks). From everywhere a written response (E-mail) or a direct connect is possible to the originator of the programme. Recent service offerings for software broadcast (Videodat® via Astra PRO7 transponder), financial information (Reuters 1000 via Astra CNN transponder) are first examples of new personal satellite data services. To join this service, subscriber have to buy dedicated decoder PC-cards, which are complete TV receivers, except the display system, with additional data decoder and descrambler. These cards, in addition, provide full TV display under Windows® or equivalent user surfaces. Similar card systems can be used in the mobile environment.

Once the transmission medium is available the growth in quantity but also in quality of programmes will create a totally new generation of broadcasters and "listeners". Listening (communicating) with almost all of your senses. Several of the above services are already offered via geostationary satellites and cable networks. The extension into the mobile environment with special services for mobile users (trucker programme, busses, ships..) will add further applications and customers.

5. Provision of Multi-media Services to Mobile Subscriber via Satellite

Archimedes System constellation

The provision of communications services to mobile subscriber requires a special space segment, particularly if high bitrate and real-time services are to be provided. Particularly for countries located in central and northern European latitudes, geostationary satellites are not considered useful for mobile applications in general due to the resulting low elevation angles.

It is, however, still an open question which elevation angle is the best for a mobile reception. "Mobile" is understood in the sense that the "terminal is on the move", i.e. not only transportable from one place to another. This terminal can be an integral part of a car, a travel bus, a train, ship, plane etc.. Under no circumstances

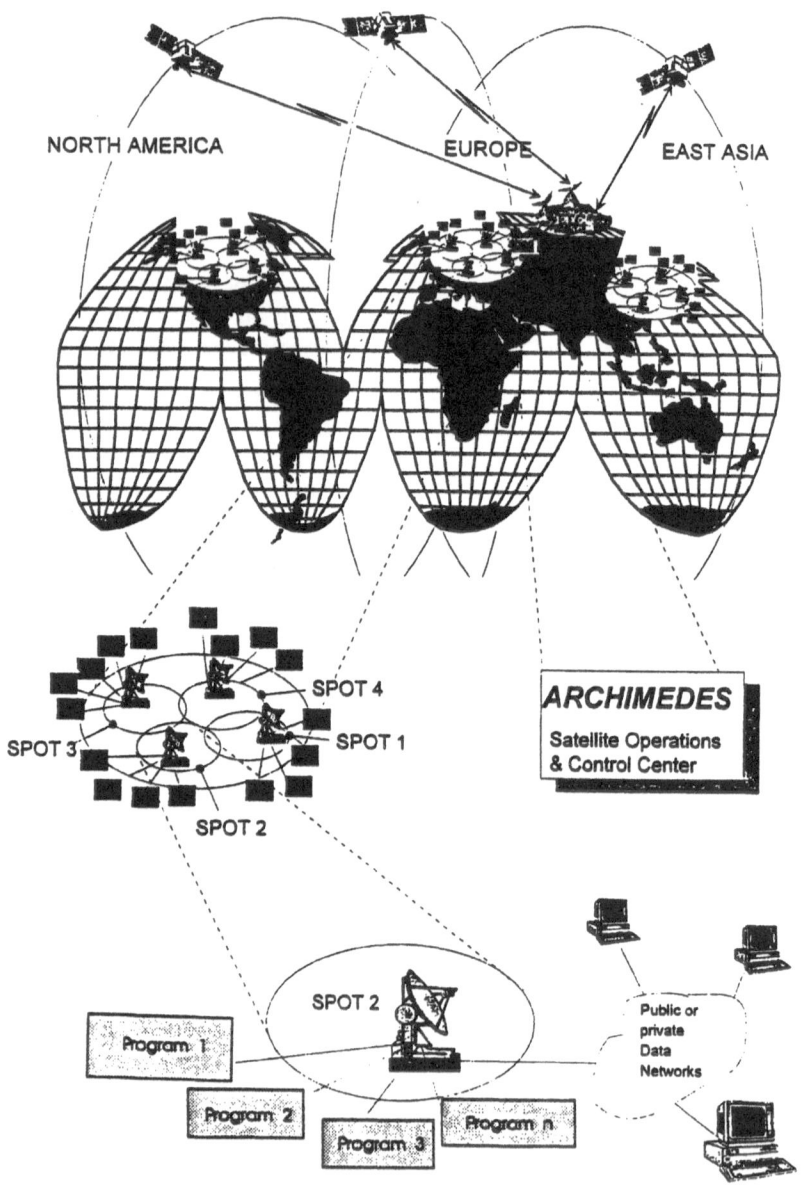

Figure 2: The **Archimedes** Audio and Data Broadcast system concept
(DASA Version)

can an availability comparable to a fixed installation be guaranteed. The actual performance of the link highly depends on many parameters which are beyond the control of a system or network operator. In the area of the Mobile Satellite Services a certain user co-operation in placing or holding the terminal is assumed, since otherwise there would be no margin sufficient to compensate all possible detrimental effects.

Even in the terrestrial GSM networks, there are still many areas with critical field strength', normally leading to an interruption of service if not corrected in time by the user. In the case of the Archimedes space segment, an elevation angle of 50° has been identified as a fair compromise between a large service area, the quality of service and considerations for efficient orbit maintenance over a ten year operating lifetime. To provide such high elevation angles continuously, a constellation of six satellites has to be implemented as a minimum constellation.

Each satellite travelling on an elliptically shaped orbital plane with an apogee of 26000km above the ground (63.4° northern latitude) and a perigee of 1000km. The orbiting period of each satellite is 8 hours, of which 4 hours are operating time during the so-called active arc, while the other 4 hours are used to re-adjust and prepare the attitude towards the new service area. After each turnaround the satellite will see the next of three service areas in sequence: Europe, East Asia, and North America. Figure 2 shows the selected system constellation for Archimedes.

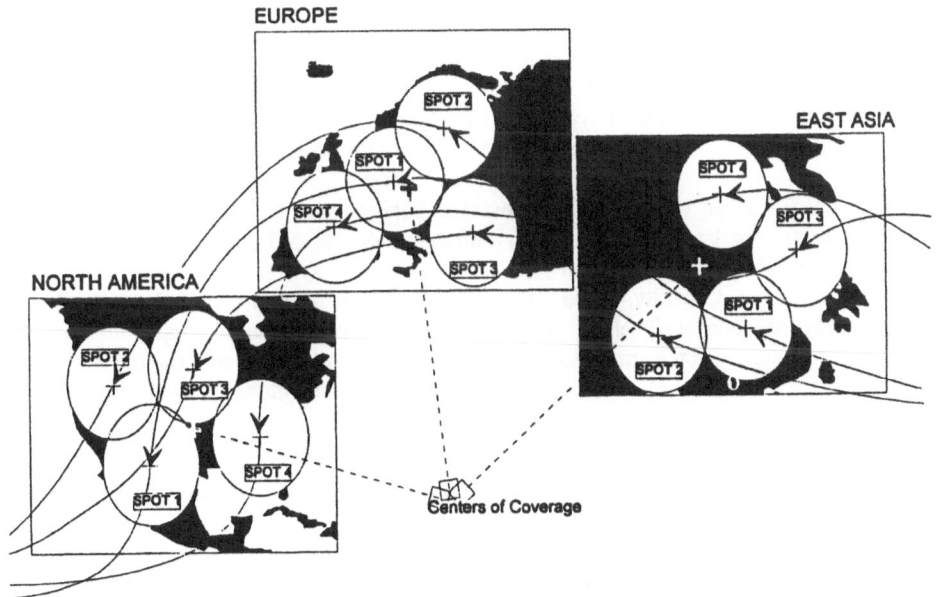

Figure 3: Moving spot beams to match different coverage requirements

The system has been designed to provide digital audio and high-rate data broadcasting services based on the DAB standard as explained before. The payload comprises four spot beams and a continental beam in the assigned frequency band 1452-1492MHz (WARC '92). The same band will be shared with the terrestrial DAB® network. One feeder station in each beam and one station in the continental beam is foreseen to create the multiplex and uplink the programmes and services which are destined for the associated service area covered by the beam. The contributions from broadcaster and service provider will be routed through the existing networks i.e. the PSTN, ISDN as well as private and public digital data networks. The control of the multiplex as well as the operation of the feeder station may be remotely controlled from a central network operation control station.

Figure 3 illustrates the need for electronically steerable beams, due to different shapes of the service areas in the three regions. The flexibility also allows to cope with future amendments or changes in the areas to be served. All satellites will be totally equal in performance to ensure a cost effective production. The planned begin of operation for a commercial Archimedes system will be summer 1999, just before the opening of the new millennium and is designed to provide the services required in the 21st century.

Terminal Equipment

The terminal equipment for DAB® services differs from traditional receiver since more sophistication and processing power is needed. The most critical VLSIs, however, required for the terminals have already been developed under the auspices of the European technology programme JESSI. As a consequence, the first field trials with near consumer type terminals (with a subsidised price) will commence end of this year with several European countries following in 1995.

Except for some very special processes at highest speed (FFT) most of the other functions can be performed under software. Hence, a Personal Computer or a derivative (e.g. lap- and palm top) may very well serve as a core for a dedicated user terminal. Capacity, speed and graphic video devices of standard devices have already reached a level of maturity that could provide a total multi-media infrastructure. Such a computer equipped with a sound device, and the mass storage facilities, including CD-ROM, can offer an entire telecommunication terminal for all services.

Instead of having a frequency display to dial the stations, DAB® terminals will have either a simple digital liquid crystal or plasma display or the more expensive an CRT display. Special versions will be integrated into cars. Here, the state-of-the-art might soon unify the on-board car computer with the telephone, radio, navigation and co-pilot devices integrated into the dashboard. All elements are interconnected over a resident optical Local Area Network. Interaction of the driver becomes possible with a set of "output" devices including video, voice messages, voice dialogue and warning tones. Figure 4 shows the trends in audio/video systems built into cars. Several of these items are already available in series cars today, most of which are sold to Japan.

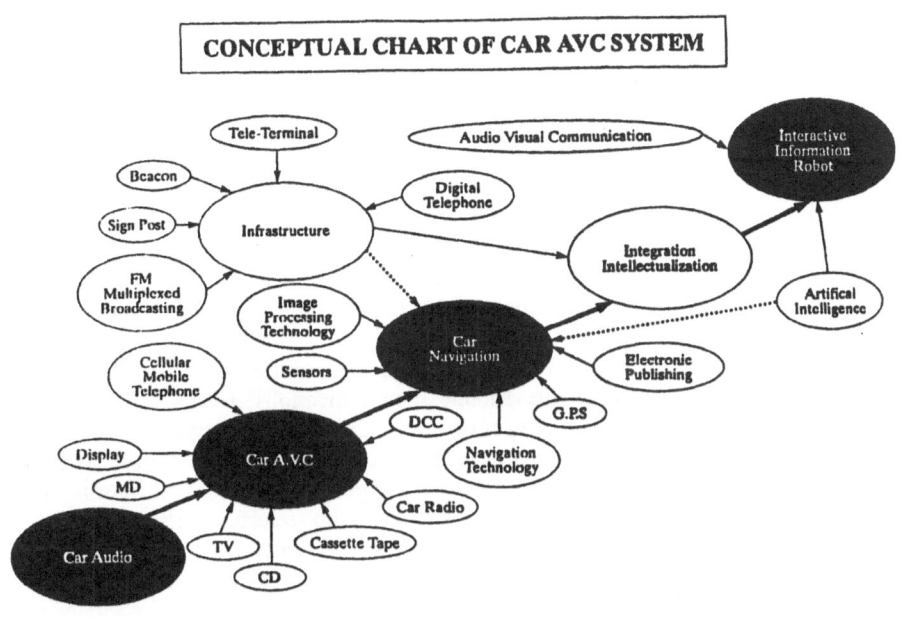

Figure 4: Trends of car audio-visual communication systems and services

Another special version of the user terminal will be the truck terminal. This terminal may in addition run special software with information for fleet management, truck status and others. Also special terminals for other vehicles (ship, trains, aeroplanes, pleasure busses etc.) are possible. Equipment will be modular with the central unit somewhere integrated into the car. An antenna unit provides antennas, LNAs and converter for all frequency ranges necessary and the functions are loaded by software to turn the computer into a telephone, a radio or any other device required.

Finally, one of the newest elements in the user terminal will be the slot for the insertion of a chipcard to provide a facility for pay services. A billing concept is presently under investigation. The key words here are pay radio and pay data services, the „dark sides" of an otherwise shiny and challenging future.

6. Conclusions and recommendations

This paper highlights the technical and business opportunities of a future satellite based telecommunications infrastructure for mobile (and fixed) services. Several systems in the personal mobile arena are seriously announcing to commence operations before the end of this decade, two of them with major European participation and contents Globalstar and Odyssey. There is a good probability that, despite some remaining critical issues in all systems one or two system eventually will survive and serve a large market.

Digital Audio Broadcasting combined with the option of a wide range of data services, particularly value added services have been identified, which give a first estimate on the potential. In conclusion it can be said, that we are presently at the beginning of multi-media services markets. The implementation of new systems and innovative services will create many opportunities for value added service provider, will therefore not only create a market for services but also a market for entrepreneurs, thus will create a great number of jobs in all three service areas.

What has not been discussed, but is of similar importance is the impact on our social, political and cultural life. Mobile communications in general, but data broadcasting in particular will bring people and nations closer together and it will also provide a very powerful means of bringing information to the people. With a data broadcast system, everybody would be in the position to access all information with a small terminal in his/her hand. But, who selects, edits, comments, intentionally or unintentionally filters the information which is then transmitted ? Probably it is worthwhile to consider for the next conference to dedicate a session for the discussion on the social impacts and its consequences for the societies in Europe and internationally ? Research should be initiated on the potential consequences and how to cope with them.

It is not so much that we can or should avoid that a system will be implemented, this will be done anyhow, if it proves to be a good business. But we should understand the mechanics to become sensitive for potential abuse. We should investigate pros and cons, dangers and challenges, to avoid that the negative impacts overcompensate the positive intentions. A broader discussion in the societies should commence including all relevant groups of a society to avoid that the great potentials in all the systems to come will not serve the evil, but will improve the quality of life for as many people as possible.

A high rate data distribution system like e.g. Archimedes providing such a wide coverage will introduce a new era of the information age on the verge to 21st century. More than Astra which, due its European wide coverage, has performed a major step towards Europe. European wide programmes, such as MTV are seen by many young people all over Europe.

7. Acknowledgements

The support of ESTEC, Alcatel Radiotelephone, DLR, DASA-ERNO, Dornier, and Space Engineering who form the team for the Archimedes study as well as the discussions I had with my colleagues at the Mercedes Benz AG are kindly appreciated.

8. References

[1] Draft European Telecommunication Standard prETS 300 401: "Radio Broadcast Systems: Digital Audio Broadcasting (DAB$^{®}$) to mobile, portable and fixed Receivers", ETSI, Sophia Antipolis, France

[2] A. Müller "Dynamic multiplexing for integrated services digital broadcasting to mobile receivers", 2nd International Symposium on DAB$^{®}$, March 14-17, 1994 Toronto, Canada

[3] F. van de Laar, N. Philips, R. Olde Dubbelink "General-Purpose and application specific design of a DAB$^{®}$ channel-decoder"; EBU Technical Review, Winter 1993

[4] John Naisbitt "Megatrends - Ten New Directions Transforming our Lives", Warner Books, USA; ISBN 0-446-35681-6

An Overview of CDMA Techniques for Mobile and Personal Satellite Communications

R. De Gaudenzi[1], T. Garde[1], F. Giannetti[2], M. Luise[2]

[1] European Space Agency
European Space Research and Technology Centre ESTEC
P.O. Box 299, 2200 AG Noordwijk, The Netherlands

[2] University of Pisa
Dipartimento di Ingegneria della Informazione
Via Diotisalvi 2, 56126 Pisa, Italy

Abstract. The recent application of Code Division Multiple Access (CDMA) to commercial terrestrial radio networks suggested the possibility of extending the same access technique to high-efficiency low-cost satellite systems. A great research and development effort has therefore started, and is still in progress, to fulfill the requirements of modern commercial satellite networks, namely: low-cost user equipments on one side and power and spectral efficiency on the other. The aim of this contribution is to address the main topics in the development of such systems for satellite-based mobile and personal communications. More specifically, we investigate first the inherent trade-off between coding and spreading, then we touch upon the techniques for the minimization of the self-noise effect, and the related issues of power control, multiuser detection and satellite diversity. As a key factor in the successful deployment of CDMA networks, we shortly review some technological aspects related to an efficient modem design via digital signal processing techniques, and the possible consequent low-cost, small-size Application-Specific Integrated Circuit (ASIC) implementation. The final part of the paper deals with some system-related issues, like the effect of the non-linear satellite transponder and the mobile multipath channel on the multiplexed user signal, and highlights the applicability of CDMA to the emerging new-generation services, such as Digital Audio Broadcasting (DAB), wideband mobile networks, and the like.

1 Why CDMA ?

"Personal Communications" has become in these days one of the keywords in the area of radio communications. The worldwide unexpected growth of traditional mobile voice communication services experienced in the last few years let us envisage a similar trend for the forthcoming Personal Communications Network (PCN), so that a tremendous research and development effort is in progress in the field [1]. It should be clear by now that such an ambitious goal calls for the integration of different heterogeneous radio networks, extending and supplementing the traditional terrestrial cellular systems [2] with ubiquitous satellite networks and indoor systems [3]. One of the key factors in attaining such a high level of integration is the unification and standardization of the radio interface, starting from the modulation and multiple access techniques. In these respect, one of the most promising modulation and access technique is (coded) Band-Limited (BL) Code Division Multiple Access (CDMA), that appears to be particularly suited both for satellite [4], [5] and for indoor/terrestrial [6], [7] networks.

The aim of this contribution is to review some of the peculiar features of BL Direct-Sequence Spread-Spectrum (DS/SS) transmission with Code Division Multiple Access for satellite-based networks, and to show how those characteristics make CDMA well suited to a number of different application areas, ranging from traditional mobile communications to broadcasting of audio programmes. After the introduction contained in this Section, we recall in Section 2 the main modulation and coding techniques currently employed in CDMA systems, and we discuss the relevant design options to attain to a maximum power and spectrum link efficiency. The "Achille's Heel" of CDMA systems, namely the need of tight power control, is dealt with in Section 3, contrasting the different requirements of terrestrial and satellite networks, while in Section 4 we tackle the pivotal issue of low-cost design and VLSI implementation of the CDMA modem. Both aspects need further research effort, since currently operating CDMA systems were mainly restricted to military and professional applications, where cost and capacity were not the main concerns. Towards system optimization, the concept of high-efficiency multiuser signal detection for asynchronous CDMA is revised in Section 5, showing how to improve the overall system performance at the expense of increased receiver computing power. The impairments induced by the distortions experienced by the transmitted signal during radio propagation and satellite transponding are discussed in Section 6, while the possibility to enhance the receiver performance through appropriate use of space/time diversity techniques is highlighted in Section 7. The customary "Conclusions" Section ends up the paper with a brief summary of the expected application areas of satellite CDMA, as resulting from the body of the previous Sections.

2 Coding, Modulation and Spreading for CDMA

The optimum coding, modulation and spreading techniques for CDMA reveals very much related to the specific system characteristics. In fact, the overall link performance turns out to be strongly dependent on three major system features, such as coherent or non-coherent signal detection, power and/or bandwidth limitations and asynchronous (A-CDMA) or synchronous (S-CDMA) access scheme, that are reviewed in the following.

2.1 Coherently Detected Asynchronous CDMA

This is the most classical scheme considered in the literature. Originally conceived to be based on binary direct-sequence spread-spectrum techniques, A-CDMA signals are generally detected by a coherent demodulator. It turns out that on the AWGN channel the spreading and despreading operations are perfectly transparent from the inner modulation and coding viewpoint, thus the very same error-protection coding techniques devised for coherent BPSK might be used. Clearly, for a given spreading sequence chip rate R_c and information bit rate R_b, the so-called *spreading factor* $M \triangleq R_c/R_s$ (representing the ratio between the spread and un-spread signal bandwidths, respectively) decreases together with the coding rate $r \triangleq R_b/R_s$, R_s being the coded symbol rate. Conversely, for a given bit rate, the actual chip rate and bandwidth occupancy will increase when r decreases, assuming that the length L of the spreading chip sequence (spreading code) is fixed and spans exactly a symbol period. The latter condition is often required to ensure that the cross-correlation properties of the user spreading codes are not affected by data modulation. This is actually achieved by imposing further synchronization constraints between the spreading sequence start epoch and the symbol clock strobe.

Assuming that the CDMA interference is generated by non-time-coordinated users, optimum coding is provided by convolutional codes (CC) [15], trellis-coded (TC) modulation of similar complexity representing a valid alternative for mild user-loading conditions only. Viterbi [11] has shown that for asynchronous CDMA systems the ultimate capacity performance can be achieved by utilizing very low rate codes generating appropriate orthogonal Walsh-Hadamard (WH) sequences. As a limit case, the spreading function could be simply an overlay PN sequence having the same rate as the coded symbols and ensuring user orthogonality. Although intriguing from the theoretical point of view, such a scheme has not found any practical application yet due to the complexity in the implementation of the decoder.

2.2 Non-Coherently Detected Asynchronous CDMA

Although widely used for system analysis and performance comparison, coherent detection is viable in a limited subset of practical situations only. CDMA satellite systems are generally used for medium-to-low data rates (say less than 64 kb/s) and often with mobile and portable terminals equipped with a small aerial. This is due to the well known high immunity to jamming and narrow-band interference and to the limited power flux density signal emission provided by CDMA systems. When operating at high carrier frequencies, such as Ku- and Ka-band, the up- and down-conversion stages in the mobile terminal are affected by high levels of phase noise. This makes the design of a low-cost equipment operating at low data rates with coherent detection a difficult task. Similar considerations hold for the inbound link of a mobile system affected by a faded modulated carrier. Phase estimation and tracking for a fast fading satellite channel results in a formidable task for the demodulator. The sparse user location in the inbound link also hinders utilization of DS/SS pilot-aided carrier phase estimation techniques. Under those conditions, robust demodulator operation requires differential or non-coherent signal detection. Restricting our attention to DS/SS systems, we can state that the considerations made by Clark and Cain in

[12] about error-protection coding for differential BPSK (D-BPSK) also apply to CDMA signals. If the carrier phase, due to fast signal fading, varies considerably over a two-bit period (thus impairing even a bit-level differential detection of the un-spread signal), it may result almost constant over a *two-chip* period, thus making a chip-level differential detection extremely effective. A further possible approach, recently proposed by Qualcomm for the Telecommunications Industry Association (TIA) Cellular CDMA Interim Standard [14], resides in utilizing an M-ary orthogonal signaling scheme with a non-coherent maximum-likelihood envelope detector [13]. A particularly efficient solution in this respect is again the use of Walsh-Hadamard (WH) sequences [18] as an orthogonal signal set [14] which can be efficiently detected by non-coherent digital techniques based on the inverse fast Hadamard transform. An overlay (long) pseudo-noise PN sequence whose chip rate is an integer multiple of the WH symbol rate is also superimposed to characterize the different users (Code Division). This approach is also being pursued for the Globalstar Common Air Interface [19]. The main advantage of that scheme lies in its enhanced bit error rate performance when compared to D-BPSK in the AWGN channel (3 dB gain at $P_b = 10^{-3}$). The inherent performance advantage is however reduced to 1 dB only in the Rayleigh fading channel. The proposed TIA CDMA cellular system mobile-to-cell link employs the optimum[3] rate 1/3, $K = 7$, convolutional code with interleaving and de-interleaving to further enhance power efficiency. This sub-optimum coding scheme provides performances very close to optimal coding techniques for M-ary WH modulation discussed in [12] with much lower complexity.

2.3 Coherently Detected Synchronous CDMA

The demand for enhanced performance over the bandwidth- and power-limited satellite channel has recently motivated the introduction of practical techniques to increase the overall efficiency of CDMA, up to the values of traditional FDMA. Several patents have been independently filed in the United States [8] and in Europe [9], [10]. The underlying idea is to exploit the inherent signal co-location in the outbound link to synchronize the spreading sequences start epoch. By so doing, the problem related to the presence of co-channel interferers can be cut straight by appropriate spreading techniques, so that perfect "isolation" among the different user signals can be achieved. Reference [8] describes a double spreading solution, now adopted by the TIA CDMA Cellular Interim Standard [14], consisting in the cascade of an inner spreading with WH orthogonal sequences, covered by an outer longer PN sequence for the randomization of the resulting cross-correlation properties when (asynchronous) offset echoes are received due to multipath reflections or co-frequency signals belonging to different satellite(s) (beams). Such an approach represents also a solution to spreading sequence acquisition and tracking, which otherwise would result difficult in the presence of WH sequences only. Reference [9] describes a simpler, but slightly sub-optimum, solution to the same problem based on the well-known Gold codes. Exact signal

---

[3] In the free-distance sense.

orthogonality is not guaranteed by this approach [4] but the resulting capacity loss is negligible [32]. This system, named Band-Limited Quasi-Synchronous CDMA (BLQS-CDMA), was successfully validated at L- and Ku-band and has been adopted by the ESA's Mobile Satellite Business Network [16] and Ku-band CDMA Network [17]. Satellite field trials at L-band with vehicular mobile terminals have demonstrated that inbound link synchronization might also be achieved. The customary interference rejection, multipath fading resistance and low power flux density emission features of S-CDMA go in this case together with the same performance and efficiency as ideal FDMA on the AWGN channel. Best signal protection against nonlinear channel distortions and possible demodulation phase offset is attained by the use of two independent I-Q spreading sequences. Since synchronous (quasi)-orthogonal CDMA systems are not interference-limited by self-noise as conventional A-CDMA, it might appear that FDMA capacity can be exceeded. Unfortunately this is not the case, because truly orthogonal codes (such as WH sequences) of length L are limited to a number of just L, and likewise the size of the preferentially-phased quasi-orthogonal Gold codebook is limited to $L + 1$. This calls for the utilization of a demand-assignment approach to share the different allowed codes among the active users. Greater PN codes families, like the extended Kasami set, provide poor mutual orthogonality properties, even when chip synchronization is used, and consequently their utilization is not convenient. The application of convolutional codes with rate $r < 1$ causes a reduction on the code period L in order to maintain the same chip rate (hence the same bandwidth occupancy). Consequently, since S-CDMA is codebook-limited, a power efficiency increase must be paid with an overall spectral efficiency reduction, similarly to coded BPSK. However, coded modulation techniques like trellis-coding of higher order constellation provide the solution to improve power efficiency without any bandwidth increase [20].

As a summary, Fig. 1 compares the overall spectral efficiency performance versus the energy per bit-to-noise density ratio (E_b/N_0) for PSK-modulated coherently-detected synchronous (S-PSK-CDMA) and asynchronous (A-PSK-CDMA) systems with either convolutional- (CC) or trellis-coding (TC). The curves refer to a spreading code period $L = 127$ and a reference $BER = 10^{-5}$. The advantage in terms of spectral efficiency of using synchronous systems is fairly apparent. Moreover, it is interesting to notice that the curves corresponding to asynchronous systems, which are interference-limited, exhibit a "smooth knee", while those corresponding to synchronous systems, which are codebook-limited, show a "sharp knee" due to the run-out of available codes. Also, the spectral efficiency attained by more conventional FDMA Single Channel Per Carrier (SCPC) multiplexing are reported as circles.

3 Power-Control Techniques

Power control techniques represent a fundamental issue in the optimization of both satellite and mobile terminal power resources, and allow to overcome the

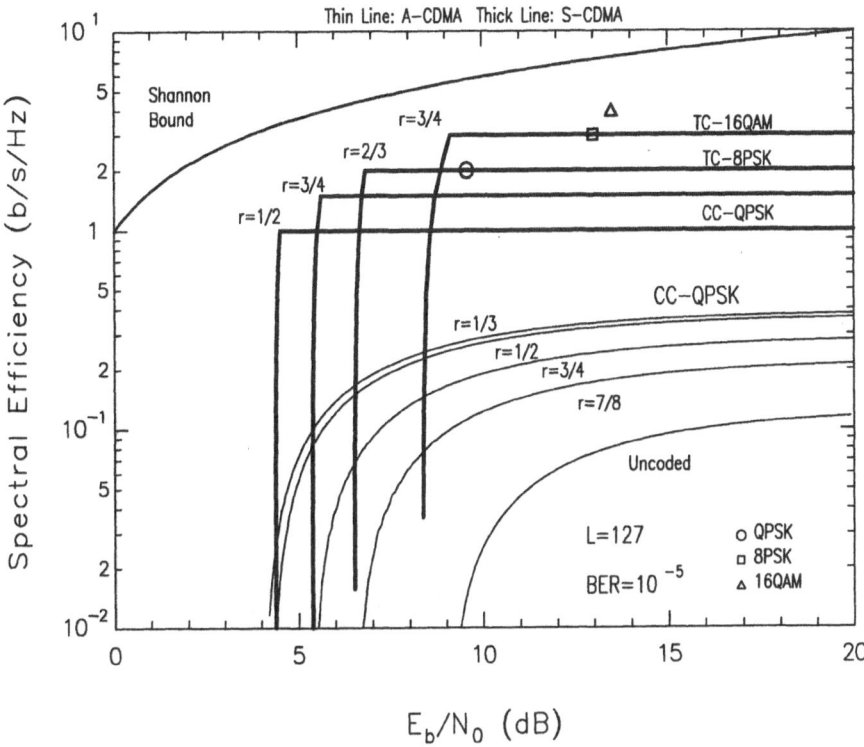

Fig. 1. System Efficiency of CDMA Networks

well known CDMA near-far effect. This latter, which is one of the main shortcomings of a conventional CDMA system, arises when the signal power received from different users is strongly unbalanced - a typical such case occurs for instance when the useful signal is shadowed and Multiple Access Interference (MAI) becomes dominant. A conventional single-user detection (SUD) scheme cannot reliably extract the information sent by a weak user under near-far conditions.

Optimum operation of any FDMA or CDMA satellite system calls for an ideally zero-margin link. This ideal operating condition can be achieved only with the aid of an adaptive power-control system for each traffic carrier. In such a way, the outbound link power utilization is maximized, and so is system efficiency, too[4]. In the outbound link for a specified receiver antenna location, the different traffic channels will be received with a signal strength inversely proportional to the actual satellite-to-user path loss (unless the satellite antenna is designed to provide an isoflux footprint). The power unbalance among CDMA signals can be much more tolerated in chip-synchronous systems with orthogonal spreading that in principle neutralize the self-noise effect. In this case, the re-

---

[4] Due to the unbalanced link characteristics, the outbound link is normally the capacity-limiting factor.

quired robust chip-timing synchronization is achieved with the aid of a powerful common DS/SS pilot reference signal. In the inbound link, power-control tries to keep continuously balanced the actual signal-to-noise ratios corresponding to the different traffic links. As a result, the transmitted power of portable terminals is optimized, thus minimizing the intra- and adjacent-carrier interference. This approach represents a satisfactory solution to the near-far effect, with no need to resort to complex (and not yet mature) multiuser detection techniques as those described in Section 5.

Similarly to a terrestrial cellular system, nominal satellite-to-mobile path losses are dependent on the actual user location with respect to the centre of the beam. Non-GEO constellations are characterized by a dynamic satellite position with respect to the user location on the Earth surface, thus making propagation losses rapidly varying with time, even in the case of a stationary user. The first cause of the variable propagation loss is related to the shape of the satellite antenna beams, providing a user-location dependent gain. Secondly, the actual satellite-to-user geometrical path loss changes depending on the user location within the coverage region of the beam. The above mentioned isoflux antenna design techniques may be used to reduce this effect. In addition, the shadowing phenomenon, caused by obstruction of the line of sight link, superimposes a random signal attenuation onto the geometrical path losses. As discussed in Section 6.2, its amplitude, generally modelled as a lognormal process, bears a standard deviation of several dBs. The last cause of signal variation is related to the fading generated by unresolved multipath components with differential delay less than one chip. Their cumulative effect can be modelled as a lognormal/Rician process affecting the line-of-sight signal. Only the "slow" shadowing components can be compensated by closed loop power control techniques. Contrarily to the terrestrial cellular environment, which is characterized by a signal dynamic range up to 90 dB, the combined satellite propagation effects cause a smaller dynamic range.

In summary, the most effective power-control strategy results in the combination of two different techniques :

- **Open-loop** power-control, to compensate for the geometrical path losses. The open-loop power-control is based on the outbound received pilot level amplitude estimation.
- **Closed-loop** power-control, to provide a fine correction of the time varying losses due to uncompensated geometrical path losses and "slow" signal shadowing. This is achieved by a double-hop Tx power control based on the actual BER or SNR estimation performed in real-time by the demodulator on the other end of the link. Corrections are then transmitted in the form of in-band signalling packets.

Clearly, the closed-loop dynamical performance is heavily dependent on the propagation and the processing delay in the modem. LEO systems have a faster geometrical dynamic evolution but also a smaller propagation delay which might ease closed-loop power-control operations.

4 DSP-Based CDMA Modems

As stated before, low-cost, small-size, low-power-consumption user terminals are key factors in the successful deployment of a satellite-based consumer communication network, independent of the particular kind of application. This basic requirement calls for state-of-the-art modem architecture design and implementation, heavily based on DSP techniques and components [21], [22]. In this respect, the continual dramatic performance improvement of both general-purpose μprocessors and ASICs is of great help, and makes the designer envisage a single chip implementation of the complete CDMA modem, directly providing the output data from a suitable IF signal (for the receiver part), and vice-versa (for the transmitter). To this aim, the cooperation of VLSI experts for the design of the chip layout, and of the telecommunication engineer for the architectural and algorithmic design reveals essential to attain to the optimum complexity/performance trade-off.

4.1 The Modulator

The basic scheme of a modulator providing the Band-Limited Direct Sequence Spread-Spectrum (BL-DS/SS) B/QPSK signal of the generic CDMA network subscriber is depicted in Fig. 2. As is apparent, the incoming binary data stream of user l, $\{d_k^l\}$, is split in two parallels streams $\{d_{p,k}^l\}$ and $\{d_{q,k}^l\}$, each one being spread with the respective chip sequence $\{c_{p,i}^l\}$ and $\{c_{q,i}^l\}$. Both resulting sequences are then shaped by a Nyquist-square-root raised-cosine filter with roll-off factor α. The filter, operating at the rate of a few samples/chip (say 4), has the purpose of band-limiting the transmitted spectrum to minimize adjacent-channel interference and to ease possible frequency re-use, and may be designed efficiently following standard optimization techniques, for instance as described in [23]. The two baseband components are then digitally quadrature-upconverted to an intermediate frequency f_{IF} and the resulting stream is input to the Digital-to-Analog (D/A) converter, followed by an analog anti-image filter, an by subsequent RF up-conversion stages, to give the l-th user transmitted signal. The processing rate of the filtering and upconverting section may be as low as 3-4 samples/chip just due to the "bandlimited" nature of the transmitted signal. By defining the following operators

$$|i|_L \triangleq i \bmod L \;\;, \;\; \lfloor i \rfloor_M \triangleq int\left\{\frac{i}{M}\right\} \tag{1}$$

we can express the baseband equivalent of such a signal as

$$\tilde{s}_T^l(t) = \sqrt{P_S^l} \sum_{i=-\infty}^{+\infty} \left(c_{p,|i|_L}^l \, d_{p,\lfloor i \rfloor_M}^l + \jmath \, c_{q,|i|_L}^l \, d_{q,\lfloor i \rfloor_M}^l \right) g_T(t - iT_c) \tag{2}$$

where P_S^l is the average power of the l-th user, $c_{p,i}^l$ and $c_{q,i}^l \in \{\pm 1\}$ are the i-th chip for the spreading sequence on the in-phase (I) and quadrature (Q) branch of the l-th transmitter, respectively, $d^l_{p,k}$ and $d^l_{q,k} \in \{\pm 1\}$ are the k-th data symbols on the I and Q branches of the l-th transmitter, L is the spreading code period, M is the spreading factor, T_c is the chip interval and finally $g_T(t)$ is the impulse response of the transmit filter. The chip interval T_c is related to the bit interval T_b through the so-called processing gain $G_p \triangleq T_b/T_c$. As is apparent from Fig. 2, two independent spreading sequences may be used for the I and Q components in order to cope best with transmission impairments (i. e. noisy carrier reference, imperfect carrier synchronization, nonlinear distortions etc.). Equation (2) fits the transmitted signal of different CDMA systems currently in use or in advanced testing status [4], [8].

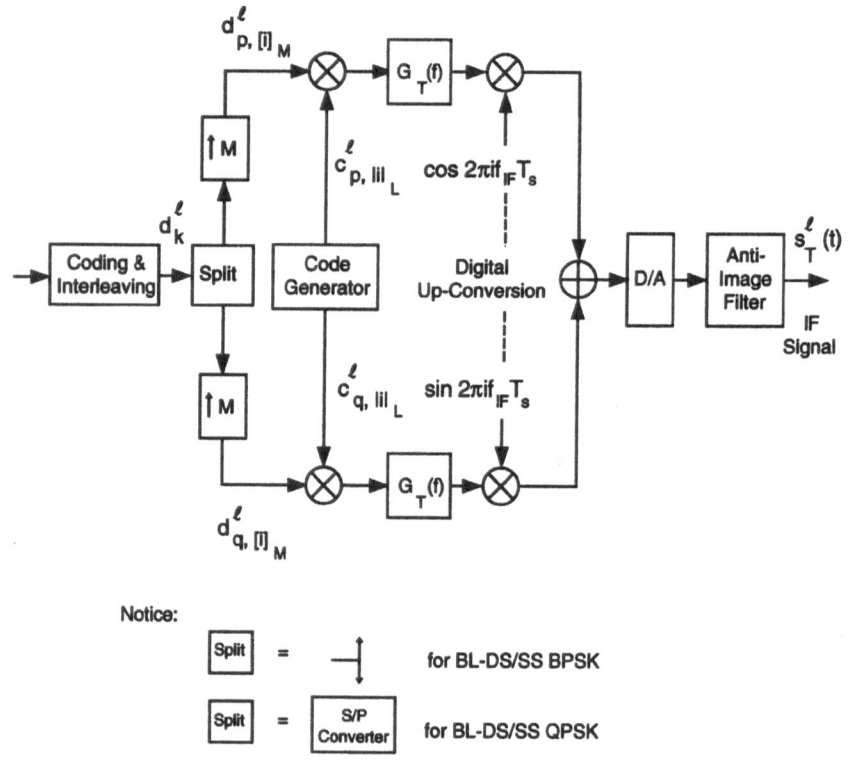

Fig. 2. Generalized B/QPSK DS/SS-CDMA modulator

4.2 The Demodulator

The simplified architecture of the companion BL-DS/SS demodulator is sketched in Fig. 3, where perfect phase, frequency and timing receiver synchronization is assumed. As will be addressed in the discussion below, several different options can be considered in the design of such a circuit, so that the one showed here should simply be regarded as a baseline for the subsequent remarks. Coming back to Fig. 3, after down-conversion form RF to f_{IF}, an automatic gain control (AGC) adjusts the signal amplitude so as to keep the total received signal power constantly equal to a reference value. This guarantees that the signal amplitude stays within the dynamic range of the subsequent IF analog-to-digital converter (ADC). The digitized bandpass signal is first downconverted to baseband, and the resulting in-phase/quadrature components are each fed to the square-root raised-cosine Chip Matched Filter (CMF) that gives intersymbol-interference-free signals for despreading and data detection. Although we assume here a baseband digital filter, the operation of matched filtering can be carried out with analog components as well (for instance with a SAW filter), directly at IF. This turns out to be particular expedient when a very low-complexity architecture of the digital section is required. This "front end" section of the demodulator bears the same processing rate (down to 4 samples/chip) as the modulator section does, while the subsequent operation of despreading simply calls for 1 sample/chip (hence the presence of the decimator). The in-phase/quadrature samples are then despread with the locally generated code replica followed by an accumulator dumped every M chip interval (i. e., a symbol period), giving the symbol-rate B/QPSK signal samples to perform data detection.

4.3   Synchronization Issues

Unfortunately, this very simple scheme is to be supplemented with appropriate signal amplitude, carrier frequency and code timing recovery subsystems and, in the case of coherent signal detection, carrier-phase loop. In this respect, it is to be said that both AGC and AFC (Automatic Frequency Control) can be effectively performed with standard analog low-complexity IF loops that can be easily integrated in the receiver analog front-end. Carrier phase correction can be also implemented at IF, with a Numerically-Controlled Oscillator (NCO) driven by an appropriate error signal derived from the baseband signal components $U_{p,i}^m$ and $U_{q,i}^m$ (hybrid loop, Fig. 4a), but the use of a non-phaselocked conversion oscillator may be favoured to get rid the NCO. In the latter case, phase error correction can be implemented digitally on the quadrature components $U_{p,i}^m$ and $U_{q,i}^m$, either using a (symbol-rate) closed-loop tracker, or an open-loop estimator (Fig. 4b) [24]. Anyway, the most crucial synchronization function the receiver has to cope with is code acquisition and tracking. In the context of BL DS/SS, the traditional function of code tracking for rectangular-chip-shaped DS/SS signals substantially reduces to that of chip timing recovery. Since the chip shape is in fact a Nyquist raised cosine function, the code tracking loop has to locate the optimum (chip-rate) signal sampling instant to yield intersymbol-interference free samples. Such a function is carried out by the so-called Digital Delay-Lock Loop (DDLL), whose scheme is sketched in Fig. 5 [25], and basically amounts to an appropriate digital version of the DLL of traditional DS/SS receivers. Operation of the DDLL is insensitive to carrier phase errors and data

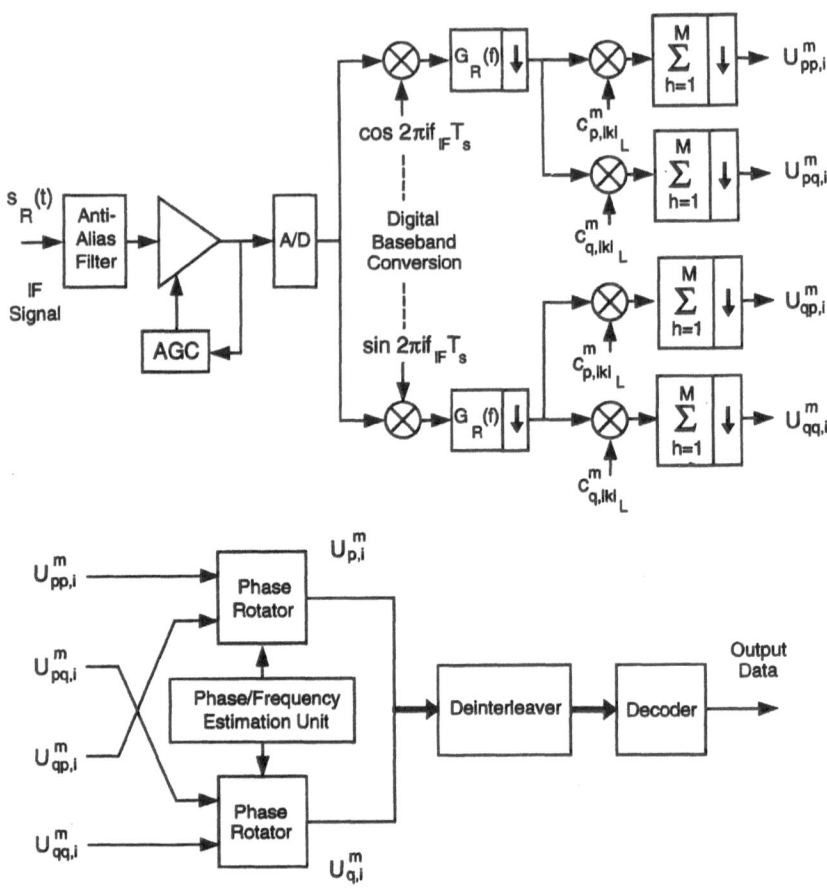

Fig. 3. Generalized B/QPSK DS/SS-CDMA demodulator

modulation due to the noncoherent nature of the relevant signal processing (notice the "squared modulus" components on each rail). Since the DDLL front-end operates at chip-rate, it may drain a non-negligible percentage of the total available computing power of the ASIC. Talking about chip timing recovery, we have two more design option in this respect: we can make the DDLL directly control the front-end A/D converters, so that the digitized signal contains the samples taken at the estimated optimum instants (synchronous sampling), or we can use a free-running A/D converter, and subsequently interpolate the digital signal to "re-synthesize" the missing signal values at the optimum instants through an appropriate interpolation algorithm or component [24]. For further detail on this aspect, we refer the interested reader to [26], [27]. The function of initial code

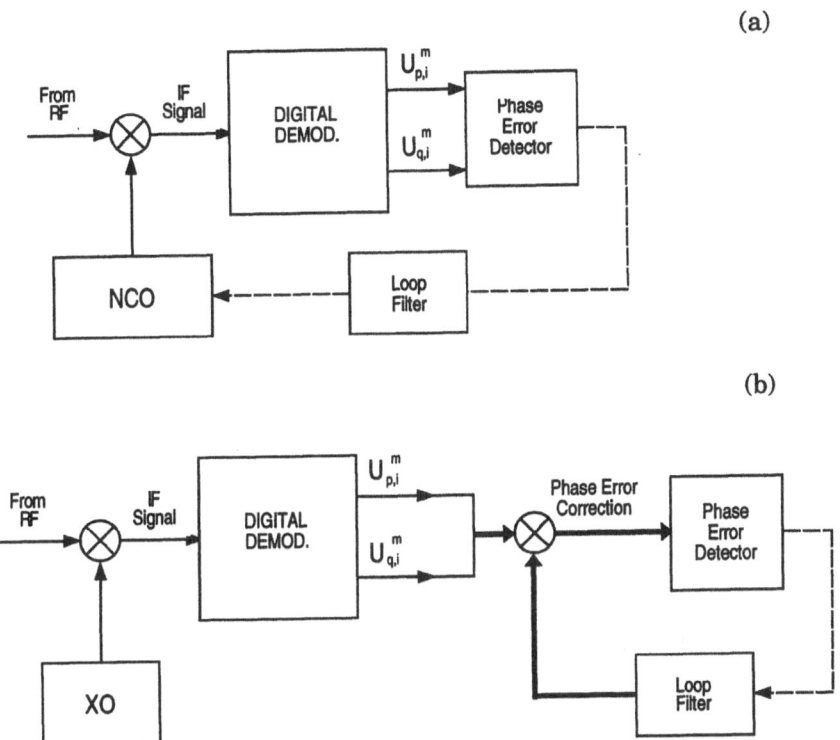

Fig. 4. Hybrid (a) vs. All-Digital (b) Carrier Phase Synchronization

acquisition at receiver start-up turns out to be still more challenging. A standard sequential acquisition subsystem [28], such that depicted in Fig. 6, is set at start-up to the so-called *search mode*, wherein chip timing tracking is frozen, and the initial epoch of the local replica code is fixed arbitrarily. The samples at the output of the chip matched filter are despread with this arbitrary-phase code, then accumulated over a bit period and passed through a squared-modulus nonlinearity. Next, W consecutive samples of the squarer output sequence are accumulated to smooth the effect of Gaussian noise, and the result is finally fed to a threshold detector. The time $T_d = WT = WMT_c$ elapsing during the accumulation of such W samples at the nonlinearity output is called *dwell time*. If no threshold crossing occurs at the end of the dwell time, the code phase employed in the despreading stage is declared *wrong*; the initial epoch of the code generator is then stepped one chip onwards and a new acquisition trial is started. Otherwise, if the accumulator output exceeds the threshold λ, the system switches to the *verification mode*, i. e. threshold crossing is tested again on a longer *verification period* $T_p = KT_d$, with K a positive integer. If threshold

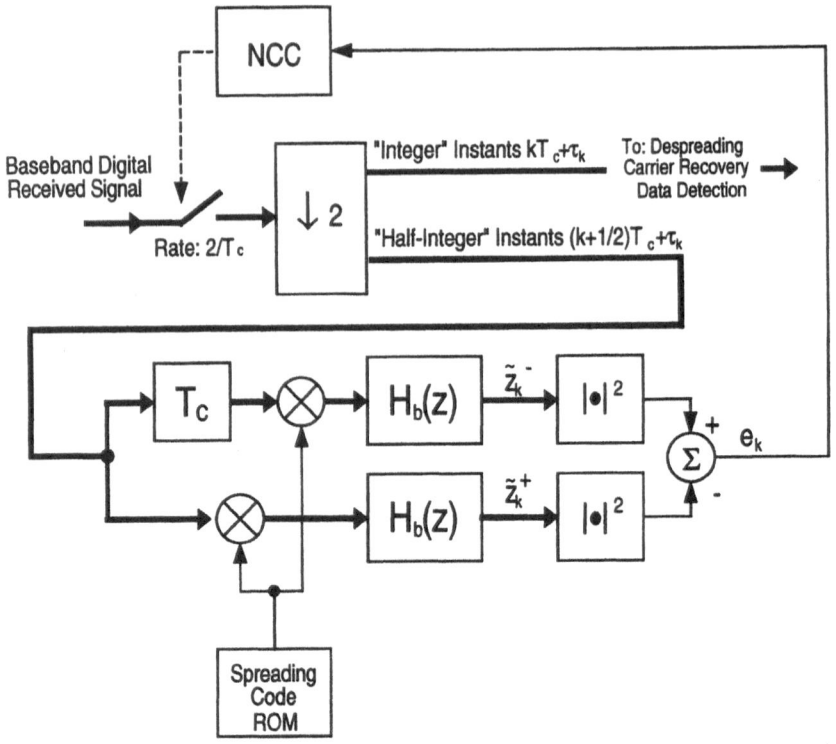

Fig. 5. Digital Delay-Lock Loop (DDLL) for fine Chip Timing Recovery

crossing occurs again, the local code phase is marked *correct*, the loop switch is closed and chip timing tracking is started. Otherwise, the code phase is marked *wrong* and the unit re-enters the *search mode*. In the latter case, the receiver has paid a *penalty time* T_p to avoid a false lock condition. The penalty coefficient K is chosen in such a way that the probability of incorrect decision at the end of the *penalty time* be vanishingly small. Double dwell techniques might also alleviate the false lock problem. A different approach to avoid the (exceedingly) long dwell time is parallel acquisition, wherein all the possible initial code phases are contemporarily investigated on a fixed observation time, and that producing the most likely output is selected [29], [30]. Of course, the algorithm is in this latter case much more simpler than the involved serial search procedure described above, but the on-chip area occupation is much larger, maybe prohibitive. When the initial carrier frequency offset due to residual Doppler or the amount of local oscillator instabilities is comparable to the inverse of the dwell time, the code acquisition turns into a two-dimensional time/frequency acquisition problem. Initial frequency uncertainty is an issue when dealing with low or medium orbiting satellites. Center-beam Doppler pre-compensation techniques will not prevent the residual Doppler shift from being larger than the symbol rate for voice communications. The carrier uncertainty problem can be solved at the expense of increased demodulator complexity through parallel acquisition in the frequency domain. The raw frequency shift estimated by the code acquisition unit is then employed to aid fine acquisition of a conventional closed-loop AFC operating in the tracking mode during receiver operations.

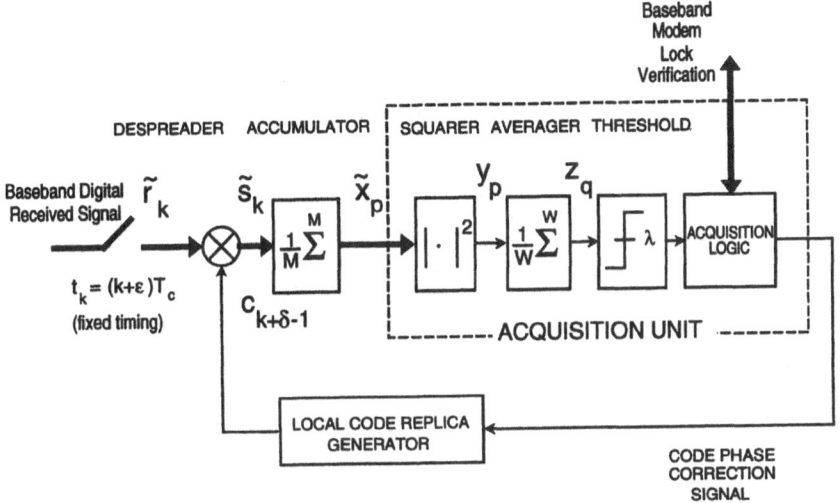

Fig. 6. Serial-Search Code Acquisition Scheme

4.4 Implementation Issues and Further Developments

A general technique to reduce further the overall implementation complexity (chip area) is the reduction of the wordlength for the internal representation of signal sample values. Some investigations have shown in fact that the overall performance degradation thus incurred may be surprisingly small. The code acquisition subsystem performance degrades for instance of about 2 dBs only when the received signal is hard-detected, i. e. quantized on one bit only [29], while it has been recently shown that the BER performance of the receiver is marginally degraded (0.5 dB only) if 4-bit quantization of signal samples is adopted [31]. On the other hand, in such a low-complexity implementation, it is mandatory to carry out some essential signal processing functions, namely CMF and carrier phase recovery, on the analog IF signal, since they turn out to be much more sensitive to quantization noise than the operation of despreading/detection. The aspect of wordlength reduction is specially attractive when a (single-chip) ASIC modem implementation is envisaged. In this context, wordlenght reduction immediately reflects into reduced chip area. Broadly speaking, the same complexity reduction due to short wordlength can not be expected when a latest-generation general-purpose digital signal processor component is employed instead. It is to be said that the current available computing power of ASICs (up to $300,000 \div 400,000$ transistors) rapidly makes viable a low-cost single chip modem with no need to resort to general-purpose chips. Once the single-chip modem is a reality, the way is paved to the rapid development of more refined detection techniques. For instance, by operating in parallel multiple receivers tracking different base stations in a cellular network or different satellites from the same constellation, it is possible to perform diversity reception (dealt with in Section 7) and/or the so-called "soft hand-over" function [50] to switch smoothly from the "setting" satellite (base station) to the "rising" one. Among the new frontiers of advanced modem design for centralized hub stations, the most promising one seems to be multi-user detection, that is the subject of next Section.

From the body of the discussion above, it should be clear that the most efficient signal detection strategies cannot help without heavy use of DSP techniques and components, that are thus going to play a key role in the development of widespread low-cost CDMA networks [33].

5 Multiuser Detectors for CDMA Signals

Operation of a conventional DS/SS demodulator may be viewed as the correlation of the received signal with a specific user signature sequence, followed by a hard-decision on the correlator output. This kind of receiver is optimized for an AWGN channel without MAI coming from the other users sharing the same bandwidth. At the hub station, one correlation/decision device is allocated for each inbound traffic channel, and the individual user streams are independently detected (see Fig. 7a). In such a conventional system, the amount of MAI immunity of each user depends on the selection of the signature sequences. For an asynchronous system very low values of the cross-correlation among signature sequences are necessary in order to achieve acceptable performance under loaded conditions. As mentioned in Section 2.3, the use of sequences that are fully or almost orthogonal to yield better interference rejection calls for user synchronization and limits the number of allowable simultaneous users. Even with (quasi-) orthogonal sequences, shadowing or multipath propagation reduces user separation (orthogonality), and the inbound link of a satellite based CDMA system results vulnerable to the near-far problem. The most common method for counteracting the CDMA near-far effect is the use of tight power-control. Some system employ clever implementations of power-control to keep the received power of the users approximately equal at the satellite side (e.g. [19]). Though mostly successful, these measures might not prove adequate due to the link propagation delay and to the limited power available at the mobile terminals. Investigation of robust receiver structures, more resilient to the near-far effect is therefore of great relevance.

5.1 The Multiuser Detector Principle

Recently, the subject of multi-user detection (MUD) has emerged as a new way of addressing the CDMA near-far problem. The basic idea behind MUD is to exploit the (side-) information available about the interfering users to cancel out the interference, and thus improve the reliability of data decisions, instead of ignoring the presence of other users in the system (see Fig. 7b), like in conventional SUD. Analysis of such MUD structures reveals that it is possible to solve effectively the previously destructive near-far problem. As a consequence, system power-control does not have to be very tight if a MUD is used. At the same time, the demand on the cross-correlation properties of the user signature sequence can be greatly relaxed - this suggests that the capacity of the CDMA system is increased by the introduction of MUD. This two-fold leap forward in the CDMA receiver performance, not surprisingly, comes at the expense of a substantial increase in the complexity (signal processing burden) of the receiver.

5.2 MUD Types

Two conceptually different approaches to MUD can be pursued. The one is *parallel detection*, wherein the receiver essentially performs a simultaneous estimation of all the user streams. The other approach is a sort of *serial detection*, also known as successive cancellation, in which the receiver processes the outputs of the matched filter bank in an iterative fashion, to estimate the information bits of each user, one at a time. Parallel MUD, similarly to multiple conventional SUD, consists of a bank of filters matched to the individual user signature waveforms. The real core of MUD lies in the elaborate post-processing of the array of matched filter outputs (sufficient statistics). The optimum AWGN MUD scheme originally proposed and analyzed by Verdu [35] uses a Viterbi algorithm to perform the parallel maximum likelihood sequence estimation. A receiver with this optimum structure has a complexity that is exponential in the number of users, and therefore it does not easily lend itself to practical implementations. Several different sub-optimum MUD structures have been proposed in the literature, all of them attempting to reduce the estimation complexity by replacing the Viterbi decoder by another device. Lupas and Verdu [36] investigated a class of linear MUD structures. Specifically the so-called decorrelating MUD performs a linear transformation on the matched filter outputs before standard hard-decisions are made. This detector has a complexity that is only linear in the number of users, but still gives a performance close to that of the optimum MUD. Xie *et al.* [37] also evaluated the performance of a family of MUD structures that perform a linear transformation on the matched filter outputs. One of those receivers is particularly reminiscent of a non-adaptive decision-feedback equalizer. A further alternative to Viterbi decoding is proposed by Xie *et al.* [38], where the use of a sequential decoder utilizing the stack algorithm with an ad-hoc decoding

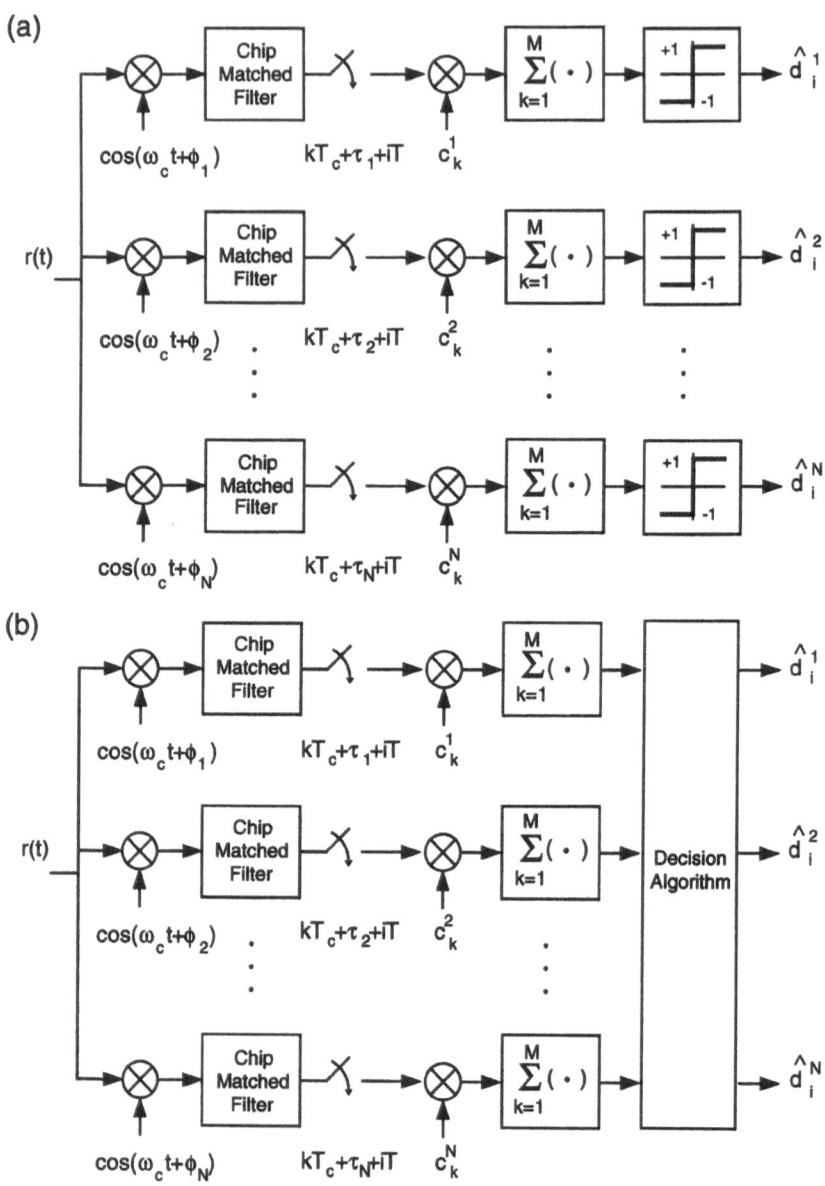

Fig. 7. Single- (a) vs. Multi-User (b) Demodulation

metric is analyzed. Again, a linear complexity in the number of users is shown to yield a performance close to that of optimum MUD. As a final example of a sub-optimum parallel MUD, Aazhang et al. [39] describe a MUD structure based on an artificial neural network, which is trained during an initial transmission of known data. Substantial performance improvements over conventional SUD are shown, but the complexity of the circuit still appears to be exponential in the number of users.

Serial MUD may be done as described by Viterbi [11], starting from the estimated most reliable user, e.g. the one with the largest matched filter output magnitude, and making a decision about the relevant transmitted bit. This bit is then re-spread with the proper signature sequence, and subtracted from the received CDMA signal (which is conveniently stored for the purpose), thus cancelling out the influence of the strongest user. The serial MUD is thus ready for the next iteration. A more implementation-oriented serial MUD structure is analyzed by Varanasi and Aazhang [40]. This scheme achieves a performance close to the optimum one, while bearing a (computational) complexity that is linear in the number of users. The serial MUD principle is inherently sub-optimum because, at each iteration stage in the cancellation procedure, decisions are made without taking into account the remaining weaker users which have not been cancelled yet.

5.3 Performance Measures for MUD

The bit-error rate (BER) performance of the different MUD structures is to be compared with two obvious reference values: firstly, an upper bound to the MUD performance is the BER of conventional SUD in the same conditions as MUD; secondly, the BER of the same conventional SUD, but in the absence of interference from other users, can be used as a lower bound on the MUD performance. In addition to these qualitative measures, two more appropriate performance parameters have become well established in the literature, namely, the. asymptotic efficiency (AE) and the near-far resistance (NFR). The AE was first introduced [35] and later elaborated further on [41] by Verdu as an indication of how fast the BER of a single user approaches zero in the presence of MAI, when we let the AWGN power approach zero. Obviously, the AE depends on the receiver structure, on the cross-correlation properties of the signature sequences and on the relative user power levels. To quantify the receiver performance under (worst-case) near-far conditions, Lupas and Verdu [42] introduced instead the NFR as the minimum AE of a single user over all possible values of the power levels of the other users. Conventional SUD is shown to have zero NFR, i. e. for a sufficiently strong interference its AE goes to zero, for under these conditions the BER of SUD is non-zero, even in the absence of noise. Optimum MUD on the other hand is near-far resistant in all but pathological cases of the signature sequence cross-correlation. This means that even with very relaxed demands on the cross-correlation, the optimum MUD attains error-free operation in the absence of noise, independently of the power of the interfering users.

5.4 MUD on Fading Channels

Vasudevan and Varanasi [47] showed that the parallel MUD structures developed for the AWGN channel do not bear optimum performance any longer in the presence of multipath propagation in an urban area. AWGN MUD is still near-far resistant (though sub-optimum) over the Ricean fading channel, but like conventional SUD it is not fading-resistant. Two modifications of the optimum and the decorrelation MUD schemes for AWGN respectively, are yet shown to be fading-resistant in addition to all of the other attractive features of AWGN MUD. Considerable improvements of serial MUD schemes on a fading channel, when compared to conventional SUD are reported by Yoon et al. [48] and Patel and Holtzman [46], even though the described detectors appear not to be fading resistant.

5.5 Applicability of MUD to the Outbound Link

Although MUD was only considered in the subsections above for a hub station, it is, in principle, also applicable to the individual mobile terminals. Receiver structures based on MUD, and optimized for the estimation of one user data only , have indeed been described by Poor and Verdu [49], both for the case where the entire set of user signature sequences is public, and for the case where network security restrictions or user privacy considerations allow only the common chip waveform to be known. These schemes promise performance gains comparable to those obtained with the optimum AWGN MUD. Despite of this, the increased hardware complexity of a MUD scheme with respect to a conventional SUD receiver still suggests considering MUD for the inbound link only.

6 Transmission Channel Peculiarities

6.1 Satellite Transponder

When dealing with a satellite-based transmission system, the assessment of the impact of the satellite transponder nonlinearity on the overall link performance is a major point. Such an issue reveals particularly critical in a multiuser system; the goal of efficient system design is, in this respect, the computation of that output back-off (OBO) of the nonlinear amplifier operating point which gives optimum on-board power utilization for a fixed target BER.

In conventional SCPC systems, the input signals are quasi-orthogonal in the frequency domain. In a CDMA system the same central carrier frequency is reused by several users and the discrimination is achieved through the overlaying signature codes. As matter of fact, the satellite nonlinearity modifies the cross-correlation properties of the code sequences causing orthogonality loss, that results particularly destructive for the case of synchronous (quasi-) orthogonal codes [32]. Recent studies on synchronous CDMA systems with PSK modulation [20], [32], [50] demonstrated that, for a 50% user loading, the overall power loss with respect to the ideal AWGN channel (no MAI), including signal distortions plus OBO, ranges from 4 to 5 dB for uncoded schemes and from 2 to 4 dB for convolutional- or trellis-coded schemes at a bit error rate of 10^{-3}.

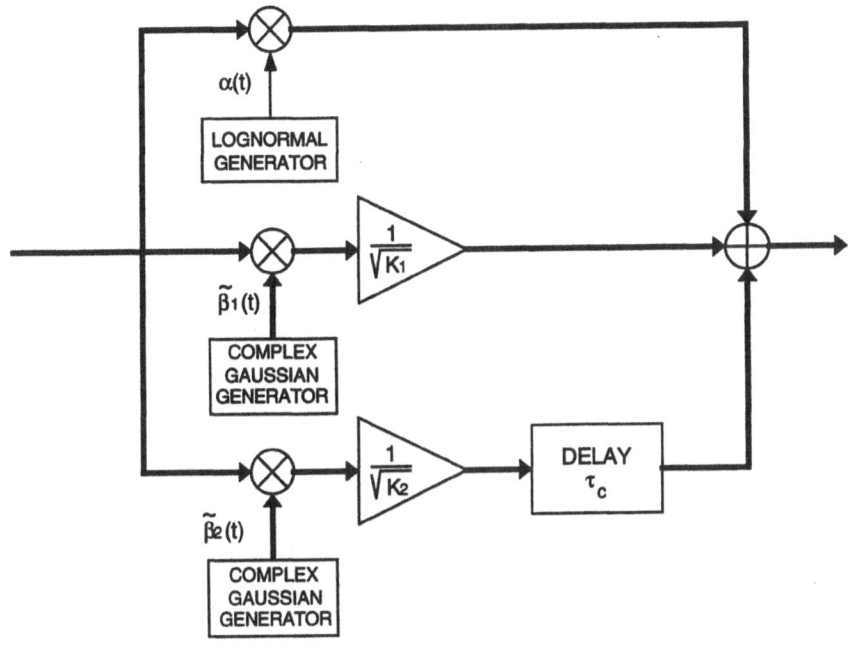

Fig. 8. Baseband Model of the CDMA Satellite Mobile Propagation Channel

6.2 Satellite-to-Mobile Fading Channel

A mobile receiver located in open areas (rural or suburban environments) has a good line-of-sight (LOS) link with the satellite available most of the time, but due to temporary shadowing caused either by man-made structures (bridges, buildings, etc.) or natural obstructions (trees, peaks, etc.) the received signal suffers from random amplitude fluctuations. Field measurements show that the LOS shadowing is well described by a multiplicative log-normal process. The lognormal process parameters clearly depend on the particular environment, the mobile speed and the satellite speed and elevation angle above the horizon. For the cases of practical interest, the received signal suffers an average attenuation ranging from 0 to 7 dB, with respect to the nominal shadowing-free LOS value, with a standard deviation of the amplitude ranges from 0 to 5 dB; the bandwidth of the amplitude fluctuations is a few Hz.

In addition to a shadowed LOS ray, the propagation channel is character-ized by a scattered Rayleigh-faded component resulting from the superposition of several reflected rays, each of them characterized by a different amplitude, phase and group delay with respect to the line-of-sight component [51], whose magnitude is largely dependent on the satellite elevation and on the mobile aerial characteristic. The rays with a small delay combine asynchronously, causing the cited zero-delay diffuse Rayleigh component that adds to the shadowed LOS ray, causing a Ricean/lognormal fading [34]. The multipath components which are delayed more than one chip will be discriminated by the SS demodulator and will simply result in an increased self-noise level (multipath rejection of the SS signal). However, such an apparently complex situation can be effectively described by the simplified channel model depicted in Fig. 8, comprising the lognormal and Rayleigh faded rays, and a third propagation path whose delay τ_c has little influence on the overall system performance (due to the multipath rejection feature) [32]. The potential DS-CDMA multipath mitigation will be of little advantage considering that in satellite links the shadowing will dominate over fading in many situations.

Unfortunately, very few experimental wideband satellite channel measure-ment results are currently available to validate extensively such a channel model. Considering the available preliminary information coming form on-going cam-paigns, we can assume as a rule of thumb that SS signals with a bandwidth up to 1 MHz will experience almost flat fading ($\tau_c = 0$), while signals with a bandwidth larger than 10 MHz take full advantage of the SS capability to discriminate the multipath components. Of course the above statement might reveal inadequate for specific operating conditions i. e. power unbalance among the signals or large operating margins.

7 Diversity Techniques

One of the claimed advantage of DS/SS CDMA is related to the inherent ca-pability of exploiting diversity signal reception to enhance link quality, and to provide soft hand-off between satellites. The feature was indeed demonstrated in the terrestrial CDMA system manufactured by Qualcomm. The large power margin available on ground systems allows to take advantage of both cell site (space) diversity and multipath propagation (time diversity). On the contrary, in a satellite system resolution of the time-diversity due to multipath propagation can not be performed because of power limitations. However, space diversity achieved by simultaneous reception of the different signals coming from two or more satellite in visibility (see Fig. 9) might considerably improve the link quality and provide seamless satellite hand-off. To this aim, the demodulator should be able to discriminate and independently demodulate the signals com-ing from the different satellites and combine them at baseband. The so-called RAKE receiver [51] (Fig. 10) performs just this operation. As is apparent, it is made up of a few independent DS/SS demodulators operating in parallel and controlled by a central processing unit. Each RAKE branch tracks a different

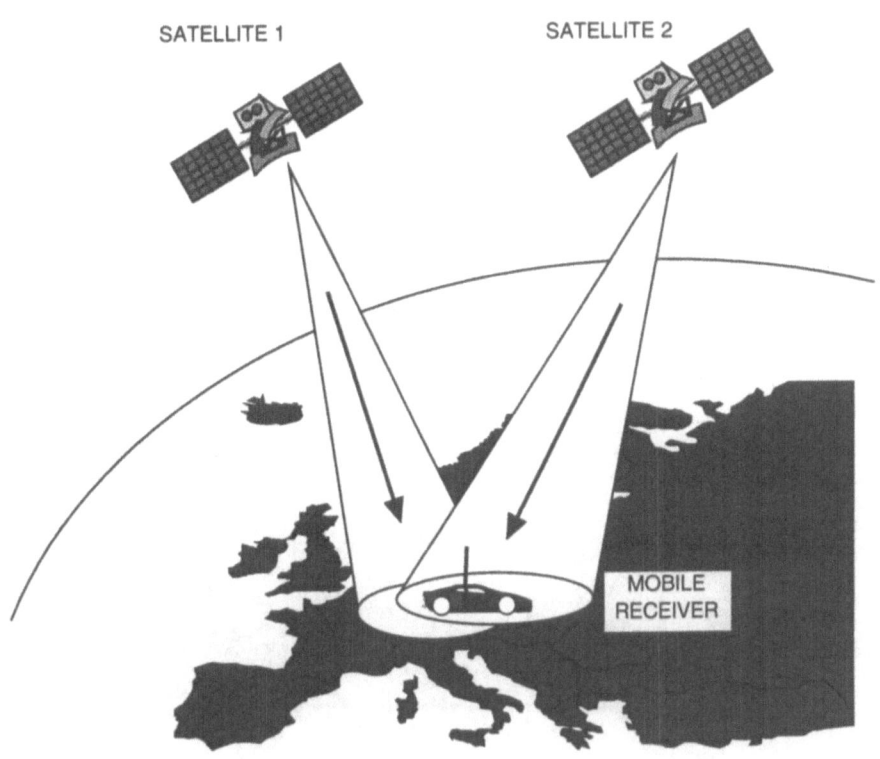

Fig. 9. Space Diversity for Satellite Networks

signal component by despreading the received signals with different time-shifted replicas of the spreading code. Baseband combining is performed after individual differential time offset compensation and appropriate amplitude weighting, so that the overall SNR is maximized. Coherent combining (i.e., separate carrier phase tracking) may also be required if coherent detection is implemented. The RAKE receiver can easily adapt to sudden channel variations due to changes in the satellite or receiver position (time-variant delay profile). The disappearance of one ray due such a change in the link geometry, does not break in fact the link, since signal reception is guaranteed by the RAKE branches tracking other non-interrupted rays.

The use of a RAKE receiver may prove effective in a mixed satellite/terrestrial mobile audio broadcasting system such that described in [50]. In this case, time-diversity due to multipath propagation can be fully exploited in the terrestrial iso-frequency gap-filler network providing urban coverage, and the RAKE receiver provides also the capability to carry out satellite to terrestrial gap-filler signal hand-off.

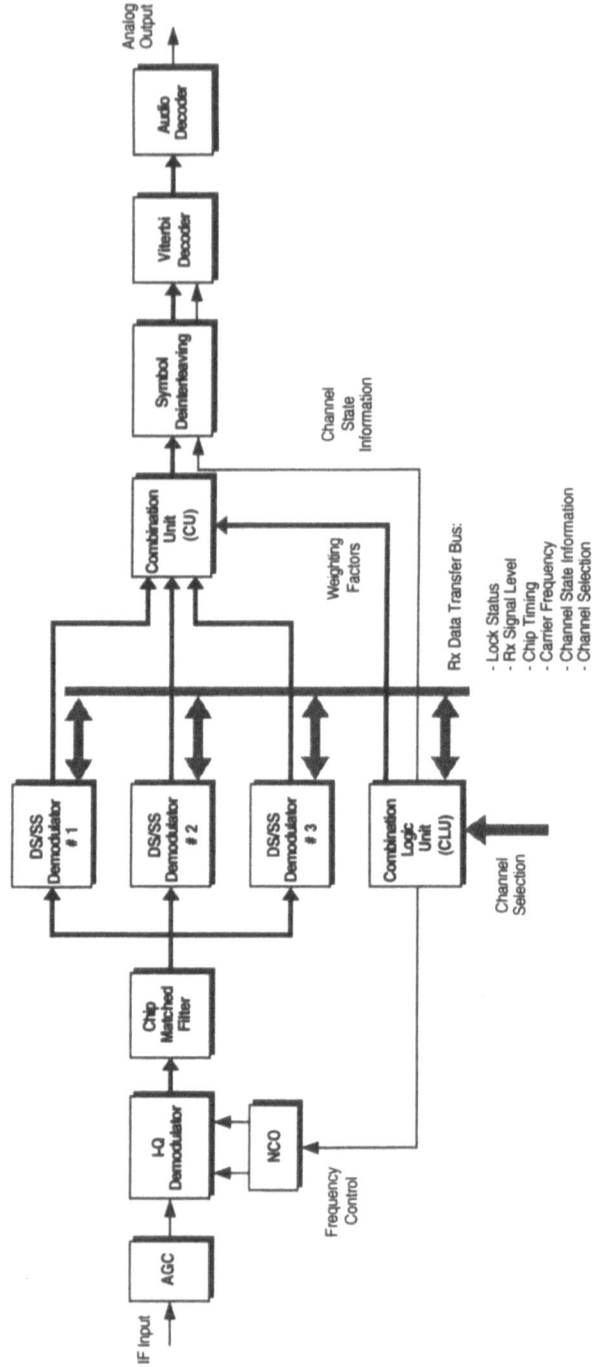

Fig. 10. Time-Diversity CDMA (RAKE) Receiver

100

8 Conclusion ?

As the question mark in the title suggests, this section should be labeled Overture rather than Conclusion, since from the collection of all of the topics examined above, it should be clear that the career of CDMA in satellite networks has just started, and looks promising yet. The increased efficiency of CDMA obtained through bandlimiting, frequency re-use, time diversity and possibly voice-activation makes it suitable for high-traffic applications, such as traditional mobile telephony. On the other hand, the inherent coding capability and the possibility of deep integration with terrestrial premises makes CDMA the ideal candidate for future wideband ISDN-grade mobile and PCN services [52]. In this respect, CDMA reveals also adequately flexible to implement variable-rate transmission capabilities, according to the momentary customer needs; that feature can be also effectively exploited in satellite-based variable-quality broadcasting of digital audio programs. The nice features listed above are to be contrasted to a somewhat higher user terminal complexity and cost, due to a less established engineering tradition. It is our hope that the synergy of clever architecture design and VLSI implementation techniques can break the last barrier in the widespread diffusion of low-cost CDMA terminals for the ubiquitous communication network of the 21st century.

References

1. *"Wireless Personal Comunications: Part I and Part II"*, IEEE Journal on Selected Areas in Communications, Vol. 11, No. 6, August-September 1993.
2. W.C.Y. Lee, *"Overview of Cellular CDMA"*, IEEE Transactions on Vehicular Technology, Vol. 40, No. 2, pp. 291-302, May 1991.
3. T.S. Rappaport, *"The wireless revolution"*, IEEE Communications Magazine, Vol. 29, No. 11, November 1991, pp. 52-71.
4. R. De Gaudenzi, C. Elia, R. Viola, *"Bandlimited Quasi-Synchronous CDMA: A Novel Satellite Access Technique for Mobile and Personal Communication Systems"*, IEEE Journal on Selected Areas in Communications, Vol. 10, No. 2, February 1992, pp. 328-343.
5. K.S Gilhousen, I.M. Jacobs, R. Padovani, L.A. Weaver, *"Increased Capacity Using CDMA for Mobile Satellite Communications"*, IEEE Journal on Selected Areas in Communications, Vol.8, No. 4, May 1990, pp. 503-514.
6. R.L. Pickoltz, L.B.Milstein, D.L. Shilling, *"Spread Spectrum for Mobile Communications"*, IEEE Transactions on Vehicular Technology, Vol. 40, No. 2, May 1991, pp. 313-322.
7. K. S. Gilhousen, R. Padovani, A. J. Viterbi, L.A. Weaver, C. E. Wheatly III *"On the Capacity of a Cellular CDMA System,"* IEEE Trans. Vehic. techn., Vol . 40, No. 2, pp. 303-312, May 1991.
8. *"System and Method for Generating Signal Waveforms in a CDMA Cellular Telephone System"*, Qualcomm Inc., Patent PCT/US91/04400 - WO 92/00639, January 9, 1992.
9. *"Code Distribution Multiple Access Communication System with User Voice Activated Carrier and Code Synchronization"*, European Space Agency, Patent PCT/EP90/01276 - WO 91/02415, February 21, 1991.

10. *"Method and Device for Multiplexing Data Signals in a Satellite Code Division Multiple Access System Based on PSK"*, European Space Agency, Patent PCT/EP92/02001 August 28, 1992.

11. A.J. Viterbi, *"Very Low Rate Convolutional Codes for Maximum Theoretical Performance of Spread-Spectrum Multiple-Access Channels"*, IEEE Journal on Selected Areas in Communications, Vol. 8, No. 4, May 1990.

12. G.C. Cain, J.B. Cain, *"Error Correcting Codes for Digital Communications"*, Plenum Press, New York, 1981.

13. J. Proakis, *"Digital Communications"*, second edition, Mc Graw Hill International Editions.

14. *"Proposed EIA/TIA Interim Standard : Wideband Spread Spectrum Digital Cellular System"*, April 21, 1992.

15. G.D. Boudreau, D.D. Falconer, S.H. Mahmoud, *"A Comparison of Trellis Coded Versus Convolutionally Coded Spread-Spectrum Multiple-Access Systems"*, IEEE Journal on Selected Areas in Communications, Vol. 8, No. 4, May 1990.

16. R. Rogard, *"LMSS: From Low Data Rate to Voice Services"*, 14th AIAA International Communications Satellite Systems Conference, Washington DC, March 1992, pp. 284-393.

17. R. De Gaudenzi, C. Elia, R. Viola, *"Band-Limited Quasi-Synchronous CDMA: a Satellite Access Technique Towards Personal Communication Systems Via Satellite"*, In the Proceedings of the 9th ICDSC, Copenhagen, Denmark, May 18-22, 1992.

18. N. Ahmed and K. R. Rao, *"Orthogonal Transforms for Digital Signal Processing"*, Springer-Verlag, 1975.

19. P. Monte, S. Carter, *"The Globalstar Air Interface"*, In the Proceedings of the AIAA 1994 Satellite Communication Conference, San Diego, California, USA, March 1994, pp. 1614-1621.

20. R. De Gaudenzi, F. Giannetti, *"Synchronous Trellis-Coded CDMA Analysis and System Performance"*, In the Proceedings of IEEE International Confernce on Communications, ICC '93, Geneve, May 1993, pp. 1444-1448.

21. A.J. Aftelak *et alii* , *"Implementation of a Spread-Spectrum Receiver and Modem for the Data-Relay System Using Digital Signal Processing"*, In the Proceedings of the 3rd ESA Workshop on DSP Techniques Applied to Space Comunications, ESA/ESTEC, Noordwijk, The Netherlands, September 1992.

22. B. Eichinger *et alii* *"VSAT Terminal Using BLQS CDMA Acces: Architecture, Performances"*, In the Proceedings of the 3rd ESA Workshop on DSP Techniques Applied to Space Comunications, ESA/ESTEC, Noordwijk, The Netherlands, September 1992.

23. N.A. D'Andrea, F. Guglielmi, U. Mengali, A. Spalvieri, *"Design of Transmit and Receive Digital Filters for Data Communications"*, IEEE Transactions on Communications, Vol. 42, No. 2/3/4, February/March/April 1994, pp. 357-359.

24. F. M. Gardner, *"Demodulator Reference Recovery Techniques Suited for Digital Implementation"*, Final Report, Gardner Research Company, Palo Alto, California,USA, August 1988, ESA/ESTEC Contract No. 6847/86/NL/DG.

25. R. de Gaudenzi, M. Luise, R. Viola, *"A Digital Chip Timing Recovery Loop for Band-Limited Direct-Sequence Spread-Spectrum Signals"*, IEEE Transactions on Communications, Vol. COM-41 No. 11, November 1993.

26. F.M. Gardner, *"Interpolation in Digital Modems-I: Fundamentals"*, IEEE Transactions on Communications, Vol. COM-41 No. 3, March 1993.

27. M. Luise, R. Roncella, *"RVLSI Implementation of a Signal Interpolator Chip for High-Speed All-Digital Modems"*, European Transactions on Telecommunications, Vol. 5, No. 4, July-August 1994.

28. A. Polydoros, C.L. Weber, *"A Unified Approach to Serial Search Spread-Spectrum Code Acquisition - Part II : A Matched Filter Receiver"*, IEEE Transactions on Communications, Vol. COM-32, No. 5, May 1984, pp. 550-560.

29. L. D. Davisson, P. G. Flikkema, *"Fast Single-Element PN Acquisition for the TDRSS MA System"*, IEEE Transactions on Communications, Vol. COM-36, No. 11, November 1988, pp. 1226-1235.

30. Y. T. Su, *"Rapid Code Acquistion Algorithms Employing PN Matched Filters"*, IEEE Transactions on Communications, Vol. COM-36, No. 6, July 1988, pp. 724-733.

31. R. De Gaudenzi, F. Giannetti, M. Luise, *"The Influence of Signal Quantization on the Performance of Digital Receivers for CDMA Radio Networks"*, In the Proceedings of IEEE GLOBECOM '94, S. Francisco CA, November-December 1994.

32. R. De Gaudenzi, T. Garde, F. Giannetti, M. Luise *"Orthogonal CDMA Transmission for Satellite-Based Mobile Communications Radio Networks"*, In the Proceedings of IEEE GLOBECOM '94, S. Francisco CA, November-December 1994.

33. D. T. Magill, F. D. Natali, G. P. Edwards, *"Spread-Spectrum Technology for Commercial Applications,"* Proceedings of the IEEE, Vol . 82, No. 4, April 1994, pp. 572-584.

34. C. Loo, *"A Statistical Model for a Land Mobile Satellite Link,"* IEEE Trans. on Vehicular Technology, VOL. VT-34, No. 3, August 1985, pp. 122-127.

35. S. Verdu, *"Minimum Probability of Error for Asynchronous Gaussian Multiple-Access Channels"*, IEEE Transactions on Information Theory, January 1986, pp. 85-96.

36. R. Lupas, S. Verdu, *"Linear Multiuser Detectors for Synchronous Code-Division Multiple-Access Channels"*, IEEE Transactions on Information Theory, January 1989, pp. 123-136,

37. Z. Xie, R.T. Short, C.K. Rushforth, *"A Family of Suboptimum Detectors for Coherent Multiuser Communications"*, IEEE Journal on Selected Areas in Communications May 1990, pp. 683-690.

38. Z. Xie, C.K. Rushforth, R.T. Short, *"Multiuser Signal Detection Using Sequential Decoding"*, IEEE Transactions on Communications, May 1990, pp. 578-5830.

39. B. Aazhang, B. Paris, G. Orsak, *"Neural Networks for Multiuser Detection in Code-Division Multiple-Access Communications"*, IEEE Transactions on Communications, July 1992, pp. 1212-1222.

40. M.K. Varanasi, B. Aazhang, *"Multistage Detection in Asynchronous Code-Division Multiple-Access Communications"*, IEEE Transactions on Communications, April 1990, pp. 509-519.

41. S. Verdu, *"Optimum Multiuser Asymptotic Efficiency"*, IEEE Transactions on Communications, September 1986, pp. 890-897.

42. R. Lupas, S. Verdu, *"Near-Far Resistance of Multiuser Detectors in Asynchronous Channels"*, IEEE Transactions on Communications, April 1990, pp. 496-508.

43. Z. Xie, C.K. Rushforth, R.T. Short, T.K. Moon, *"Joint Signal Detection and Parameter Estimation in Multiuser Communications"*, IEEE Transactions on Communications, August 1993, pp. 1208-1216.

44. A. Radovic, B. Aazhang, *"Iterative Algorithms for Joint Data Detection and Delay Estimation for Code Division Multiple Access Communication Systems"*, In the Proceedings of the 31st Annual Allerton Conference on Communications, Control

and Computing, University of Illinois at Urbana-Champaign, September-October 1993.

45. M.K. Varanasi, *"Noncoherent Detection in Asynchronous Multiuser Channels"*, IEEE Transactions on Information Theory, January 1993, pp. 157-176.

46. P.R. Patel, J.M. Holtzman, *"Analysis of Successive Interference Cancellation in M-ary Orthogonal DS-CDMA System with Single Path Rayleigh Fading"*, In the Proceedings of the 1994 International Zurich Seminar on Digital Communications, Zurich, Switzerland, March 1994, Springer Verlag, pp. 150-161.

47. S. Vasudevan, M.K. Varanasi, *"Fading Resistant Multiuser Detection over CDMA Fading Channels"*, In the Proceedings of IEEE International Conference on Communications, ICC '93, Geneve, Switzerland, May 1993, pp. 137-141.

48. Y.C. Yoon, R. Kohno, H. Imai, *"A Spread-Spectrum Multiaccess System with Cochannel Interference Cancellation for Multipath Fading Channels"*, IEEE Journal on Selected Areas in Communications, September 1993, pp. 1067-1075.

49. H.V. Poor, S. Verdu, *"Single-User Detectors for Multiuser Channels"*, IEEE Transactions on Communications, January 1988, pp. 50-60.

50. R. De Gaudenzi, F. Giannetti, *"Analysis of an Advanced Satellite Digital Audio Broadcasting System and Complementary Terrestrial Gap-Filler Single-Frequency Network"*, IEEE Transactions on Vehicular Technology, Vol. 42, No. 2, May 1994, pp. 1-12.

51. G.L. Turin, *"Introduction to Spread-Spectrum Antimultipath Techniques and Their Application to Urban Digital Radio"*, Proceedings of the IEEE, Vol. 68, No. 3, March 1980, pp. 328-353.

52. A. Baier et al. *"Design Study for a CDMA-Based Third Generation Mobile Radio System,"* IEEE Journ. on Sel. Areas in Comm. Vol. 12, No. 4, May 1994, pp. 733-743.

Payload Design Alternatives for Geostationary Personal Communications Satellites

M. Lisi, M. Piccinni
Alenia Spazio, via Saccomuro 24
I-00131 Rome Italy

A. Vernucci
Space Engineering, via dei Berio 91
I-00155 Rome Italy

Abstract

GeoStationary-Orbit (GSO) satellites are nowadays being used for providing worldwide communications services to mobile users equipped with terminals utilizing a *reasonably low size* antenna (e.g. Inmarsat-M, Inmarsat-C). New GSO systems (e.g. AMSC, EMS) will shortly be implemented to increase the service capacity available at regional level, and especially to face the increasing demand for voice services.

In the present flourishing of proposals for all kinds satellite systems offering service to hand-held Mobile Stations (MSs) utilizing *almost omni-directional* antennas (the so-called *personal* communications services), the European Space Agency (ESA) has promoted study activities aimed to investigate the *suitability* of GSO satellites to this kind of application and to assess and compare different *implementation approaches*, encompassing both analog and digital solutions.

In consideration of the high EIRP and G/T values to be provided for serving hand-held MSs, it was immediately recognized that the elements deserving particular consideration are the *antenna*, the *RF power amplifier* and the *routing processor*.

This paper summarizes the main outcomes of the above mentioned study activities, by presenting the architectural payload configurations which were considered the most attractive for each implementation approach and the relevant quantitative assessments, in terms of payload mass and power consumption.

1 Introduction

Satellite systems based upon "unconventional" orbital configurations (namely LEO, Low-Earth Orbit, ICO - Intermediate Circular Orbit, M-HEO - Multiregional Highly Elliptical Orbit) are currently being proposed and pursued, with the objective to make ubiquitous personal communications services available to users equipped with an hand-held MS, with tariffs, quality and availability scores fairly comparable to services rendered by terrestrial Public Land Mobile Networks.

Nevertheless, the utilization of a more "traditional" GSO-satellite constellation for the provision of this services category may still be regarded as a potentially attractive solution (to be perhaps complemented by a low-capacity LEO- or ICO-system for polar regions coverage), because of the following main reasons:

— potential *synergies*, in terms of both *space-* and *ground-segment*, with the traditional mobile satellite systems, which are, or are going to be, all implemented by means of GSO spacecrafts;

— remarkable simplification of *mission* operations (small total number of spacecrafts to be controlled) and of *communications network* operations (no need for complex link-management protocols, such as satellite hand-offs, dynamic re-routings, etc.);

— simpler overall *network configuration* (e.g. any MS can unrestrictedly get access to any GateWay Station (GWS) falling within the satellite coverage);

— easier *frequency coordination* among systems operating in the same spectrum.range. On the other hand GSO-satellite systems have to face important challenges, e.g.:

— the need for a *large on-board antenna*, offering a gain high enough to contrast the large free-space loss typical of GSO-systems, thus allowing to yield a G/T consistent with the low hand-held MS antenna gain;

— the capability of *routing*, in a flexible manner, the individual carriers to the numerous spots, such as to be able matching, as closely as possible, the unavoidably variable traffic distribution across the service area;

— the necessity to generate quite a *large total RF power* and to flexibly distribute it to the various spots, consistently with the current carrier-to-spot assignment pattern.This, in conjunction with the remarkable antenna gain, will result in an earth-flux allowing sufficient margin to offer dependable system operation.

It has to be noted, however, that the last two issues also represent an important challenge for all LEO-systems currently being proposed or developed.

2. Objectives and requirements

The assumed service / performance objectives and system / access requirements for the GSO personal satellite system were determined by elaborating certain guidelines provided by ESA, which reflected the indications currently emerging in the fora where mobile systems are studied and / or planned and the results of the attendant technical investigations.

Among the most important system design guidelines we here cite:

— the *user channel* information rate is 4 Kbit/s;

— the Forward-Link *FDMA* access comprises 32-Kbit/s *TDM* carriers (32-Kbit/s *TDMA* carriers in the Return-Link), each one supporting eight user channels. With QPSK modulation, the nominal TDM, or TDMA, carrier spacing is 30 KHz;

— the TDM / TDMA time plans are arranged such that the MS can adopt an *RF switch* instead of a *diplexer*, thus simplifying the MS design and reducing RF losses;

— MSs operate at *L-band* (up-link) or *S-band* (down-link), while GWSs operate at *C-band*. The maximum User-Link bandwidth (S- or L-band) is 10 MHz. The maximum Feeder-Link bandwidth (C-band) is 50 MHz (25 MHz per polarization);

— the required satellite receive *G/T* is 7 dB/K at Edge-Of-Coverage (EOC). The corresponding maximum spot aperture is 1.9 deg. @ the 4-dB crossover, with an EOC gain of 33.5 dBi (1.55 deg. @ 3 dB). For an hemispherical coverage, a total of 109 spots would then be needed;

— under the requirement of limiting coverage to land masses and adjacent coastal waters, the *number of spots* to be actually generated by a satellite was found to vary with the orbital position, for a worst-case of 41 spots. Furthermore, it was possible to find a unique pattern of 64 spots which allows to generate the required coverage at any of the specified orbital positions, by simply de-activating some of the spots:

- total nominal *system capacity* is 5,000 user channels (corresponding to 625 TDMs), each spot being required to support up to 20% of this figure (i.e. 1,000 channels or 125 TDMs, corresponding to 3.75 MHz). Such a traffic re-routing flexibility has an impact on both the on-board routing processor and the transmit front-end design. The minimum spot capacity is 1 TDM;

- the resulting average *frequency re-use* factor of 1.875 is accommodated by means of a 7-cell frequency re-use cluster (assumed isolation requirement of 21 dB);

- the above capacity figure results in challenging requirements for the RF transmit front-end (total *EIRP* of 60.7 dBW);

- dynamic *level-control* (Forward-Link) and *power control* (Return-Link) is adopted not to impose the maximum margin (7 dB) on all channels simultaneously. Voice activation is also used to reduce the total on-board EIRP requirement (a 40% activity factor was assumed);

- a unique *reconfigurable satellite design* is to be employed for operation at different orbital positions (each serving an oceanic area). Furthermore, each satellite shall be in-flight reconfigurable to act as a spare satellite for any of the others;

- for compatibility with an *host spacecraft* to be launched with Ariane-V, the payload shall remain within a total mass envelope of 3,500 Kg (at launch) and a power envelope of 6400 W (EOL at equinox).

3 Implementation approaches

The above referred to study activity has been focussed upon the assessment and trade-off among three main payload implementation approaches, i.e.:

- *Conventional* payloads, i.e. "transparent" payloads which utilize traditional filtering devices (e.g. SAW filters) to realize the required channel-to-spot routing functionality. Flexible RF power allocation to spots is ensured by means of appropriate Analog Beam-Forming (ABF) antenna solutions, exploiting Multi-Port Amplifiers[1] (MPAs) or active design concepts (e.g. imaging or multi-matrix arrangements).

- *Processing* payloads, with particular reference to transparent approaches which utilize "unconventional" devices, built around DSP techniques, to achieve the above mentioned spot-routing function, while still relying upon the same ABF antenna concept suitable for the conventional payloads category.

- *Digital Beam-Forming* (DBF) payloads, which, taking advantage of the potential synergy between the digital repeater implementation techniques and novel digital spot-formation techniques, may result in an integrated architecture not only capable of satisfying requirements in a flexible and efficient manner, but of also offering additional system-level features not otherwise attainable with less advanced antenna design approaches.

Payloads representative of the three categories are discussed in sects. 5, 6 and 7. Before this, the ABF antenna design and implementation aspects are presented in sect. 4, this antenna concept being common to both the *Conventional* and the *Processing* payload categories.

4 Analog beam-forming antennas

Three antenna architectures[2] were considered particularly feasible for the GSO personal mission, namely:
— focussed reflector with single-feed-per-beam;
— focussed reflector with triple-feed-per-beam;
— Direct-Radiating Array (DRA).

The last approach was however early discarded because of the technology criticality, the excessive size, the complex deployment and cable routing.

Separate antennas are used for transmit (S-band) and receive (L-band), this solution largely simplifying the antenna design and allowing to control the passive intermodulation products problem, which may otherwise be hard to manage for a payload generating a large amount of RF power. Fig. 1 shows the satellite with the 5-m S-band unfurlable reflector and the 8-m L-band unfurlable reflector.

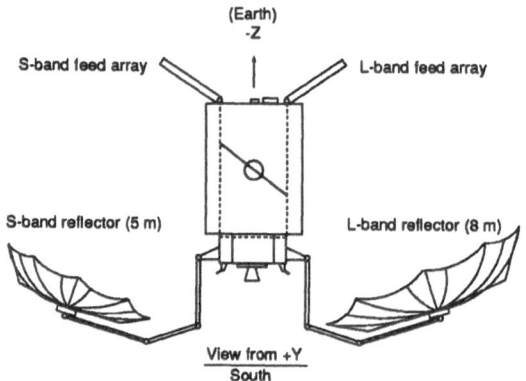

Fig. 1 GSO satellite pictorial

For the transmit antenna a trade-off was made between a single-feed-per-beam configuration based upon eight 8-by-8 MPAs, and configurations based upon beam-synthesis, either adopting five 16-by-16 MPAs or ten 8-by-8 MPAs. The main results of the trade-off are summarized in tab. 1, in which also the power consumption of the RF power amplifier has been duly taken into account.

Configuration	8 8-by-8 MPAs	5 16-by-16 MPAs	10 8-by-8 MPAs
Beam synthesis	1 feed per beam	≥ 3 feeds per beam	≥ 3 feeds per beam
Mass (Kg)	169	214	206
DC power (W) for EIRP = 60.7 dBW	2700	2000	2000
Traffic support flexibility	12.5 % of total power in a spot	≥ 60 % of total power in a spot	≥ 30 % of total power in a spot
Development risk	low/medium	high	medium

Tab. 1 Transmit antennas trade-off (ABF)

The configuration featuring ten 8-by-8 MPAs was adopted as baseline. Fig. 2 shows the transmit S-band antenna layout, comprising a feed array of 80 elements. The 8-by-8 MPAs in the output section allow the realization of the spot-synthesis approach (3 feed-per-beam was assumed), while complying with the traffic reconfigurability requirement (a maximum of 30% ot total power can be delivered into a single spot).

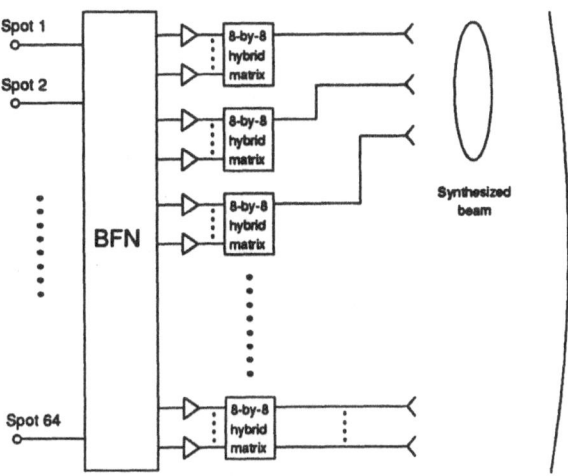

Fig. 2 Transmit antenna architecture

At to the L-band receive antenna, a similar approach is proposed, still with an 80-element feed array. This arrangement is shown in fig. 3.

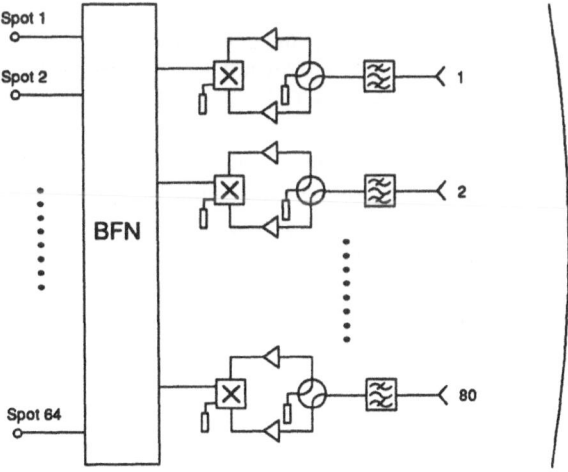

Fig. 3 Receive antenna architecture

State-of-the art technologies are deemed necessary to meet the requirements: as a matter of example, use of HEMT-based low-noise amplifiers and square coaxial technology for hybrid matrices should be considered.

A key component of both the transmit and the receive active antennas is the passive Beam Forming Network (BFN) which implements the spot synthesis function.

5 Conventional payloads

This category of payloads is in principle attractive mainly because of the minimization of the inevitable risks associated with new technologies.

With the assumed use of fixed-bandwidth SAW-filters, only a discrete number of different bandwidths can be provided, this resulting in a generalized oversizing of bandwidth assignments to spots with respect to the minimum values required for supporting the current traffic levels. In particular the following problems are experienced:

– loss in User-Link bandwidth utilization efficiency, caused by the quantization of bandwidth allocations to spots. As a consequence, it is not possible to optimally herd the band portions being really utilized;
– loss in Feeder-Link bandwidth utilization efficiency, significantly impaired by the fact that the cumulative bandwidth of all the on-board filters is larger than needed;
– the power-robbing effect, whereby some of the on-board RF power is wasted by up-link noise being retransmitted into the down-link (this effect may be particularly important for the Return-Link).

In addition to the above issues, it was shown that these type of repeaters can hardly meet the specifications, while they do not allow for significantly simpler realizations. For this reason the study of *conventional* repeaters was limited to what needed for establishing a *benchmark* configuration, against which to measure the complexity and performance of the competing approaches.

The Forward-Link (FL) section of the *conventional* repeater is subdivided into two identical branches, one per up-link (C-band) polarization. Each branch feeds four IF Processors, each serving 8 down-link (S-band) spots, for a total of 64 spots.

Each IF Processor has the ability to allocate fixed-bandwidth chunks to each spot and to appropriately position the said chunks in the down-link frequency plan, such as to achieve a certain flexibility in meeting variable traffic distributions

The Return-Link (RL) section is structured in symmetrical manner.

Tab. 2 presents the power and mass budgets applicable to the analyzed *conventional* payload configuration. As expected, budgets are dominated by the S-band active antenna, while a non-negligible impact is given by the frequency generation subsystem.

6 Processing payloads

Despite certain advantages which would take place with on-board regeneration on the FL, it was decided to propose the adoption of a *transparent* FL repeater architecture, this solution offering maximum operational flexibility. On the other hand, a regenerative RL architecture would make little sense, as the increased RL repeater complexity may hardly be compensated for by system-level advantages.

	Mass (Kg)	Power consumption (W)	Power dissipation (W)
C-band receiver	5	16	16
FL IF Processor	48	104	104
S-band active antenna	206	2000	1400
L-band active antenna	139	36	36
RL IF Processor	48	104	104
C-band transmitter	19	272	152
FL Local Oscillators	56	336	336
RL Local Oscillators	24	80	80
Total	545	2948	2228

Tab. 2 Conventional payload budgets

The FL repeater block diagram is shown in fig. 4. An *all-digital* approach was proposed, with variable-bandwidth Digital DeMultipleXers (DDMXs), one per up-link polarization, separating variable-bandwidth groups of TDMs to be routed to different spots. In the proposed design, the fine position of individual TDMs in the output frequency plan is adjusted at digital level, while the gross position is set at analog level: doing so, the resulting increase in digital processing power is more than compensated for by the simplification occurring in the analog up-conversion arrangement following Digital-to-Analog (D/A) conversion.

Among the possible solutions to perform the demultiplexing function, the use of the so-called *Bulk FFT*[3] technique was proposed, as it allows for full flexibility in terms of TDM grouping and output frequency plan, a fact which also turns out in minimal waste of up- and down-link bandwidth. *Bulk-FFT* utilizes fast convolution through overlap-save (or alternatively overlap-add) FFT processing. A DDMX based upon *Bulk-FFT* performs the following functions:

— transformation in the frequency domain of the FDMA signal by FFT processing;

— frequency-domain filtering to separate the 32 output TDM groups;

— transformation in the time domain of each of the 32 TDM groups, via inverse FFT processing.

Bulk-FFT does not result, for this application, in excessive power consumption; furthermore it has to be noted that ASICs performing the elementary functions required for implementing a *Bulk FFT*-based DDMX are currently being developed by Alenia Spazio, with Space Engineering support, in the context of an Intelsat R&D contract.

This all-digital repeater architecture was preferred to an alternative one based upon *Chirp Fourier Tansform* (CFT) processing, because the complexity of an all-digital approach appears to be manageable even with today's technology, while CFT technologies are not yet mature for space applications. Furthermore the power consumption of

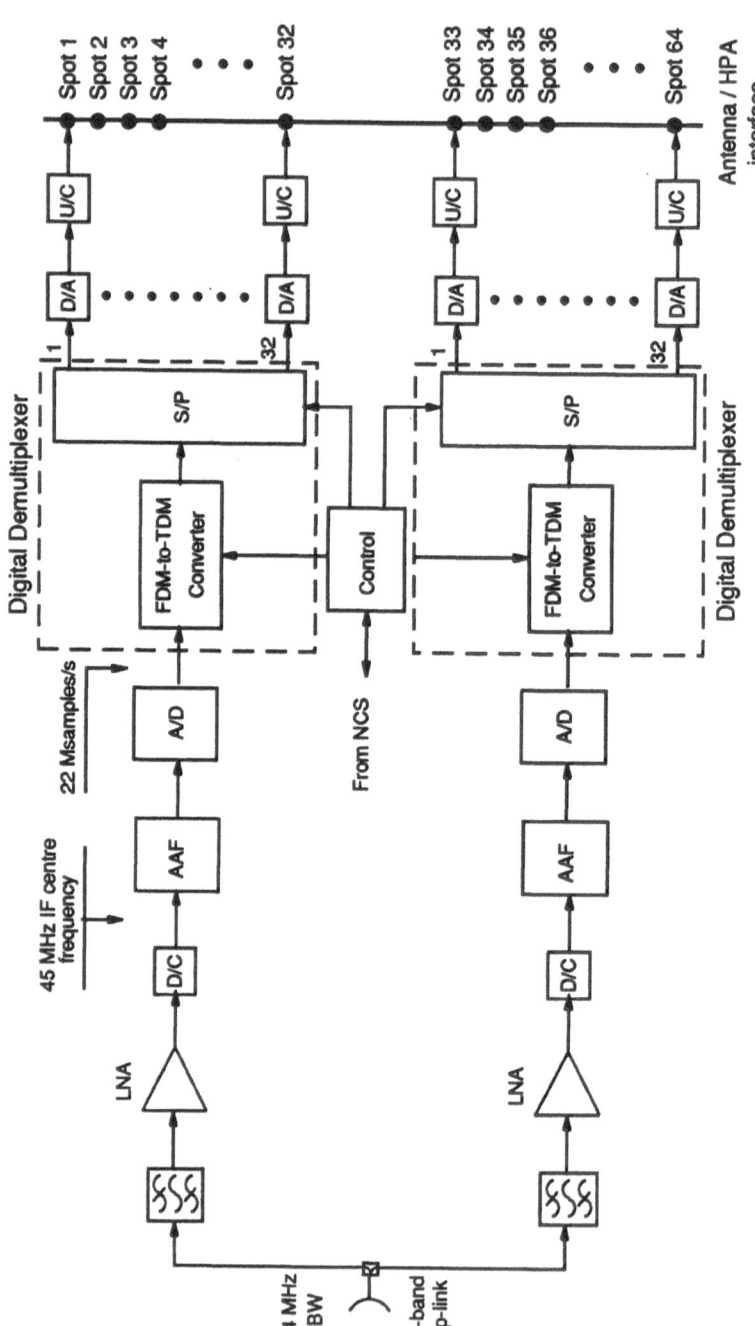

Fig. 4 Processing repeater block diagram (Forward-Link)

digital technology is expected to decrease significantly in the future, especially when the supply voltage of space-qualified ASICs will be lowered.

The RL repeater block diagram is shown in fig. 5. Again an all-digital architecture was proposed, with Analog-to-Digital (A/D) conversion of variable-bandwidth TDM groups (one per spot). With the aim to utilize a unique Anti-Aliasing Filter (AAF) bandwidth, thus simplifying the analog repeater front-end, TDM groups are all sampled at the same rate (the highest required). A digital multiplexer, also based upon the above referred to *Bulk-FFT* technique, is then used to assemble the down-link FDMA arrangement (one per polarization).

A different configuration would have to be adopted should one want to be able adjusting, on-board, the level of individual TDMs, so that the down-link power associated to each TDM can remain the same independently of up-link level variations. This approach, not further pursued, would require to firstly demultiplex the individual TDMs, and then to re-multiplex them into the down-link, after level adjustment (a fast-attack AGC would be required).

Tab. 3 shows the power and mass budgets for the analyzed *processing* payload configuration.

	Mass (Kg)	Power consumption (W)	Power dissipation (W)
C-band receiver	7	20	20
FL Digital Processor + U/C	23	90	90
S-band active antenna	206	2000	1400
L-band active antenna	139	36	36
D/C + RL Digital Processor	25	84	84
C-band transmitter	15	259	139
FL Local Oscillators	3	128	128
RL Local Oscillators	2	40	40
Total	420	2657	1937

Tab. 3 Processing payload budgets

This budget shows that savings of the order of 22% in terms of mass and 10% in terms of power consumption can be achieved, while allowing an almost ideal traffic routing flexibility and minimizing all C-band, L-band and S-band bandwidth wastes.

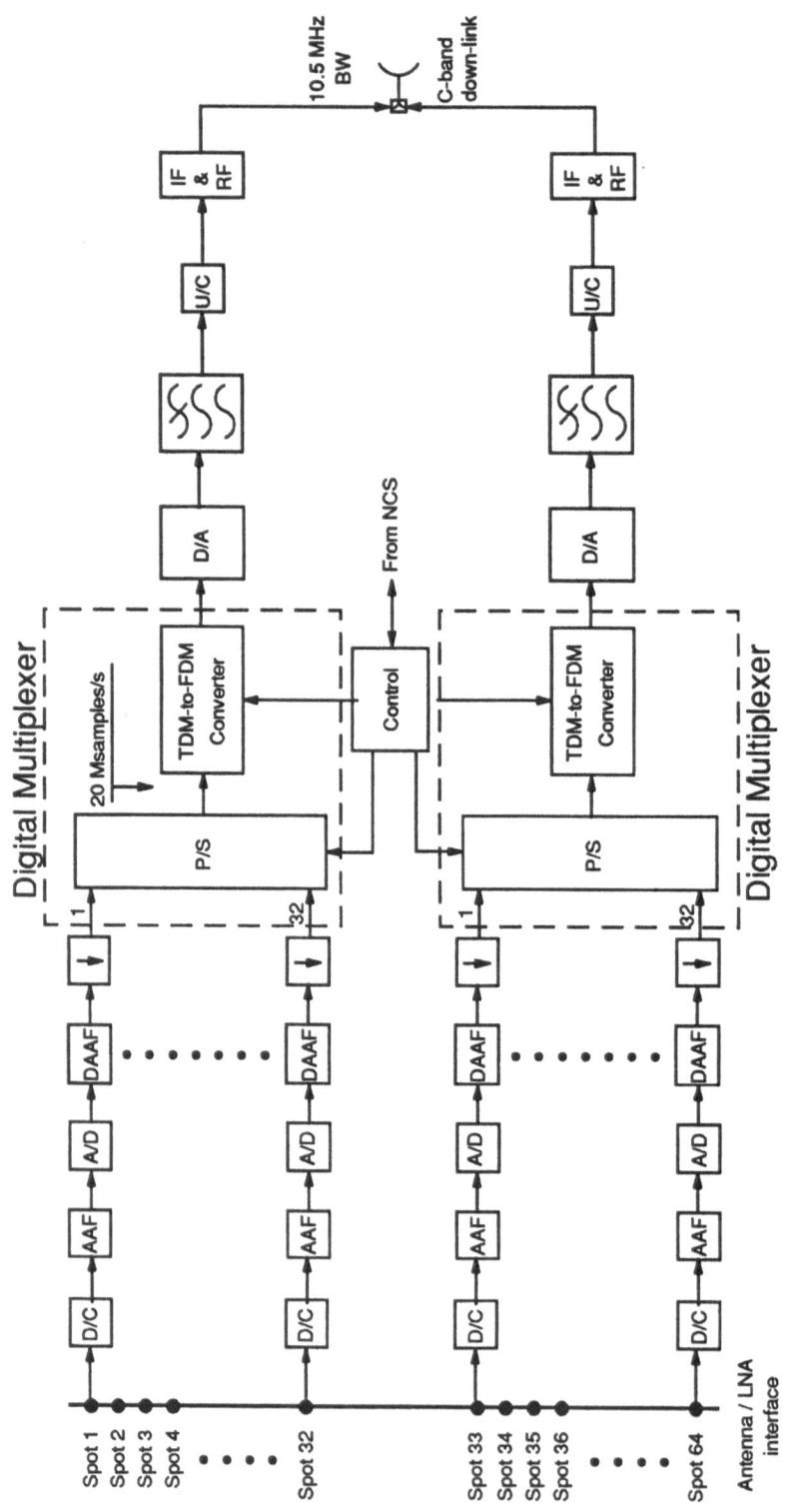

Fig. 5 Processing repeater block diagram (Return-Link)

114

7 Digital beam-forming payloads

The increase of complexity, versus the number of spots, for a payload based upon a *DBF* antenna is not as fast as with an *ABF* antenna. Due to the above fact, use of *DBF* techniques for mobile communications has often been proposed to generate a very large number of spots, in the limit one per mobile user, with the aim to recover the 3-dB loss typically experienced, with a traditional coverage, by users located at spot border.

In our case, *DBF* ha been examined under a different perspective, i.e. that of implementing a coverage concept largely different from that assumed for the *conventional* and *processing* payloads cases. Under this new concept the service area would be covered by quite a large number of fixed spots (up to about 300), overlapping at the -1 dB contour. With this solution, the link budget for worst-case users located at spot borders gets improved by almost 2 dB, this allowing to yield important savings in terms of on-board power. The flexibility offered by the *DBF* concept is also exploited to reconfigure the Earth coverage while in-flight, thus more easily permitting to utilize a single spare satellite for all orbital positions.

With such an heavy overlapping, a different frequency re-use cluster had to be adopted. Early investigations led to conclude that a 9-cell frequency re-use would be adequate. Although such a re-use cluster inherently tends to reduce the frequency re-use performance, the achieved frequency re-use factor did not drop as much, due to the compensation effect deriving from the increase in the total number of spots.

The block diagram of the FL DBF payload is shown in fig. 6. In this case the two branches corresponding to the two up-link polarizations are merged prior to beam forming as, with *DBF*, the use of a single BFN results in a lower payload complexity. DDMXs have here to operate at *single-TDM* level instead of at *TDM-group* level.

After the digital BFN, Frequency-Reuse (FR) Combiners merge samples of TDMs operating on the same down-link frequency in different spots (frequency re-use).

As far as multiplexing is concerned, the performed trade-offs led to conclude that, under the given challenging requirements, a fully-digital implementation may have to be delayed until when less power-hungry digital technologies will be available. Recourse was then proposed to a two-step *hybrid* approach[4] whereby digital technology is used for performing fine TDM multiplexing (up to a bandwidth of some hundred KHz), while analog technology, namely CFT devices, are instead adopted for final multiplexing (see fig. 6).

In line with the above strategy, the Multi-Carrier (MC) Multiplexers following the Combiners perform a first-level digital multiplexing, whereby several TDMs are frequency-division multiplexed into band segments assumed to all have the same bandwidth of a few hundred KHz.

After D/A conversion, Inverse-CFT (ICFT) devices act as fixed-bandwidth variable-center-frequency reconstruction filters multiplexing, on different frequencies, the various band segments.

The antenna architecture assumed for the *ABF* cases (80 feeds, spot synthesis by 3 feeds) was determined to be well applicable to the *DBF* case too (clearly disregarding the BFN, which is now implemented in digital form), and was therefore adopted. It has to be noted that the advantage provided by the *DBF* approach is more important for the FL, whereby each BFN input is distributed, with appropriate phases, to 24 BFN outputs, than for the RL, whereby only 3 BFN inputs are to be distributed to each BFN output.

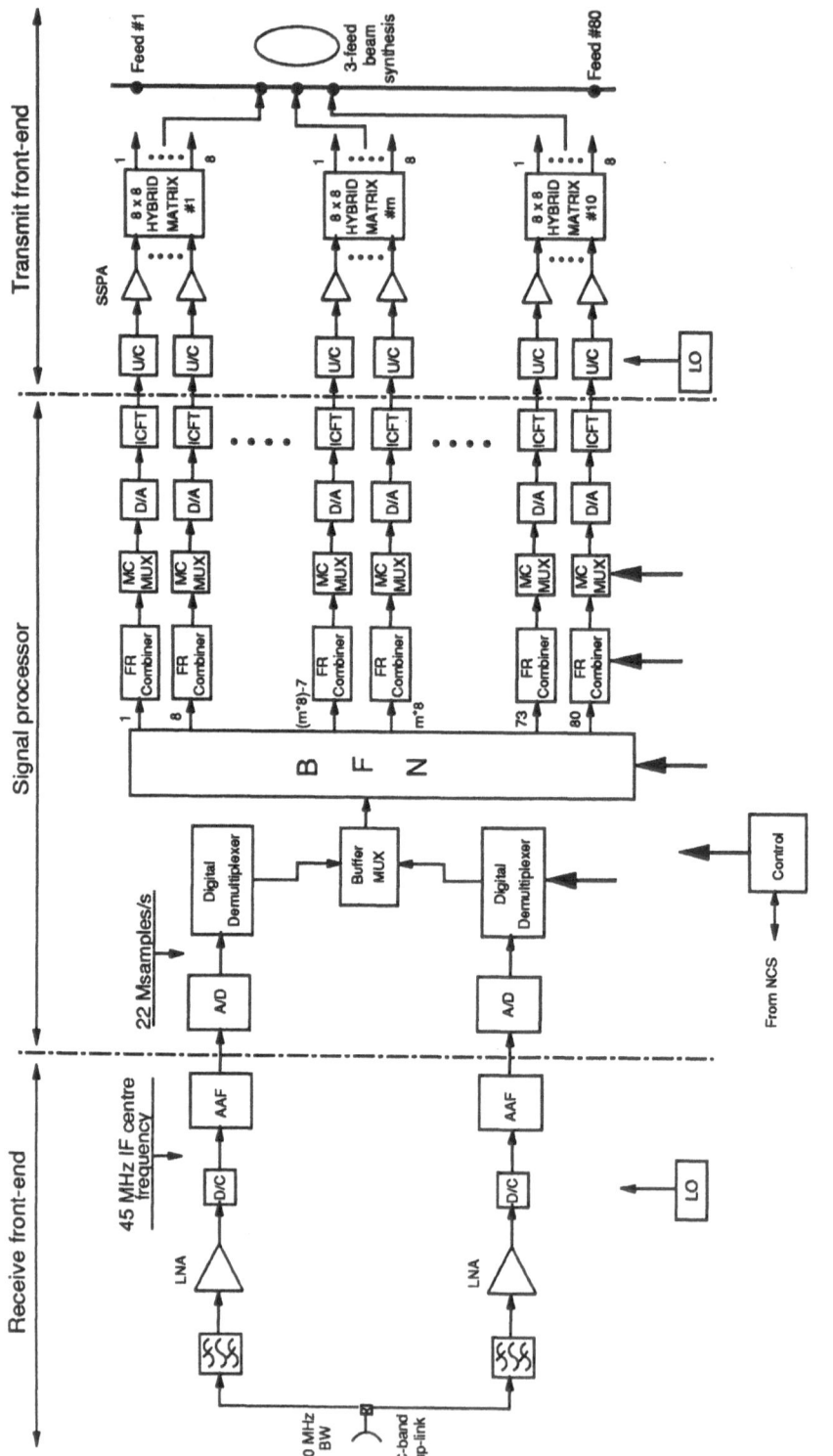

Fig. 6 *Digital Beam Forming payload block diagram (Forward-Link)*

Nevertheless, as the complexity of a *DBF* arrangement increases almost linearly with the number of feeds, further studies should be devoted to verifying whether a lower number of feeds may be sufficient to achieve the desired performance over the coverage area, possibly synthesizing each spot with a greater number of feeds, each one excited with a different amplitude and phase.

The block diagram of the RL *DBF* payload is shown in fig. 7. Considerations reciprocal to those presented for the FL case were made to support the adoption of CFT devices for implementing fixed-bandwidth variable-center-frequency anti-aliasing filters, which implicitly demultiplex the input TDM groups. Similarly to the FL payload, the digital multiplexers following the digital BFN operate at *single-TDM* level instead of *TDM-group* level (as in the *processing* payload case). This last feature may be exploited to implement an on-board level-control function on individual channels (as already hinted in sect. 6), also by means of digital techniques.

Tab. 4 shows the power & mass budgets for the analyzed *DBF* payload configuration. Despite the slightly lower saving (17%) in terms of mass now achieved with respect to the *conventional* payload, the power saving (34%) appears to be very significant. This saving mainly arises from the EIRP reduction made possible by a coverage with spot-overlapping @ 1-dB contour, notwithstanding the higher power consumption of the hybrid processor.

	Mass (Kg)	Power consumption (W)	Power dissipation (W)
C-band receiver	7	20	20
FL Hybrid Processor + U/C	46	116	116
S-band active antenna	184	1250	875
L-band active antenna	124	36	36
D/C + RL Hybrid Processor	69	243	243
C-band transmitter	15	259	139
FL Local Oscillators	3	16	16
RL Local Oscillators	2	16	16
Total	450	1956	1461

Tab. 4 Digital Beam-Forming payload budgets

8 Conclusions

A satellite personal communications system based upon GSO spacecrafts certainly poses formidable challenges to system and hardware designers. In particular, the high

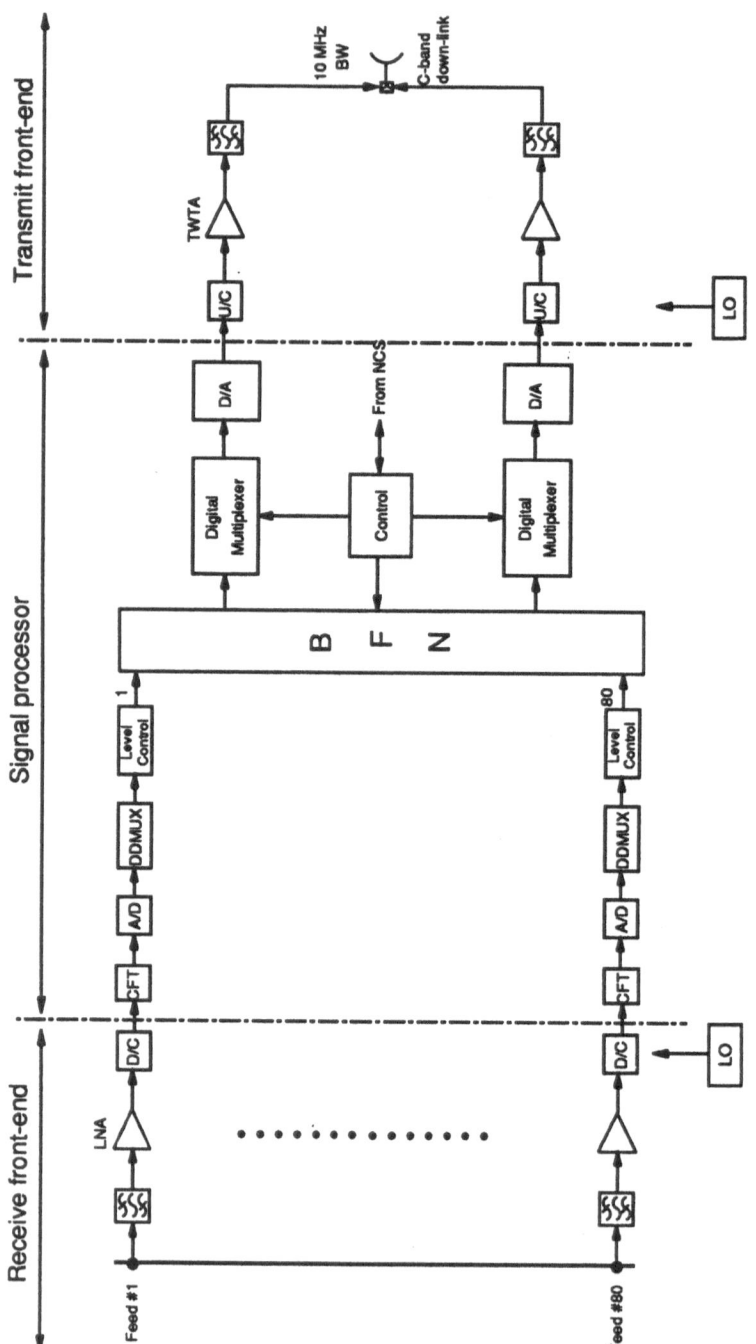

Fig. 7 Digital Beam Forming payload block diagram (Return-Link)

G/T and EIRP values to be attained for compatibility with hand-held terminal operation turn out into the need for large reflector antennas and huge on-board power amplifiers. Furthermore, with the necessarily high number of spots, provisions must exist for on-board channel routing and flexible RF power allocation, such as to efficiently match different and changing traffic distributions across the service area.

Both the above goals can be reached by means of active antenna designs, whereby each spot is synthesized by several feeds. Clearly, the repeater must also incorporate the demultiplexing / multiplexing functions needed for assigning bandwidth to spots, as required.

The use of a *conventional* transparent repeater design, in which the required filtering functions are implemented by means of traditional devices (e.g. SAW filters), is in principle attractive, because recourse need not be made to new and possibly unqualified technologies. The herein referred to study concluded that such a payload category may not be able meeting the given stringent flexibility requirements, unless at the expense of unreasonable complexity. A moderate-flexibility solution was then analyzed, which on the other hand did not prove to be either simple or particularly attractive in terms of mass and power consumption. Additionally, bandwidth utilization could not be optimized, both on the feeder- and user-link, not to make the payload architecture too complex.

The next step was that of investigating the suitability of a *processing* repeater, still fully transparent, whereby the required filtering functions are performed by means of fully-digital techniques (in particular use of the *Bulk-FFT* approach was suggested). With such an architecture, some reduction in mass (22%) and power consumption (10%) could be achieved, but the most important advantage certainly was the very high flexibility occurring when trying to match different traffic distribution. This fact also turned out in the virtual elimination of bandwidth wastes.

The last solution investigated was that of integrating a *Digital Beam-Former* into the digital processor, with the aim to generate an Earth coverage featuring a very high number (up to about 300) of overlapping spots (@ -1dB contour), which would otherwise be virtually unfeasible with analog beam forming techniques. This coverage concept allows saving a significant amount of RF power for a given total EIRP, while also resulting in the simplification of certain operational procedures (e.g. power control).

The transparent *DBF* processor, which with today's technology may still have to be implemented with a blend of digital and analog (namely CFT / ICFT) devices, obviously displayed a higher power requirement. This disadvantage was however more than compensated for by the signifcant EIRP reduction allowed by the highly over-lapped coverage. As a result, the *DBF* payload allowed sparing 34% of the power consumed by the *conventional* payload, while offering an unlimited flexibility, also in terms of in-flight coverage reconfigurations.

In conclusion, several solutions exist for implementing GSO payloads suitable to supporting personal communications services. Clearly, their actual feasibility will also depend on the efforts which will be paid on the development of critical technologies, mainly the reflectors, space-qualified 1.2 μm ASICs, solid-state power amplifiers, local oscillators, etc.

Finally, although systems based upon different constellations (LEO, HEO, M-HEO) may appear to be less demanding in terms of technology developments, they will certainly result to be much more complex from the mission operations and network management standpoints.

REFERENCES

[1] M. Lisi - "An Introduction to Multi-Port Amplifiers" - Microwave Engineering Europe, February 1993.

[2] G. Codispoti, M. Lisi, G. Panariello, V. Santachiara - "Alenia Spazio Activities on Active Antennas" - XXVII Scientific-Technical Conference on Antenna Theory and Technology, Moscow August 23-25 1994.

[3] "Study of a 2 Mbps Multi-carrier Demodulator and its Associated Front-End" - ESA contract No. 7195/87.

[4] "Study on Digital Beam-forming Networks" - ESA contract 8087/88.

Part III

Processing and network aspects

Interworking procedures between cellular networks and satellite systems

E. Del Re, P. Iannucci, F. Argenti
Dip. di Ingegneria Elettronica
Università di Firenze
Firenze - Italy

B. Evans, W. Zhao, R. Tafazolli
Centre for Satellite Eng. Research
University of Surrey
Guildford - England

Abstract

In the near future personal mobile communications will be provided either by cellular networks and by satellite systems. In terms of capacity and coverage features, the two types of systems are complementary and integration between them could assure noticeable benefits. The paper addresses an efficient management strategy of the RF resources of an integrated system compound by a cellular component and a satellite component and highlights procedures that permit interchange of traffic load between these two parts of the system.

1   Introduction

Future mobile communication systems will be probably compound by a cellular component and a satellite component. It has been well recognized, in fact, that a satellite system can offer a major chance to cover zones where the deployment of cellular infrastructures is not cost effective or not possible.

The benefits that may be achieved in areas where the satellite coverage overlaps to the cellular coverage have been less investigated but, in our opinion, they could have relevant impact on the quality of service offered to the users. The first part of this paper (sections 2 and 3) addresses the role of a satellite system in areas where same mobile services are assured by cells also. The second part (sections 4, 5, 6 and 7) focuses on the "interworking procedures", that is those procedures that allow interchange of traffic load between the two components of the mobile network. In these sections, where particular emphasis will be assigned to the handover, GSM protocols will be taken as reference. It has been assumed that the two systems are interconnected and that they share the same network infrastructures (e.g. the database for the localization of the mobile terminals).

2   Role of the satellite system

It is a well accepted idea that a satellite system can play a significant role to complete and enlarge the cellular service area, covering zones where terrestrial infrastructures are unable or not effective to supply mobile services. There are three situations that may be mentioned to this regard:

- *areas characterized by low density of users.* A cellular network does not cover the whole territory. Sparsely populated areas (rural and mountain) are left out of the coverage because the investments for the deployment and the maintenance of cellular infrastructures are not cost effective;

- *holes inside the cellular coverage.* A typical difficulty encountered in the implementation of a cellular network is to assure a uniform coverage inside the service area: because of natural obstacles and man made structures it is very demanding to place a base station so as to assure a suitable reception quality wherever in the cell.

- *ships and aircraft.* Cellular networks can satisfy land users. Mobile services aboard of vehicles which does not move on the land surface have to be managed necessarily by means of satellite systems.

Another promising application of satellite systems in mobile communication context is to provide a back up option when the cellular network is out of use because of failure arising in base stations or other cellular infrastructures.

All these topics refer to the possibility of using the satellite system to replace either permanently or occasionally the cellular network. However, a satellite system can offer significant benefits even in areas where cellular coverage is available, since it provides further traffic channels. In these areas, a mobile terminal, at most for a restrict set of services, may choose between the cellular and the satellite link. The choice criterion should optimize the performance of the whole system, taking the maximum advantage from the integration.

The simplest approach is to consider the spot beam resources as purely additional channels in the double coverage areas. The consequence of this assumption is that, in the zones where the satellite signal is stronger than the one from the cellular base stations, satellite resources are preferred, even if cellular channels with acceptable quality and power are available: in these areas the satellite is the first choice and mobile stations try to use the cellular component only when all spot beam channels are busy. Cellular and satellite resources play reverse roles in areas where the signal from the base stations is better received.

This strategy does not seem very convenient, if the following aspects are considered:

- the capacity of a satellite system is much more limited than that of a cellular network. This effect is essentially due to the width of the spot footprints: even if a wider frequency band than the one of a cell is allocated to the spot beam, the frequency re-use distance is much higher because of the extension of the spot footprint radius. If the choice between satellite and terrestrial RF band is based only on link quality parameter, satellite resources will be saturated by the mobile terminals roaming in areas where satellite signal is stronger, even if cellular coverage is available. In these areas, which may be supposed limited in size, a lower blocking probability is experienced, but in the remaining cellular coverage the satellite presence is not noticeable;

- use of satellite resources, probably, will result in higher charging for the user. It is not reasonable to use a more expensive option, when a cheaper one is available.
- power consumption at the mobile terminal is greater during satellite communications.

These considerations suggest the following conclusions:
- from the network management point of view, the satellite channels should be regarded as a more valuable resource;
- from the user point of view, they represent a more expensive option.

Both these items lead to a more reasonable resource management strategy. It is based on the principle that the mobile stations make access to the satellite band only when all the cell channels are completely busy: a satellite channel should not be allocated when at least one cellular channel is available. In other words, the cellular channels should be always the first choice, provided that they guarantee a satisfactory quality level, even in the areas where satellite signal is better received.

2.1 Blocking probability in an integrated system

One of the most relevant parameter that characterizes the quality of service of a communication system is the blocking probability experienced by the users: the lower this parameter, the better the service assured. A satellite system superimposed on the cellular coverage adds further traffic channels and permits to reduce the blocking probability experienced by the users during the access to the radio resources. However, if the second approach is assumed this reduction is more noticeable. This conclusion is suggested from the results of a simulation that has been carried out to compare the efficiency of the two approaches. The simulation performed considers a simplified case since the mobility of the terminals is disregarded. As a consequence there are not handovers from cell to cell.

Other hypotheses are:
- 80 traffic channels per cell;
- 37 cells inside a satellite spot footprint;
- exponential service time distribution, with mean 3 min.;
- the interarrival time of the new calls is supposed exponentially distributed, with mean $1/l$; l has been chosen so that the cell blocking probability is 0.03;

The aim of the simulation was to evaluate the blocking probability, versus the number of spot traffic channels, in the following two cases:
- *Selective choice.* In this approach the mobile terminals try to use the satellite resources only if all the cell channel are busy.
- *Non-selective choice.* In this strategy the mobile terminals use the best channel available, disregarding if it belongs to the satellite or cellular pool. It is assumed that a fixed percentage (3%) of terminals, inside each cell, perceives the satellite signal better that the cellular ones. These terminals try the set-up on cellular resources only when the spot frequency band is completely busy. The remaining number of terminals, on the contrary, make an access to cellular channels firstly.

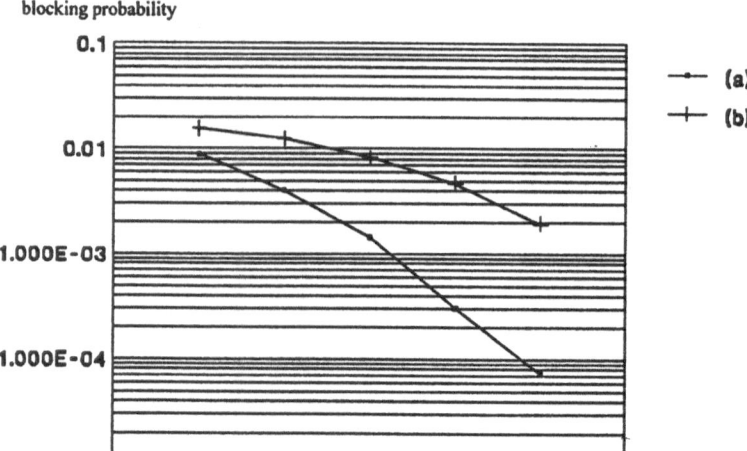

Figure 1 - Blocking probability: a) selective choice; b) non selective choice

The results obtained are shown in figure 1. It is evident the increased efficiency of the selective strategy.

Reduction of the blocking probability - The reduction, on average, of the blocking probability may be employed in two way to enhance the performance of the whole system:

- *design of the cells with a lower number of channels*; the cellular network has to cope with sudden concentrated overload, due to the mobility of the customers and channel assignment is usually tailored to traffic peak conditions. If there is a satellite band that collects overflow traffic, it is possible to design the cellular sites according to looser requirements;

- *delay of the investments*; the constant growth of customers requires additional investments to assure the same quality of service. With satellite overflow channels these investments can be postponed, prolonging the network life.

3   Selective choice and interworking procedures

The previous sections suggest that the satellite frequency band has to represent a sparing possibility with respect the cellular coverage, replacing the terrestrial channels when and where, for whatever reason, they are not available. In the following sections we will consider the "interworking procedures", that is those procedures by which a Mobile Station (MS) performs the "switch" between the satellite component and the terrestrial component of the integrated system. The switch can be carried out both when the mobile terminal is in idle mode and when it is communicating. As interworking procedures we can consider the following ones:

- selection and reselection in the cellular band and in the satellite band;
- handover between cell and spot beam or vice versa.

A mobile terminal operating in a cellular network is able, with the selection and reselection procedures, to select the Base Station (BS) most suitable for communicating over the radio path. With the handover the mobile terminal is assigned a new traffic channel during the communication. In a cellular network, both these procedures are part of the mobility management which makes it possible to keep track of the roaming of the terminal within the network, both when the terminal is idle mode and it is communicating.

In an integrated system, the selection and the handover make possible the switching between the two components of the integrated network.

4 Selection and reselection

In a cellular network, the mobile terminal performs the selection activity when it is idle. Aim of this activity is to determine the BS most suitable for the communication, that is, in short, the cell where the terminal is located. It is possible to distinguish between procedures of selection and reselection. The selection is carried out when the terminal does not have information of the cellular environment: this situation arises after the switch-on or the re-entering into the coverage area of the network. The reselection is a routine procedure performed when the terminal is on, has selected a cell and moves inside the radio coverage.

The cell choice is based on the evaluation of some parameters, related to the strength and the quality of the signal received by the mobile terminal on the Broadcast Control CHannel (GSM terminology), broadcast in each cell. During the set-up, the selected BS is the access physical point to the network: if the MS is the calling side, it requests a dedicated channel on the Random Access Control CHannel (RACCH) of the selected cell; if the terminal is the called side, it replies to the paging message sent by the base station of the selected cell.

If a satellite component is superimposed to the cellular component, the terminal should carry out a "selective selection", firstly by searching for a suitable terrestrial cell. Only if it not possible to find a cell satisfying the minimum threshold related to the intensity of the received signal (*selection* or *reselection failure*), the MS tries to select a spot beam, switching to the satellite band (such a situation occurs in the holes of the cellular component or outside the cellular coverage).

Selection and reselection in the satellite frequency band could be performed with a specific procedure, different from the selection arrangement of the cellular component. However what is important for the switching process is that in every spot a "beacon" signal be present, to be checked by the terminal in order to ascertain the availability of satellite resources [2].

Suppose the MS selects a spot beam because it is outside the cellular network. It is possible that it re-enters in the cellular coverage. It is convenient, to avoid the unnecessary occupation of satellite resources, that the MS, when it has selected a spot, starts a selection procedure periodically to verify the availability of the cellular coverage. Figure 2 shows the logical diagram of the selective selection and reselection.

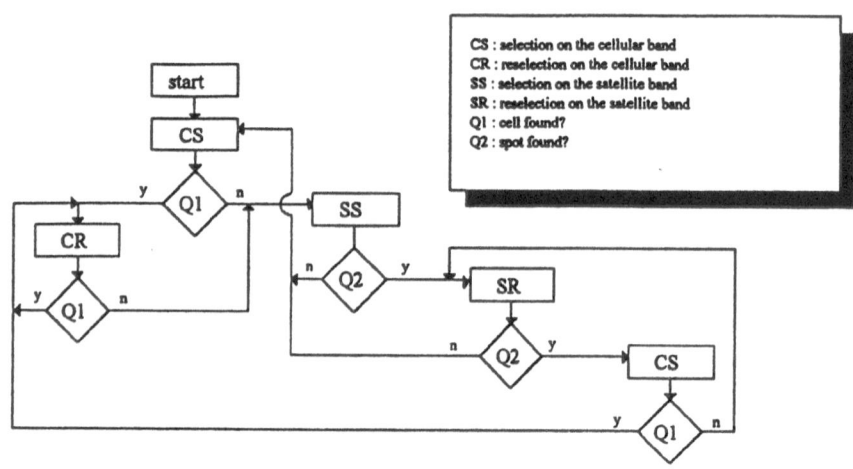

Figure 2 - Selection and reselection in an integrated system

Overflow detection - With selection and reselection procedures a terminal is able to detect when it is outside the cellular network. It can ascertain this situation if it is not possible to find a radio channel satisfying the prescribed quality and power level. In a similar way, if the mobile station is inside the cellular coverage, it should be able to detect traffic overload condition. When this situation arises the MS should start searching for a satellite resource, so that the satellite acts as a buffer collecting the overflow of the cells.

Traffic congestion of RF cellular links might be signaled from the base stations by a suitable parameter (*barred access* parameter) that should be checked by the mobile terminals during the selection and reselection procedures. When the parameter is set, the access to the network from this cellular site is denied. If all the cells satisfying signal strength requirements have barred access, the terminal tries to select a spot beam. After a spot selection the terminal has to perform the registration on the satellite sub-network.

Actually, this approach works only if some refinements are adopted. In fact, as soon as the barred access parameter is set, a lot of mobile terminals must presumably perform the location updating on the satellite link. To avoid congestion on satellite service channels, some modifications, with respect to the previous mentioned method, can be proposed:

- all the terminals could be shared in different *access classes*, which can be distinguished by a number;
- when the BS sets the barred access parameter, it broadcasts also a series of access class numbers: only the terminals belonging to this classes have denied access;

- the barred access parameter must be set when a certain number of cellular traffic channels is still available;
- the MS with barred access selects a spot beam and starts the registration procedure to be registered in the appropriate satellite paging area.

A further modification may be introduced to avoid the use of satellite service channels for registering the mobiles in the satellite paging area:
- when the barred access parameter is set, also a series of access classes and the indication of one satellite selection channel (satellite BCCH) are broadcast. The satellite selection channel is referred to the spot beam where the cells with barred access are included;
- the involved mobile stations tune to the satellite selection channel and send an acknowledgment;
- the network deletes the involved terminals from the terrestrial paging areas and registers them in the appropriate satellite paging area. In this way it is unnecessary for the MSs to start a location updating procedure.

5 Handover between cell and spot footprint

In a cellular context, the handover avoids a forced termination of the call when the terminal is leaving the serving cell and entering the next one; however it can also be carried out for other causes, for instance traffic management reasons: the cellular network, monitoring the occupancy of the cell channels, may decide to redistribute the traffic load preventing congestion situations. The passage of a call between the two components of an integrated system may happen in two direction:
- from satellite footprint to cell;
- from cell to satellite footprint.

The two cases must be examined separately, since they present specific problems.

Handover from cell to spot - The handover from a cell to a satellite footprint might be useful in the following circumstances:
- when the mobile terminal is in the boundary cells of the cellular network and it is leaving the cellular coverage, but still inside a the spot coverage. In this case, the handover from a cell to a spot may be regarded as an extension of the usual inter-cell handover, as it permits the communication to be continued in spite of the moving of the user outside of a cell.
- when traffic congestion arises in the cellular network and the zone involved is covered by the satellite footprint. In this way, teletraffic that would be refused by cells is collected by the satellite component.

So the implementation of the handover from the terrestrial component to the satellite one might be limited only to specific cells, that is:
- boundary cells of the terrestrial coverage area with a noticeable outgoing traffic form the cellular network;
- cells inside the cellular coverage and affected by occasional peak teletraffic.

In the first case the cellular network can detect that the mobile terminal is leaving the radio coverage if the received signal is weak and do not exist other base stations which receive the signal sufficiently strong. Then the handover procedure toward the satellite link can be activated. In the second case the cellular network may hand over the communication to the satellite link if the cells where the call would be passed to do not have available channels. In both the cases, the destination spot should be easily determined by the satellite sub-network, on the basis of the position of the cell.

Handover from a spot to a cell - The handover from a spot footprint to a cell can be viewed as a interesting management chance to reduce the occupancy of satellite resources: this possibility enhances the performance of the whole system making more available the satellite channels.

The implementation of this procedure presents specific key issues:

- it must be possible for the satellite system to identify if the user is inside a cell that can perform this kind of handover. The problem is related to the locating of a mobile terminal by the satellite component. The terminal position information should be updated during a via satellite communication and this information should permit to spot the cell where the MS is at the moment, accurately enough;

- the terminal has to evaluate the quality of the signal received from the base station and, possibly, to check some operative parameters on the RF cellular link;

- before beginning the procedure, the satellite component has to know if traffic channels are available in the zone of interest, that is in the cell where the communication should be handed over. This information could be supplied either by the terrestrial sub-network or reported by the mobile terminal: in the latter case the busy condition on the RF cellular link must be communicated by the base station.

A possible way to simplify these issues is to restrict the implementation of the procedure to a limited number of cells which have to be carefully individuated. Recall that a communication is handed over to the satellite when the user is approaching a cell completely busy and, then, the cells provided with this capability should be the cells affected by occasional peak load. When the user is leaving this cell, it is convenient that the communication comes back to the cellular channels. Then, the cells able to perform a handover from the satellite system to the terrestrial environment should be adjacent to those cells able to carry out the reverse switch.

It should be highlighted that the condition that has determined the access to the satellite component could affect the feasibility of the procedure:

- if the set-up on the satellite link happens because of traffic congestion on the cellular one, the MS has at the beginning, "good knowledge" of the local cellular environment;

- if the access to the satellite occurs because the MS is outside the cellular network, it has no information of the cellular environment.

6 Handover procedures

In this section, possible handover techniques between cell and spot footprint are presented and discussed. It is supposed that in the satellite system there are two types of Earth Stations:

- *Primary Earth Station Controllers (PES/Cs)*; PES/Cs represent point of attachment of the mobile terminal to the satellite system. The access to satellite resources by a terminal happens always trough a PES/C, both during the call set-up and the handover from cellular component to the satellite component. After the access, the PES/C assign the terminal a dedicated channel;

- Primary Earth Stations (PESs); PESs do not have direct control of satellite resources: its traffic and associated signaling channels are allocated by a PES/C. All the functions of a PES are included also in a PES/C.

The reference architecture of the satellite system is shown in figure 3.

Two types of handover may be considered:

- backward handover; in this handover the access of the terminal to a new base station is performed after the fixed network has established a new link toward this base station;

- forward handover; the access to the new base station is carried out before the link is established.

In the following, the GSM handover procedures will be taken as reference. It is supposed that the access to the satellite link is of TDMA type.

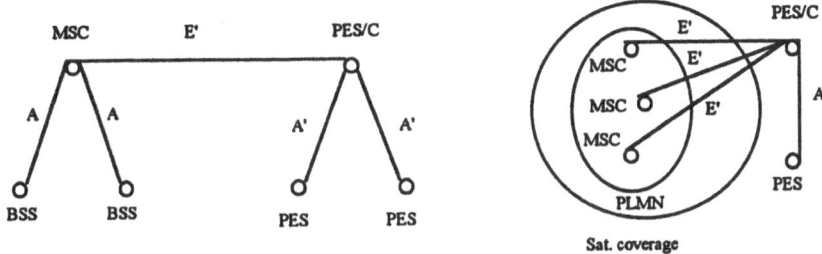

Figure 3 - Selected integration architecture

6.1 Backward handover

The GSM handover procedure is a backward handover. In this section some problems related to the application of such procedure to an integrated system are analyzed.

Handover from cell to spot beam - Figure 4 shows a possible procedure of handover from cell to spot: it is a GSM-like procedure, with some changes. This procedure present two main drawbacks:

- access to the satellite link. In the GSM, the serving BS communicates the new channel by means of the message *handover command*. In this message frequency and timeslot of the new channel are indicated. The MS sends *handover access* bursts to the new BS at the beginning of the assigned timeslot. Note that the instant of the beginning of the slot is detected by the MS after a propagation delay T_{BM}, dependent on the distance between BS and MS. In turn, the handover access burst reaches the BS delayed by T_{BM}. Due to the two delays, the time interval, at the BS, between the beginning of the slot reserved for the access and the arrive of an access burst is equal to $2T_{BM}$. However, in the GSM the access burst is sufficiently short so that it does not overlaps to the next slot, even if the MS is at the maximum distance from the BS. After receiving this burst, the BS calculates the single trip delay from the MS to its own position and converts this delay to the *Timing Advance* (TA) parameter. The TA is communicated to the MS, which,

131

from now, transmits its bursts with an advance of TA, with respect to the beginning of the allocated timeslot.

On the satellite link, this protocol must be modified. In fact, the delays involved are much wider, as a double hop FES-satellite-MS occurs. Moreover, also the range of the delay variation is much wider, since it depends on the position of the satellite on the sky and the position of the MS inside the spot. Using the GSM techniques the duration of an access slot should be $2(T_{FM})_{MAX}+t_{ab}$, where T_{FM} is the single trip delay and t_{ab} is the duration on an access burst. This timeslot may be reduced if the BS is able to communicate to the MS a roughly TA before the terminal send out the access burst. This TA should be equal to the minimum time advance required, with respect to the all possible positions of the satellite and of the terminal inside that cell. The MS will send the access burst using this TA. The PES evaluates the correction to be added to the minimum TA and communicates it back to the terminal.

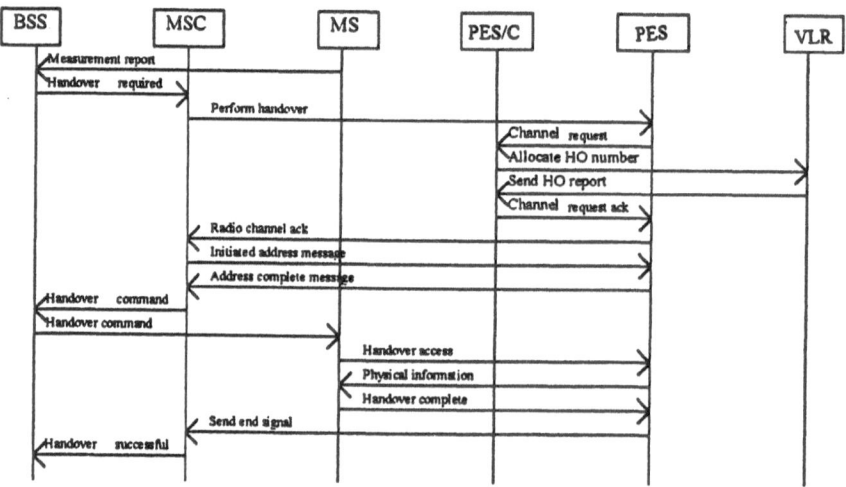

Figure 4 - Handover from GSM cell to spot footprint

- *total interruption delay*. In the procedure of Figure 4, after the MS send out a handover access burst, the PES/C allocates the TA through a *physical information* message. Receiving this message, MS send out *handover complete* to PES and begins to transmit messages on the assigned traffic channel by using this TA value. Note that, in this process, only delay related to this exchange of messages takes up to three single trip delay. If we consider the packet transmission delay, signaling processing delay and transcoding as well, total interruption duration can be calculated according to different satellite constellations. Table 1 shows different interruption times, T_d, for different satellite constellations. For LEO, MEO and GEO constellations, proposed systems are considered; for HEO, supposed figures are used. In this table H is the satellite height, a is the minimum elevation angle of a MS in degree, b is the minimum elevation angle of PES/C. T_d includes three parts: single trip propagation delay, packet transmission delay and signaling processing delay in the switching machine. Packet length is the same as in GSM

(240 bits) and radio channel gross bit rate is 9600 bit/s. Also for processing delay in the switching machine, the GSM figure is taken (150 ms).

From this table, it is evident that handover from cell to spot with exact GSM protocols will produce a significant interruption. This degrades the call quality dramatically.

Table 1 Handover interruption duration

	H (Km)	a (deg)	b (deg)	T_d (ms)
IRIDIUM	780	8.2	5	227.55
MAGSS-14	10350	28.5	5	451.30
HEO	25000	30	5	753.86
INMARSAT-M	36000	30	5	976.98

Handover from spot beam to cell - A possible procedure, suggested from GSM protocols, is shown in figure 5. Also in this direction problems arise:
- *identification of the candidate cells*. In the GSM during a call, the MS reports information about the strength and quality of signals from neighboring cells. These measurements are communicated to the BS by means of the message *measurement report* , sent periodically by the terminal. Each cell is identified through the BCCH carrier frequency and the Base Station Identification Code (BSIC), that permits to distinguish all the cells being around the serving one. To identify uniquely all the cells adjacent to the serving cell, it is necessary for them to be characterized by a different couple BCCH-BSIC. If we apply the same method to the handover from spot to cell, it must be noted that cells with the same couple BCCH-BSIC might be inside the spot beam. The problem to identify uniquely the candidate cells may be resolved if we use the *Cell Global Identification* (CGI) code. The CGI differentiates each cell of the GSM coverage and is broadcast on the BCCH. In the message measurement report, the identity of at least one cell, among those reported by the mobile, must be indicated with the CGI. It is suitable that the specified cell is the one from which the received signal strength is the highest.

Note that the PES which passes the call sends the message *handover required* to the satellite MSC. In GSM, this message may be generated as reply to the *handover candidate inquiry* sent by the MSC and includes a list of candidate cells. Also in this list, the couple BSIC-BCCH, used in the GSM to identify each cell must be replaced by the CGI.

Another aspect deserves attention. During the call, in a cellular network, the MS has to monitor the BCCH carrier of the neighboring cells. The frequencies to be monitored are communicated to the terminal by the serving BS through the BCCH allocation list. In the same way, when the MS communicates through the satellite, in this list the PES should indicate the frequencies that are BCCH carriers in the area where the MS is at this time. In other words, the PES should adapt the transmitted list according to the position of the terminal in the spot footprint. This problem may be solved in the following way. Suppose that the terminal knows the CGI of the cell where it is immediately before the set-up on the satellite link. Soon after the set-up, it can communicate this information to the PES, using the message *measurement report* only for transmitting this indication. Knowing the MS position, the PES sends a BCCH allocation list, where only the frequencies that are BCCH carriers in that zone are indicated. During the communication the list will vary according to the MS position, which is determined by the PES on the basis of the cell explicitly indicated in the *measurement report*. Note that if the cell indication is missing at the beginning of the communication, the procedure cannot start. This condition excludes a handover from spot to cell when the set-up occurs outside the cellular coverage. Instead the MS can specify the cell when the access to the satellite happens because of traffic overload in the cellular network;

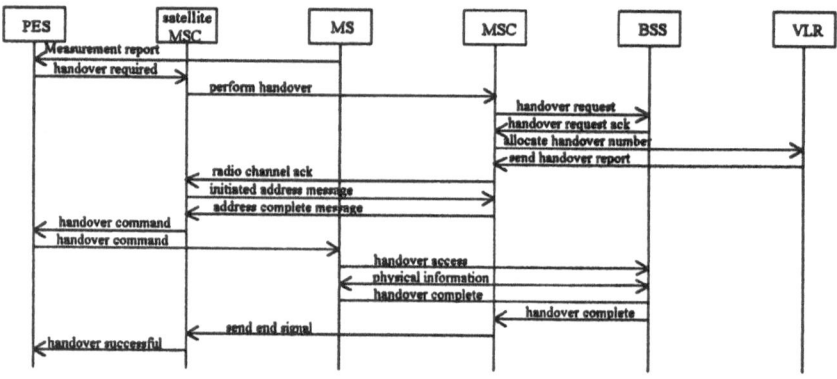

Figure 5 - Handover from spot footprint to GSM cell

- *packets overlap during handover*. Handover from satellite to GSM may cause some messages to overlap their packets during handover process. This will happen both on uplink and downlink. In order to explain this problem, we choose uplink transmission as an example and whole process has been shown in Figure 6.

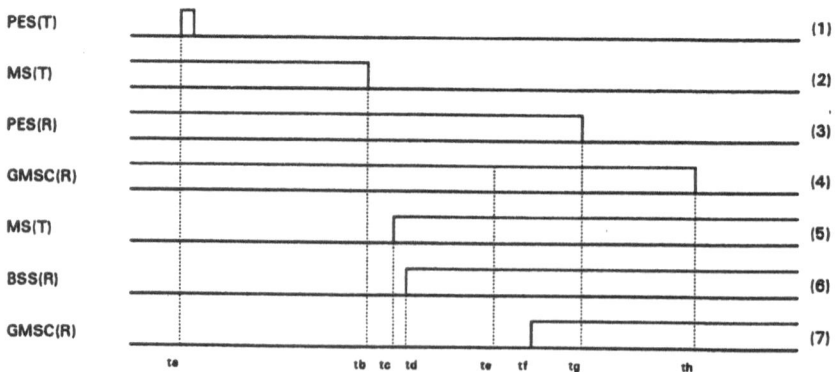

Figure 6 - Packet overlap during handover

134

In this figure, the first four lines represent message transmission in the satellite system during handover, the last three lines relate to message transmission in GSM. As mentioned, after satellite network has decided a handover to GSM is necessary, a *handover command* is sent out from PES to MS at time 'ta' (line 1). Receiving this command, MS will stop any transmission and prepare to contact GSM BSS at time 'tb'. (tb-ta) is one way propagation delay in satellite network. After transcoding and rate adaptation(th-tg), those transmitted packets by MS between 'ta' and 'tb' will finally arrive in GMSC between 'te' and 'th' (shown in line 4). Now we look at other packets which arrive in GMSC through GSM network. After MS contacts its target BSS and gets its timing advance ('tb' to 'tc' in line 5), it begins to transmit its following packets from time 'tc' (line 5), and the first packet will arrive in GMSC in 'tf' after propagation delay (td-tc) and transcoding delay (tf-td) which is shown in line 7. Comparing line 4 with line 7, we find those packets transmitted through GSM will overlap with those coming from satellite by (th-tf).

Table 2 Overlapped packets in different constellations

	IRIDIUM	MAGSS-14	HEO	INMARSAT-M
(th - tf) (ms)	17.02	91.61	192.60	266.98
Packet Number	3.70	19.85	41.80	57.85
(th - tc) (ms)	106.52	181.11	282.10	356.48

Time (th-tf) varies depending on different satellite constellation. Table 2 has depicted different value of (th-tf) and their corresponding packet number according to different satellite constellation (supposed figures of transcoding delay, frame length and time slot number in satellite system are the same as GSM ones). From this table we can see, some constellation will result in significant packet overlap. The main reason for this overlap is due to satellite propagation delay, and it's difficult to eliminate by modifying the handover protocol. There are two possible solutions: either in GMSC we buffer those packets coming from GSM and wait until the satellite packet transmission has been finished, or we discard those overlapped satellite packets during 'tf' and 'th' in line 4. The first option will produce a significant delay for GSM transmission as is shown in Table-2(th-tc). The second one will result in a detectable speech loss. According to the accepted GSM service quality, maximum one way delay should not exceed 90 ms, the second solution seems more practical since it will not introduce significant delay to GSM.

Packet overlap also happens in the downlink during handover process. In this case, overlapped packets will be received by MS but not GSM gateway. The figures are exactly the same as in Table 2. As a conclusion, handover from satellite to GSM will produce some overlapped packets in both uplink and downlink. It is inevitable for the GSM system to accept some quality degradation.

6.2 Forward handover

It is obvious to see the old signalling channel is not reliable to exchange handover information because it could fail unexpectedly. One solution to overcome this problem is to establish a new signalling channel between the MS and target base station. Handover information which is stored in the MS can be transferred through this new signalling channel. Finally, through the fixed network, the new base station will instruct the previous base station to release radio and network resources. This handover approach is known as forward handover. Figure 7 is a forward handover signalling protocol which is based on network structure shown in Figure 3. Several problems has been identified if forward handover is used in the GSM to satellite handover.

For a forward handover, MS should make handover decision. When it find a handover to satellite is necessary, MS contact a PES/C which has a better power level. Then handover parameters can be sent to PES/C through *HO parameter* signalling. PES/C is responsible to allocate traffic channel to PES, obtain handover number and finally inform MS these information. E'-interface between MSC and PES is only be used when new channel has been established and PES inform original MSC to release old channel (by *Send end signal*).

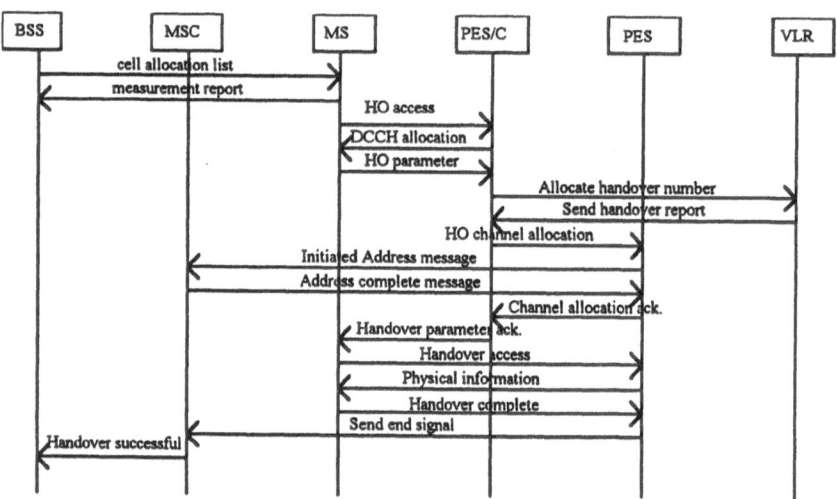

Figure 7 Forward handover from GSM cell to spot footprint

7 Comparing between backward and forward handover

In this section we compare the quality between backward and forward handover, several factors have been considered to be significant to represent the final result. Only handover from GSM cell to spot beam is discussed, forward handover in another direction is not suggested because significant signalling modification is necessary on GSM side.

7.1 GSM signalling modification

Signalling modifications will concentrate on A-interface, E-interface and radio interface. Following is a comparing between backward handover and forward handover as signalling modification on GSM is concerned.

Backward handover - Some modifications required to the GSM protocols have been discussed in section 6.1. Further problems arise if the MS has to perform measurements on the satellite environment also. According to GSM protocols, receiving *Cell allocation list* from BS, MS perform measurement for both current BS and surrounding BS, finally send them back through *Measurement report*. In the integrated system, if the MS has to determine target spot beam, parameters relating to satellite channel should be included in the *Measurement report*. Basically, *Cell allocation list* should also be changed to include satellite BCCH number, but this is difficult because of its structure which is designed specifically for GSM. In order for the MS to obtain its surrounding satellite broadcast channel number without modifying *Cell allocation list*, one possible solution is the MS scans satellite frequency band and find out those acceptable ones. Because scanning satellite band will take longer time, measurement report about satellite diversity and their power level can not be reported to BS very frequently.

Another reason to modify this radio channel signalling is target PES determination.. If MS provide PES with its satellite diversity together with spot beam diversity, MSC should be able to find out possible target PES because satellite motion is regular. The only way to provide MSC with satellite information is through *Measurement report*.

Forward handover - From Figure 7, related GSM signalling interface is A-interface and E'-interface. In a forward handover, handover parameters are not exchanged through A-interface and E'-interface. These interfaces are only used to inform original BSS to release old channel. As a result, only a subset of these interface is used. Modification to these interface is not necessary. On the contrary, minor modification on the radio link is considered to be necessary. Handover function between different sub-network is not allocated to each GSM cell, thus the MS should always be notified the possibility to switch to spot beam from a cell during a call. This information can be sent from BSS to MS either through BCCH or through SACCH.

7.2 Handover break

As shown in Figure 8, the handover break in a backward handover is significant. This is because before MS begins to contact with target PES, communication with the original BSS has already stopped. In a forward handover, communication activity with the original BSS will not stop until a new channel has been established (if original channel doesn't deteriorate too quickly). In this case, handover break is only limited to single trip delay time from MS to target PES. Typical handover break in a forward handover for different satellite constellation is shown in Figure 9.

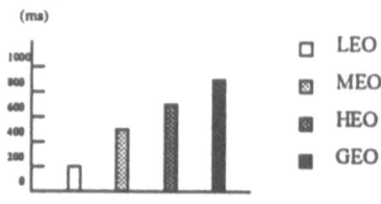

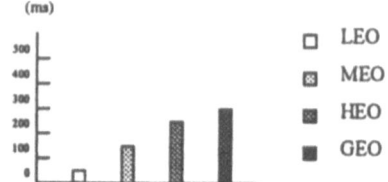

Figure 8 - Handover break   **Figure 9** - Handover break

7.3 Forced termination

Backward handover - A backward handover uses its original signalling channel to exchange handover information with the target base station. In our integrated system, target base station is a PES. According to Figure 4, the network, after it has detected a handover is necessary, contacts the target PES and get some handover parameters from the PES, finally informs MS through *handover command*. Getting those parameters can be very time consuming. If current channel corrupts before MS obtains those parameters, a call is released. The time T from the beginning of a handover is necessary (power level or BER is under threshold in uplink or downlink) till MS receives *handover command* can be calculated in following way:

$T = T(average)/2 + T(transmit) + T(signal)$

T(average): averaging time for those received measurement. It represent how long it takes for the network to detect a handover is necessary. In GSM, it is 15 s (32 samples, each of them takes 480 ms).

T(transmit): transmission time for the signalling *Measurement report*. This will prolong the handover detection time since every sample has to be transmitted back to BSS. It uses SACCH channel which bit rate is 382 bit/s. If the signalling include one satellite signal information (diversity is not considered), this figure can take up to 544 ms.

T(signal): signalling exchange delay. This is the time interval from the moment a handover is detected to be necessary until MS is informed to switch to a new channel (by *handover command*). Switching process delay is supposed to be 150 ms. In calculating signalling transmission delay, parameters from [5] are used, supposed average distance between MSC, PES is 200 km, MSC and PES to PES/C is 1000 km. The resulted figure is around 1.5 s.

Finally, the figure T is around 9.544 s. During this interval, if current channel doesn't corrupt, handover can be successful. Otherwise the call has to be released.

Forward handover - Comparing with backward handover, result coming from a forward handover can be different. In forward handover, the only reason for a call to be released is that there is no satellite visibility or there is congestion on the target satellite channel. Handover successful rate has nothing to do with current channel condition. A poor radio channel condition will only result in a handover break. Considering the handover time from a handover is necessary (power level or BER begin to reach handover threshold) until handover is successful(new channel is established by the time *Send end signal* comes out from PES), this figure can be calculated in the following way:

$T(HO) = T(average)/2 + T(signal)$

T(average) and T(signal) can be calculated in the same way as backward handover, finally, resulted handover time in a forward handover is shown in Figure 10.

Generally this figure is around 9-10 s depending on different satellite constellations since MS exchange every signalling through satellite channel (considering random access is successful at the first time). This figure is very similar to the figure calculated in a backward handover which determines if a forced termination will happen or not. Within 9-10 s, if current channel remain a constant change, handover will be finished with very little break which has been discussed before. If channel becomes corrupted during this time, handover still can be finished but with a longer break. This break can be well accepted if we consider an non interrupted call is more important than handover break problem.

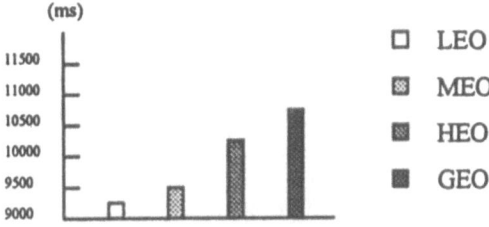

Figure 10 - Handover time in F-HO

7.4 Double channel occupation

During handover process, a traffic channel has to be established between target PES and controlling GSM MSC through terrestrial link. This is finished by signalling *Initiated address message* and *Address complete message* on E'-interface which is supported by SS#7. For a backward handover, from Figure 4, old traffic channel will not be released until current BSS has received *Handover successful* signalling. As a result, two terrestrial channels exist during handover process at the same time and one of them is redundant. Figure 11 shows double channel occupation time for different satellite constellations during a backward handover from GSM to satellite sub-system. Now we look at forward handover. From Figure 7, before a PES inform the PES/C *Channel allocation ack*, it has to ensure terrestrial traffic channel is available and obtain this channel by exchanging signalling *Initial address message* and *Address complete message* with controlling MSC. This channel is established afterwards. Double channel occupation time is shown in Figure 12. Comparing with Figure 11, it takes longer during a forward handover.

If a GSM cell has a heavy traffic density, many handovers can be found at any time. Because each of them takes two terrestrial channels for a little while, this will result in low efficiency for the terrestrial link is concerned. This interval should be as little as possible.

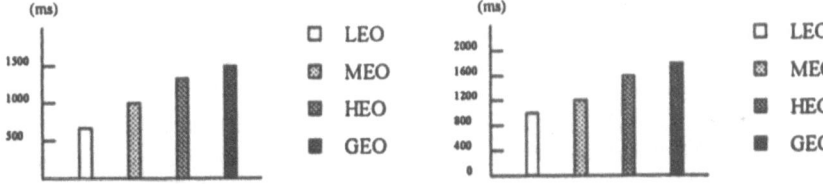

Figure 11 - Double channel occupation   Figure 12 - Double channel occupation

7.5 Reliability of the handover decision

In order to realise a forward handover, handover decision should be made by MS. MS has to get enough information to make this decision. Two kinds of handover strategies have been foreseen to support this handover. One is mobile controlled handover (MCHO), another is network assistant handover (NAHO). In a NAHO, handover information will be based on power level and BER from both uplink and downlink (like DECT system) because uplink and downlink may have different quality since they work on different frequencies. Current channel measurement is make by MS and BS, BS will downlink measurement result to MS. In the integrated system, a NAHO is difficult to realise because GSM BS never downloads those information to MS and modification to the GSM protocol is not suggested. Possible solution is MCHO. In MCHO, handover decision is only based on downlink information. Shadowing condition can be detected with this scheme but co-channel interference in the uplink can't be avoided. In a MCHO, a MS will measure power level and BER from the serving BS, and power level from its surrounding BSs. Because the decision is based on incomplete information, resulted handover will not be reliable and unnecessary handover probability must be higher than GSM handover. This problem can be relieved by the fact that frequency hopping is used in GSM, so that any co-channel interference can only exist on the uplink channel of the same MS temporarily. Concerning about this, handover reliability is thought will not deceased too much. Handover quality relating to this problem need to be discussed further.

8 Conclusion

In the first part of the paper, comparison between two management strategies of the RF resources of an integrated system has been carried out. Preliminary results support the idea to assign the satellite channel a lower priority in areas where the service is assured by the cellular network also.

The second part of the paper focused on the interworking procedures and considered the state of the terminal either in idle mode and during a communication. Particular emphasis has been assigned to the handover between the two component of the system and comparison (in the direction from cellular network to the satellite system) between backward handover and forward handover has been carried out. The main features of backward and forward handover are listed in Table 3. From this table, we have following conclusions:

1. A forward handover can avoid great signalling modification on GSM subsystem in the handover from GSM to satellite sub-system.

2. In an evolutionary change condition of current channel, forward handover is helpful in reducing handover break.

3. Result of a backward handover depends on the current channel condition. If received signal quality remain constant during handover for at least more than 9-10 s, backward handover can be successful but with a great break. On the other hand, forward handover work very well on this circumstance which only produce a very short break. When current channel condition becomes poor (signal quality goes under threshold in no more than 9-10 s), backward handover doesn't work anymore because lack of reliable channel. In this case forward handover can be

used but we have to accept a longer break (several seconds). Handover quality depending on different channel condition is shown in Table 4.

4. Forward handover overcomes problems coming from backward handover in the price of low terrestrial traffic channel efficiency and low handover reliability. In choosing handover scenarios in the integrated system, trade-off between these two sides should be well balanced.

Table 3 HO scenario comparison

	B-HO	F-HO
GSM signalling modify	A and radio interface	
handover break	long	short
termination probability	high	low
double channel time	short	long
HO reliability	high	low

Table 4 Handover quality in different channel quality

	B-HO	F-HO
Good channel quality	HO is successful. with a long break.	HO is successful. with a short break.
Poor channel quality	HO is failed.	HO is successful. with a long break.

References

1 R. Menolascino, E. Del Re, P. Iannucci, F. Settimo, *Traffic management strategies in an integrated satellite/cellular network for mobile services*, 2nd International Conference on Universal Personal Communications, Ottawa, October 1993

2 Multiple Access Communications Ltd, Southampton, England, *The Evolution of Second Generation Mobile Networks to Third Generation PCN*, A report for the Commission of the European Communities, DGXIII, April 1993

3 E. Del Re, F. Delli Priscoli, P. Iannucci, R. Menolascino, F. Settimo, *Architectures and Protocols for an Integrated Satellite-Terrestrial Mobile System*, 3rd International Congress on Mobile Communications, JPL Pasadena, June 1993

4 W. Zhao, R. Tafazolli, B. Evans, *Handover signalling analysis in the GSM and satellite integrated system*, RACE Mobile Telecommunications Workshop, Amsterdam, May 1994

5 W. Zhao, R. Tafazolli, B. Evans, *Handover scenarios in the GSM and satellite integrated system*, COST 227, 1994

Network Aspects on the Integration between the UMTS Network and Satellite Systems

Francesco Delli Priscoli

Dip. di Informatica e Sistemistica, University of Rome "La Sapienza", Rome, Italy.

Abstract

The integration between satellite systems and the third generation of mobile systems (referred to as Universal Mobile Telecommunication System (UMTS) by the European Telecommunication Standard Institute) is a largely debated issue.

This paper outlines the UMTS network architecture and identifies some basic issues for extending such architecture to an integrated system including a satellite. The possible options for carrying out the aforesaid integration are examined together with the relevant trade-offs.

The paper is based upon the work performed by the author for TELESPAZIO (Rome) in the framework of the RACE Project: "Satellite Integration in the Future Mobile Network (SAINT)". The opinions herewith reported are not necessarily those of the SAINT project as a whole.

1 Introduction

The third generation of mobile systems is referred to as Universal Mobile Telecommunication System (UMTS) by the European Telecommunication Standard Institute (ETSI) .

The integration between the UMTS and satellite systems is a very promising issue since it permits to immediately provide areas lacking in terrestrial facilities with the radio services offered by cellular networks [1],...,[4].

In general, satellite systems can provide a limited capacity with respect to terrestrial networks, nevertheless they are particularly suited in order to cover large terrestrial areas offering a scarce amount of traffic, since in these areas it is not convenient to implement the cellular network infrastructure.

Moreover, satellite systems can be profitably used in order to cope with contingent situations of unavailability of the terrestrial carriers and/or in order to shorten the terrestrial tails (a mobile and a fixed user involved in a call can communicate, via satellite, through the base station closest to the fixed user).

Finally, an interesting application is the use of the satellite spot-beams as they were *umbrella* cells. This means that satellite systems can be used for absorbing the traffic peaks caused by rapid and unpredictable traffic variations in the cellular network cells.

In the light of above, a likely scenario foresees several (GEO or non GEO, regenerative or transparent) satellite systems, each one tailored to particular requirements, integrated with the UMTS network. Common characteristics of these satellite systems should be a spot-beam arrangement (for coping with the limited performance of mobile terminals), the operation in a multi-Fixed Earth Station (FES) environment (for shortening terrestrial tails, for satisfying political requirements, etc.), some on-board switching capabilities (for increasing the space segment flexibility).

In the integrated network single-mode Mobile Terminal (MT) only equipped with a terrestrial transmitter are expected to coexist with dual-mode MTs also equipped with a satellite transmitter. In this paper we only refer to this last type of terminals.

In the above perspective, the present paper instead of proposing a single integration scenario, identifies the key characteristics (hereinafter, referred to as *attributes*) of the integrated systems, highlighting, for each of these attributes, the possible implementation alternatives and the trade-offs underlying the alternative selections. Clearly, the aforesaid selections are strictly dependent on the assumed *boundary conditions,* i.e. the target service area, the type of services to be provided, the target quality performance (e.g. in terms of BER, link availability, blocking and dropping probability), the target space segment complexity (e.g. in terms of payload capacity, payload switching capabilities,....), the technological state-of-art when the system is implemented, the selected access technique (either a Time Division Multiple Access oriented technique or a Code Division Multiple Access oriented technique), the status of the terrestrial network when the system is implemented and so on.

Section 2 describes some key characteristics of the UMTS which must be taken into account for its integration with satellite systems. In Section 3 some basic integrated system attributes are identified, highlighting their possible implementation alternatives and related trade-offs.

2 UMTS Architecture Basic Characteristics

At present, the UMTS architecture definition is at a very preliminary stage; so, the concepts exposed in this section are subject to variations.

2.1 Network Entities

In the framework of the MONET study [5], [6], in order to define the UMTS network architecture, a set of Functional Entities (according to [5] a Functional Entity is "a grouping of service providing functions in a single location") have been identified and have been mapped into some Network Entities (according to [5] a Network Entity is "a grouping of Functional Entities and constitutes the atomic unit of implementation"). A hierarchy is defined among the Network Entities (hereinafter simply referred to as *entities*): an entity at a given hierarchical level is *associated* with (in the sense that it is physically interfaced with and includes functionalities which have interest for) one or more entities at the hierarchical level immediately lower. Hereinafter, the entities are listed in increasing hierarchical order:

- the *Base Transceiver Stations* (BTSs): they include the functionalities to adapt the network signalling and user traffic to the radio interface; moreover, the BTS is charged for performing the measurements of link quality (e.g. in terms of C/I) to be used, together with the similar measurements performed by the MTs, in order to perform the handovers;
- the *Cell Site Switches* (CSSs): they represent the basic switch functionality and they contain the functionalities to control the communications between the MSCP and the one hand and one or more associated BTSs on the other hand;
- the *Mobility and Service Control Points* (MSCPs): they include the functionalities to guide the associated CSSs as concerns the control mobility procedures and service operation;
- the *Mobility and Service Data Points* (MSDPs): they contain the functionalities of a node of a distributed database; a MSDP stores the data relevant to the associated MSCPs.

Moreover, the following additional entities have been identified:

- *Local Exchange* (LE); it is responsible for local switching and should include the functionalities of an IBCN LE;
- *Transit Exchange* (TX): it is responsible for transit connections and related switching functions and should include the functionalities of an IBCN TX.

A distinction is performed between the *access network* (i.e. the part of the network *closer* to the radio interface which includes *local* functionalities) and the *core network* (which includes *global* functionalities). In general, access network entities can only be associated with access network lower hierarchical entities, whilst core network entities can be associated with both access and core network lower hierarchical entities. LEs and TXs are only present in the core network which, for the sake of simplicity, will be no longer considered in the following.

Fig. 1 shows an example of a network architecture for UMTS which highlights the possible relationships between the UMTS entities. The dotted line represents the boundary between the core network (included into the dotted line) and the access network. For instance, MSCP 1 can be associated with CSSs 2 and 3; MSDP 4 can be associated with MSCPs 5 and 6; MSCP 7 can be associated with (via LEs 10, 11 and TXs 12, 13) CSSs 2, 3 and 8.

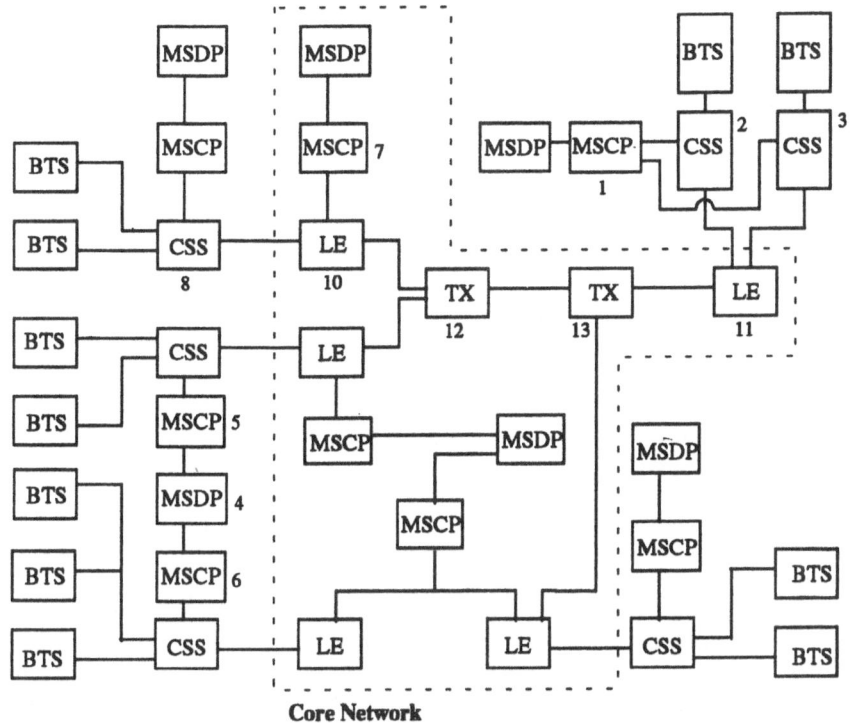

Fig. 1 - Example of network architecture for UMTS

2.2 Functional Areas

A *cell* is the locus where a *beacon* broadcast by a BTS is received with an *acceptable quality*. In a given point P, at a certain time t, we say that a given beacon is received with acceptable quality if and only if the received C/I ratio in P, at time t, is greater than the minimum one which still guarantees the target quality requirements, e.g. in terms of Bit Error Rate (BER). Note that, according to the previous definition, a given point P can belong to more than one cell.

Each beacon contains a *logical cell identification*. A one-to-one association exists between logical cell identifications and beacons.

Each BTS only broadcasts a single beacon. Therefore, a one-to-one association exists between cells and BTSs and a cell is univocally identified by its logical cell identification.

An other basic UMTS functional area is the *location area*, i.e. the area in which a MT may roam freely without updating the location information; location information update is required whenever a switched-on MT crosses the location area boundaries.

The following functional areas are related to the entities introduced in Sect. 3.1 [7]:

- *CSS area:* the CSS area corresponding to a given CSS includes all the cells served by BTSs associated with the CSS;
- *MSCP area:* the MSCP area corresponding to a given MSCP includes the CSS areas corresponding to the CSSs associated with the MSCP;
- *MSDP area:* the MSDP area corresponding to a given MSDP includes the MSCP areas corresponding to the MSCPs associated with the MSDP.

3 Basic Characteristics of the Integrated System

An integration scenario can be characterized through a certain number of characteristics, hereinafter referred to as *attributes*. For each attribute various alternatives are recognized. A scenario is identified by selecting, for each attribute, one alternative among the possible ones.

The attributes examined in this paper should be fundamental for a preliminary definition of the architecture and/or the procedures of an integrated system.

The attributes which will be examined in the following seven sections are hereinafter listed. For each attribute, some of the most promising alternatives will be outlined and the basic trade-offs to be considered for the alternative selections will be highlighted. Specific implementation and/or procedural problems are outside the scope of this paper.

- The configuration and the switching capabilities of the space segment (see Sections 3.1 and 3.2).
- The satellite counterparts of the UMTS entities and functional areas addressed in Sect. 2.1 and 2.2 and their integration level with the terrestrial segment (see Sections from 3.3 to 3.6).
- Cell selection/reselection and handover (see Sect. 3.7).

It is important remarking that the selection of a certain alternative for a given attribute is not independent of the selection of the other alternatives for the other attributes. So, the problem of identifying the integration scenario can not be simply decoupled in a number of independent problems equal to the number of identified attributes.

In the following, we assume that the satellite segment includes N FESs, M spot-beams in the mobile link (i.e. in the link between the payload and the MTs) and S spot-beams in the feeder link (i.e. in the link between the payload and the FESs).

In the following, for the sake of brevity, whenever we mention "spot-beam", without specifying whether we refer to the feeder or the mobile link, we always mean "spot-beam in the mobile link".

3.1 Space Segment Configuration

This attribute concerns the parameters which describe the space segment (e.g. number of satellites, satellite orbit, number and shape of spot-beams, link budget related parameters...)

The various alternatives have to be selected in order to fit the required coverage area (e.g. global, regional, rural, urban,...) and the required performance (e.g. in terms of capacity and quality).

The discussion of the specific technological trade-offs underlying this attribute and of the various alternatives does not concern network aspects and therefore is outside the scope of this paper.

3.2 The Space Segment Switching Capabilities

This attribute concerns the ability of the satellite segment to connect, on a real time basis, the feeder link spot-beams with the mobile link spot-beams.

Consider a parameter $C_{ij}(t)$ $(i=1,...,N)$, $(j=1,...,M)$ equal to 1 or 0 according to whether, at time t, a connection, via satellite, between the i-th FES and the j-th spot-beam is or is not possible. Obviously, such connection is possible if and only if the satellite system arrangement, at time t, permits the connection between the feeder link spot-beam where the FES i is placed and the mobile link spot-beam j (it should be noted that, in general, the above parameters depend on time since some connections could or could not be possible depending on the present on-board switch configuration).

Clearly,

$$C_j(t) = \sum_{i=1}^{N} C_{ij}(t) \quad , \quad C_i(t) = \sum_{j=1}^{M} C_{ij}(t)$$

represent the number of FESs which, at time t, can be connected with the j-th spot-beam and the number of spot-beams which can be connected with the i-th FES, respectively.

Obviously, for this attribute the following two contrasting requirements are to be traded off. By improving the on-board switching capabilities (i.e., on the average, by increasing $C_j(t)$ and/or $C_i(t)$), we gain in space segment flexibility. On the other hand, payload complexity increases.

As a matter of fact, also depending on the space segment configuration, for achieving the target switching capabilities, (in increasing complexity order) Intermediate Frequency on-board processors, baseband on-board processor, inter-satellite links could be required.

Clearly, the maximum flexibility degree corresponds to the implementation of switching capabilities guaranteeing, at any time t, full FESs-to-beams connection, i.e. $C_j(t) = N$ and $C_i(t) = M$.

3.3 Satellite Cell Definition

The definition of the satellite cell is one of the key issues for the identification of the integration scenario, since all the UMTS procedures will refer to the cell concept.

By reminding the one-to-one association between beacons and cells (see Sect. 2.2), it can be noted that the number of *satellite cells* $M_i(t)$ served, at time t, by FES i is equal to the number of beacons broadcast, at time t, by the FES i towards different spot-beams. According to the concepts introduced in the previous section, $M_i(t) \leq C_i(t)$.

The number of FESs $N_j(t)$ which, at time t, broadcast a beacon towards the spot-beam j is equal to the number of satellite cells whose coverage, at time t, coincide with the spot-beam j coverage. Note that these $N_j(t)$ satellite cells have the same coverage, but are served by different FESs. Clearly, $N_j(t) \leq C_j(t)$

Consider the following positions :

- $T_{ij}(t)$ is a parameter equal to 1 if, at time t, a beacon is broadcast by the FES i towards the spot-beam j and is equal to 0 in the opposite case;
- $C_S(t)$ is number of satellite cells at time t.

Then, it can be easily deduced:

$$N_j(t) = \sum_{i=1}^{M} T_{ij}(t) \;,\; C_S(t) = \sum_{i=1}^{N} M_i(t)$$

The number of beacons to be broadcast by each FES has to be selected by taking into account the following trade-off. By increasing the number of broadcast beacons, the number of spot-beams the FESs can be simultaneously connected with increases: $M_i(t)$ spot-beams can be simultaneously connected with FES i. Moreover, the degrees of freedom in the choice of the serving FES from the MT increase (a MT roaming in spot-beam j can choice, at time t, its serving FES in a set of $N_j(t)$ FESs), which can result in an optimization with respect to a proper criterion (see Sect. 3.7).

On the other hand, the FES complexity increases and more radio resources (possibly, not well exploited if the beacon is broadcast in a spot-beam which rarely needs to be connected with the FES) have to be assigned to each FES.

One extreme alternative (applicable in the case of full FESs-to-beams connection) is that all FESs continuously broadcast a different beacon in every spot-beam; in other words, $M_i(t) = M$ ($i=1,...,N$), $N_j(t) = N$ ($j=1,...,M$), $C_S(t) = MN$. Fig. 2 exemplifies this alternative. The satellite system sketched in the figure includes three FESs and two spot-beams (i.e. $N=3$ and $M=2$); hence, six satellite cells are present.

The other extreme alternative is that all FESs only broadcast a single beacon (such alternatives requires a minimum on board switching capability), i.e.

$M_i(t) = 1$ ($i=1,...,N$), $C_s(t) = N$. By referring to the satellite system introduced in the previous example, Fig. 3 shows this alternative.

Intermediate alternatives include the case in which the number of beacons broadcast by a given FES varies as time varies (e.g. in order to follow traffic fluctuations). Note that, even if this number is constant, the set of spot-beams towards which a given FES broadcast its beacons can vary as time varies (e.g. again in order to follow traffic fluctuations).

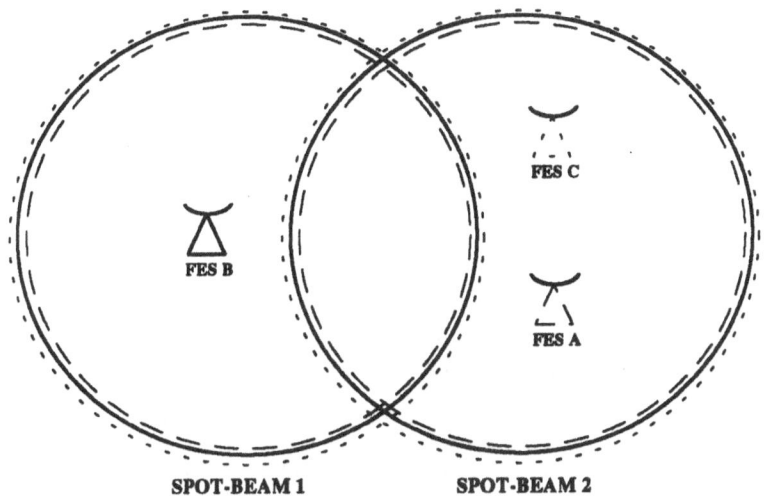

SPOT-BEAM 1　　　　**SPOT-BEAM 2**

The integrated system includes six satellite cells identified by the couples: (FES A, Spot-beam 1), (FES A, Spot-beam 2), (FES B, Spot-beam 1), (FES B, Spot-beam 2), (FES C, Spot-beam 1), (FES C, Spot-beam 2).

In the figure the coverages of the six satellite cells (i.e. the areas covered by the associated spots) are highlighted:
- the coverages of the satellite cells (FES A, Spot-beam 1) and (FES A, Spot-beam 2) are sketched with a dashed line;
- the coverages of the satellite cells (FES B, Spot-beam 1) and (FES B, Spot-beam 2) are sketched with a continuous line;
- the coverages of the satellite cells (FES C, Spot-beam 1) and (FES C, Spot-beam 2) are sketched with a dotted line.

Fig. 2 - Each FES transmits a different beacon towards each spot-beam

3.4 Satellite Location Area Definition

The satellite location area is the area where a MT, served by the space segment, can freely move without updating the location registers.

This attribute entails a trade-off which is typical of terrestrial cellular systems as well. By increasing the dimension of satellite location areas, the overhead caused by the signalling exchanges necessary for location register updates is reduced.

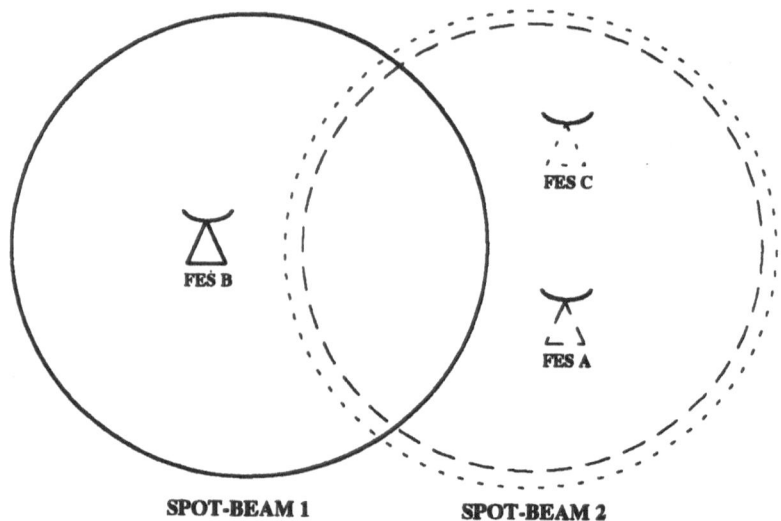

SPOT-BEAM 1   **SPOT-BEAM 2**

The integrated system includes three satellite cells identified by the couples:
(FES A, Spot-beam 2), (FES B, Spot-beam 1), (FES C, Spot-beam 2).

In the figure the coverages of the three satellite cells (i.e. the areas covered by
the associated spots) are highlighted:
- the coverage of the satellite cell (FES A, Spot-beam 2) is sketched with a dashed line;
- the coverage of the satellite cell (FES B, Spot-beam 1)is sketched with a continuous line;
- the coverage of the satellite cell (FES C, Spot-beam 2) is sketched with a dotted line;

Fig. 3 - Each FES transmits a single beacon

On the other hand, since the accuracy the fixed network is aware of the MT
position decreases, the paging overhead in case of fixed network originated call
set-ups increases.

In the following, the total number of location areas in the space segment, at
time t, is referred to as $L(t)$.

A possible alternative is to regard each satellite cell as a satellite location
area, i.e. $L(t) = C_S(t)$. this alternative suits GEO satellite systems where the
satellite cells are rather large and, consequently, a low satellite cell reselection
frequency is expected. In the case introduced in Fig. 2 we have six satellite
location areas coinciding with six satellite cells.

An alternative solution is that each satellite location area includes all
satellite cells served by a same FES, i.e. $L(t) = N$. This alternative suits non GEO
satellite systems where the frequency of spot-beam changes is high, so that the
reduction of location register update frequency is more important than paging
overhead reduction. Fig. 4 shows this alternative for the satellite system
introduced in Fig. 2: three satellite location areas, corresponding to the three
FESs, are present.

Inter-segment location areas (i.e. location areas including both satellite and terrestrial cells) can be convenient for including, for instance, a satellite cell together with a terrestrial cell from which MTs very frequently leave the terrestrial coverage (e.g. because a road is present going out of the terrestrial coverage).

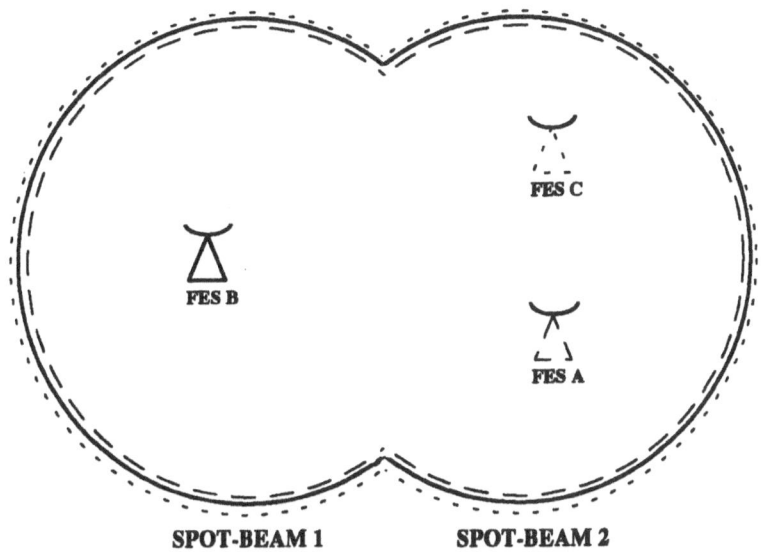

SPOT-BEAM 1   **SPOT-BEAM 2**

The integrated system includes three satellite location areas.

In the figure the coverages of the three satellite location areas are highligthed:
- the coverage of the satellite location area relevant to FES A is sketched with a dashed line;
- the coverage of the satellite location area relevant to FES B is sketched with a continuous line;
- the coverage of the satellite location area relevant to FES C is sketched with a dotted line;

Fig. 4 - Each FES is associated with a single satellite location area

3.5 Fixed Earth Station (FES) Configuration

For defining FES configuration, it is necessary to identify the access network UMTS entities to be included in the FES.

Four possible alternatives can be recognized: (i) the FES only includes the functionalities of the lowest hierarchical level entity, i.e. the BTS functionalities, (ii) the FES includes the BTS and the CSS functionalities, but not the MSCP functionalities, (iii) the FES includes the BTS, the CSS and the MSCP functionalities, but not the MSDP functionalities, (iv) the FES includes the functionalities of the highest hierarchical level entity, i.e. the MSDP (and, hence, also the functionalities of the entities at a lower hierarchical level).

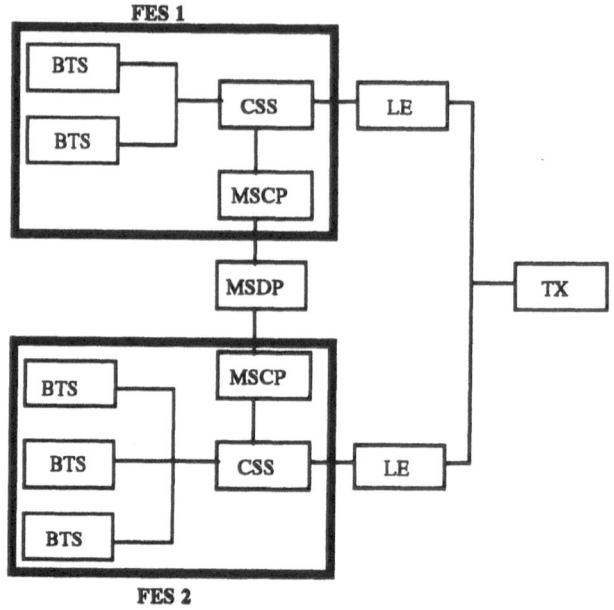

FES 1

FES 2

Fig. 5 - FES configuration example

Fig. 5 shows an example of alternative (iii). By reminding the one-to-one association between BTSs and (satellite) cells, from Fig. 5 it can be deduced that FES 1 and FES 2 can simultaneously serve 2 and 3 satellite cells, respectively; nevertheless, BTSs in a same FES can share most of hardware and software.

The following trade-off has to be taken into account in order to select the most appropriate alternative. The more the functionalities included in the FESs are, the least is the signalling overhead to be sustained by the fixed network; as a matter of fact, the number of signalling exchanges which require the interrogation of entities outside the FES, decreases.

On the other hand, this prevents a simple sharing among the FESs of the entities and of the resources handled by these entities.

3.6 Integration Level between Terrestrial and Space Segment

The integration level between a satellite system and the terrestrial UMTS entities is determined by the hierarchical level at which common terrestrial-satellite entities are allowed. So, four alternatives can be recognized according to whether the integrated system avail of (i) separated entities, (ii) common MSDPs (but separated MSCPs and CSSs), (iii) common MSDPs and MSCPs (but separated CSSs), (iv) common MSDPs, MSCPs and CSSs; note that, in any case, the BTS functionalities are separated.

We have *pure satellite* CSS (or MSCP, or MSDP) areas in case the CSSs (or the MSCPs, or the MSDPs) relevant to the space segment are separated from the ones relevant to the terrestrial segment; on the contrary, we have *mixed terrestrial-satellite* CSS (or MSCP, or MSDP) areas in case the CSSs (or the MSCPs, or the MSDPs) are shared among the two segments. For instance, alternative (ii) entails mixed terrestrial-satellite MSDP areas and pure satellite MSCP and CSS areas.

For instance, a pure satellite CSS area corresponding to a given CSS includes the satellite cells served by the satellite BTSs associated with the CSS; a mixed terrestrial-satellite CSS area corresponding to a given CSS includes the satellite and the terrestrial cells served by the satellite and the terrestrial BTSs associated with the CSS.

Basically, for this attribute, the following trade-off can be identified. By allowing the presence of common entities and functional areas even at low hierarchical levels, the various entities can be shared (with the consequent hardware saving) between terrestrial and space segment and the resources handled by these entities can be shared, in a simple way, among the two segments.

On the other hand, the number of signalling exchanges which requires an interaction with the terrestrial segment increases; this entails a decrease in the possibility of tailoring the procedures involving the satellite to the peculiar space segment needs.

In the example in Fig. 5, the CSSs and the MSCPs are included in the FES and hence they are pure satellite entities; the relevant CSS and MSCP areas are pure satellite as well. Conversely, the MSDP can be either pure satellite (if, as in Fig. 5, it only controls pure satellite MSCPs) or mixed terrestrial/satellite (if it also controls one or more terrestrial MSCPs). In this last case the relevant MSDP area is a mixed terrestrial/satellite MSDP area.

3.7 Cell Selection/Reselection and Handover

By sophisticating cell selection/reselection and handover procedures, it is possible to improve the integrated system efficiency and flexibility, but this, in general, entails an increase in the integrated system complexity.

In a fully integrated system both intra-space segment satellite cell reselections (occurring when a stand-by MT passes from a satellite cell to another satellite cell) and inter-segment cell reselections (occurring when a stand-by MT passes from a satellite cell to a terrestrial cell and vice versa) must be implemented.

It should be remarked that the most critical of the aforesaid cell reselections is the one entailing the passage from a satellite cell to a terrestrial cell, since it implies the monitoring of a high number of beacons. As a matter of fact, in a cellular system, a stand-by MT perceives it is changing cell by comparing the power level of the channel it is tuned on, with the power level of the beacons broadcast in the surrounding cells. When dealing with an integrated system, the concept of surrounding cell is well applicable in case of intra-satellite cell reselections (the satellite cells adjacent to a given satellite cell covering a certain spot-beam are the ones covering spot-beams adjacent to the first spot-beam), or even in case of cell reselections entailing the passage from a terrestrial to a satellite cell (one or more satellite cells covering a given terrestrial cell can be considered adjacent to this cell); conversely, in case of cell reselections entailing the passage from a satellite to a terrestrial cell, in general, no particular adjacent terrestrial cell(s) can be associated with a given satellite cell and therefore all the terrestrial segment beacons must be monitored by a MT served by a satellite cell.

Important advantages can be achieved by allowing call-by-call cell reselections. As a matter of fact, since, in general, a MT is covered by more than one satellite cell (see Sect. 3.3), at each call set-up, a MT can select the most convenient satellite cell according to a given criterion. Possible criteria are the selection of a satellite cell relevant to the FES closest to the fixed user involved in the call (such criterion aims at the shortening of the terrestrial tails), or the selection of a satellite cell relevant to the idlest FES (such criterion aims at the reduction of the blocking probability) [8], [9].

As for the handover facilities, the intra-space segment handover is fundamental if a non-GEO satellite is envisaged. On the contrary, in a GEO satellite system case, this kind of handover has marginal utility since the graceful degradation of the spot-beam coverage, the expected MT speed and the expected call duration suggest that, in most cases, a call initiated in a given spot-beam can terminate in the same spot-beam.

The inter-segment handover is useful, but not fundamental [9]. As for the handover from a terrestrial cell to a satellite cell, it is useful in case an in-call MT leaves the terrestrial coverage, but still remains in the satellite coverage. The handover from a satellite cell to a terrestrial cell is useful in order to immediately free a satellite channel in case an in-call MT, served by a satellite cell, goes into the GSM coverage (in general, a satellite channel is a more valuable resource than a terrestrial channel).

It should be noted that, in both the above cases, after call termination, the cell reselection procedure provides for reselecting a satellite (in the first case) or a terrestrial (in the second case) cell, even though the above-mentioned handovers are not foreseen.

For similar reasons as the ones described for cell reselection, the handover from the space to the terrestrial segment is more complicated than the one in the opposite direction.

4 Conclusions

This paper has described several key attributes which characterize an integration scenario between the UMTS network and a satellite system. In particular, the on-board switching capabilities, the satellite counterparts of the entities and functional areas defined in the UMTS framework and the cell selection/reselection-handover procedures have been dealt with.

For each attribute, the paper has outlined some of the most promising alternatives and has discussed the underlying trade-offs.

Once for each attribute an alternative is chosen, an integration scenario is outlined. Since a likely integration scenario foresees several satellite systems, each one tailored to particular requirements, integrated with the UMTS network, these choices not only depends on the aforesaid trade-offs, but also on the target the considered satellite system aims at.

Acknowledgments. The author wishes to thank Prof. Fulvio Ananasso (TELESPAZIO) for its helpful suggestions in the framework of the study.

References

[1]   E. Del Re: "Overview of COST 227 PROJECT: Integrated Space/Terrestrial Mobile Networks".

[2]   F. Ananasso, M. Carosi: "Architecture and Networking Issues in Satellite Systems for Personal Communications", International Journal of Satellite Communications, Special Issue on Personal Communications Via Satellite, 1994.

[3]   F. Delli Priscoli, E. Del Re, P. Iannucci, R. Menolascino, F. Settimo: "Architectures and Protocols for an Integrated Satellite-Terrestrial Mobile System", 3rd International Mobile Satellite Conference and Exhibition. Pasadena (USA), June 1993.

[4]   F. Delli Priscoli, F. Muratore: "Assessment of a Public Mobile Satellite System Compatible with the GSM Cellular Network", International Journal of Satellite Communications, Special Issue on Personal Communications Via Satellite, 1994.

[5]   MONET Project, "UMTS Network Architecture", Document MoNet/FACE/GA3/22 Issue 5.1, June 1994.

[6]   MONET Project, "UMTS Functional Model", Document MoNet/FACE/GA3/DS/P/029/b1, February 1993.

[7]   MONET Project, "Location Areas, Paging Areas and the Network Architecture", Document MoNet/PPTNL/MF1/057 Issue 1.2, April 1992.

[8]   F. Delli Priscoli: "Network Aspects relevant to the Integration between the GSM System and a Satellite System", 2nd International Conference on Universal Personal Communications (ICUPC '93), Ottawa (Canada), October 1993.

[9]   F. Delli Priscoli: "Procedures for a Fully Integrated System Including a Cellular Network and a GEO Satellite System", European Transaction on Telecommunication (ETT), Sep-Oct 1994.

Gateway Earth Stations for Future LEO Communications Satellite Systems

Paolo Capodieci   Mauro Caucci   Renato Del Ricco   Antonio Vernucci
        Alenia Spazio   Italspazio (c/o ALS)   Space Engineering
                      Via Cannizzaro 83   Via dei Berio 91
                      00156 Rome, Italy   00155 Rome, Italy

Abstract

The most important technical and operational challenges for new mobile satellite services (MSSs) in low and medium earth orbits (LEOs, MEOs) currently being proposed and expected to be operational by the end of the decade, are often referred to the Space Segment, i.e. to the need for launching quite a large number of satellites, maintaining them in orbit and replacing failed units. Somewhat less emphasis is usually placed on the Ground Segment, and in particular on the fixed earth stations acting as "gateways" to the terrestrial network, them implicitly being considered to involve more conventional techniques and technologies. In reality, these gateway stations deserve careful consideration because of their potentially significant impact on the overall system economics. With reference to the specific case of Globalstar, this paper is addressed to defining the main guidelines for a gateway design optimization. After a brief description of the main characteristics of the Globalstar system, this paper describes the most important tradeoffs affecting the integration into the terrestrial network, and the design of the RF subsystem of the gateway.

1   Introduction

Mobile Satellite Services (MSSs) will represent one of the fastest-growing segment of future satellite communications systems. Besides the existing networks using geostationary satellite orbit (GSO) satellites, that currently provide voice and data communications for a variety of land, maritime and aeronautical applications, new mobile satellite services are just around the corner. These new systems, employing multiple satellite constellations in low and medium earth orbits (LEOs, MEOs), are aiming to deliver a new level of service, i.e. world-wide communications via handheld/rural telephones. These systems have the potential not only to operate as a complement to existing terrestrial communications systems, but also to provide communication services to parts of the world which are grossly underserved. Among the almost limitless applications of these new systems are cellular-like mobile services, radio determination satellite services (RDSS), search and rescue communications, disaster management communications, environmental monitoring, paging services, facsimile transmission services, cargo tracking, and industrial monitoring and control. Domestically, this service will help meet the demand for a seamless, nationwide communications system that can offer a wide range of voice and data telecommunication services. In addition to enhancing the market for cellular-like service in those areas served by cellular providers, these new mobile satellite systems will offer services in rural areas where no phone service exists. LEO/MEO MSSs can be seen as a way to rapidly increase the capacity and the coverage to meet the demand for basic and mobile communications. In some cases, due to geographic and terrain difficulties, it may not be economically feasible to provide in the near future basic public switched telephone network (PSTN) services to rural areas with wireline networks or terrestrial cellular networks.

LEO/MEO satellite system configurations can potentially extend these benefits throughout the world, and can provide those countries that have not been able to develop a nationwide communication service an "instant" global telecommunications infrastructure at minimal cost for them.

Typically, LEO/MEO systems consist of three major segments: the Space Segment, the Ground Segment, and the User Segment The Space Segment comprises the constellation of satellites. The Ground Segment consists of gateway stations; the Telemetry, Tracking, and Command (TT&C) stations; the Satellite Operation Control Center (SOCC); and the Ground Operation Control Center (GOCC). The User Segment includes different kinds of user terminals, such as vehicle-mounted units, hand-held units, and RDSS-only units.

Various LEO/MEO satellite mobile communications systems have been proposed by several applicants. Three of the more mature proposed LEO/MEO systems are the Globalstar, Iridium, and Odyssey systems. Table 1 summarizes, as a comparison, the major technical characteristics of these three systems.

This paper is addressed to one of the major system entities constituting the "ground segment", i.e. the "gateway" earth stations. In fact, one of the main characteristics common to all future LEO/MEO MSS systems, is that to have, in addition to the user links between satellites and mobile users, one or more fixed earth stations (called gateways) which complete the transmission paths by processing the information being transmitted, and providing interconnections with terrestrial communications networks. Taking as a reference the specific case of Globalstar, one of the most important LEO systems being currently implemented, the paper defines the guidelines for gateway design optimization, dwelling in particular on the following main topics:

- The functions required to implement the interface between the gateway and the terrestrial network, e.g. the local Public Land Mobile Network (PLMN) where this exists, in the perspective to take advantage of the roaming facilities already implemented in this type of networks thus avoiding unnecessarily duplicated functions.

- The possibility to directly interface the GW to the PSTN, needed to support the implementation of rural networks for serving regions where no PLMN facilities exist.

157

Table 1 - Technical Characteristics of LEO/MEO MSS Systems

		Globalstar	Iridium	Odyssey
Constellation:				
No. of Satellites		48	66	12
Orbit/Inclination		Circ/52°	Circ/86.4°	Circ/55°
No. of Planes		8	11	3
Satellites per Plane		6	6	4
Spacing of Planes		45°	31.6° (planes 2-5) 22° (planes 1 & 6)	120°
Sat. Spacing within Plane		60.°	32.7	90°
Sat. Phas. between Planes		7.5°		120°
Altitude (Km)		1,414	780	10,370
Frequency (GHz):				
User-Link:	- Up	1.610-1.6265	1.616-1.6265	1.610-1.6265
	- Down	2.4835-2.500	1.616-1.6265	2.4835-2.500
Feeder-Link:	- Up	5.025-5.225	29.1-29.3	29.5-30.0
	- Down	6.875-7.075	19.4-19.6	19.7-20.2
Intersatellite-Link		na	22.55-23.55	na
Bandwidth (MHz):				
User-Link		16.5	10.5	16.5
Feeder-Link		200.	200.	101.
Intersatellite-Link		na	200.	na
Polarization:				
User-Link		LHCP	RHCP	Circular
Feeder-Link		LH/RHCP	RHCP	Circular
Intersatellite-Link			Vertical	
Connectivity:				
Max No. of Circuits/Sat		2400	3840	2800
Connection Time		10-12 min	9 min	2 hours
Min Operat. Elev. Angle		10°	8.3°	15°
Satellite Designs:				
Stabilization		3-axis	3-axis	3-axis
Transponder		Bent pipe	Processing	Bent pipe
Mission Life (yrs)		7.5	5	15
Dry Mass (kg)		<400	<650	<1100
User-Link Antenna		16-beam phased array	3 16-beam phased array	Rigidly mount. 37/32 u/d ant.
Feeder-Link Antenna		Non-steerable	4 steer. beams	Steerable
Intersatellite-Link		No	Yes; 4 crosslinks	No
Transmission Parameters:				
Multiplexing		FDMA/CDMA	FDMA/TDMA	FDMA/ CDMA
Modulation		QPSK	QPSK	QPSK
FEC		Convolutional (r=1/2;K=9)	Convolutional (r=1/2;K=7)	Convolutional (r=1/3;K=7)
BER		1E-3/1E-6 V/D	1E-2/1E-5 V/D	1E-3/1E-6 V/D
Data Rate (Kbps)		1.2-9.6 V & D	4.8/2.4 V/D	4.8/1.2-9.6 V/D
Coded Data Rate (Mbps)		1.2288	.05/6.25/5 ul/fl/isl	<4.833
BW per Channel (MHz)		1.230	.315/4.375/17.5 ul/fl/isl	5.500

Additionally, the considerable number of functions to be supported in conjunction with the inherent logistic spreading of the gateway layout (a gateway avails of up to four separate antennas) leads to the need for optimizing the apportionment of gateway functions among the various gateway sites, with the main objectives of: 1) reusing off-the-shelf hardware as much as possible, thus leading to development risks reduction; 2) minimizing the total equipment count; and 3) adopting the most cost-effective approaches.

2   Globalstar System Overview

The Globalstar system has the capability of providing communications (mobile and fixed voice and data services) from any point on the earth surface that is served by gateway stations, exclusive of the polar regions. The satellite orbits are optimized to provide highest link availability in the area between 70° South Latitude and 70° North Latitude. Service is feasible in higher latitudes with decreased link availabi- ity. It also has the capability of providing position location to the user.

The Globalstar system consists of a constellation of satellites and terrestrial gateway stations, user terminals (including hand sets, mobile sets, and fixed sets), ground operation control and accounting facilities and satellite operation control facilities as shown in Figure 1.

The primary service consists of continuous near-toll quality full duplex voice. Data services up to 4.8 kbps is also supported. Higher data rates may be supportable with specialized user terminals.

The Globalstar Space segment consists of 48 satellites in 1414 Km Low Earth Orbits. The low orbits permit low power hand sets similar to cellular phones. These satellites are distributed in 8 orbital planes with 6 equally spaced satellites per orbital plane. Satellites complete an orbit every 114 minutes. User terminals in a particular location on the surface of the earth can be served by a satellite 10 to 15 minutes out of each orbit.

An handover process between satellites provides continuous communications for the users. The orbital planes are inclined at 52°. This provides full earth coverage with at least two satellites in simultaneous view, providing space diversity, over most of the area. There is some small sacrifice in multiple satellite coverage at the equator. Coverage is maximized in the temperate areas.

Both the forward link (gateway terminal to user) and the reverse link (user to gateway terminal) use spread spectrum direct sequence PN coding in conjunction with Forward Error Correction Coding (FECC) and interleaving to provide Code Division Multiple Access (CDMA) for interference and noise rejection. Voice links employ near toll-quality vocoders to reduce the average bit rate of a voice channel. Power control and diversity combining is employed to mitigate the effects of fading, and permit operation of the subscriber units at low average power.

The system is designed so that the user-to-satellite and satellite-to-user antenna beams can accomodate 13 channels, each with 1.23 MHz bandwidth. Each of these beams supports up to 128 CDMA circuits allocated between traffic and control functions. There are 16 satellite beams for user access on both transmit and receive bands which together fill a subtended solid angle of 108° centered on the satellite nadir. Every beam is capable of reusing the 16.5 MHz bandwidth allocated for user to satellite and satellite to user. The satellite translates the frequency band from each beam into adjacent 16.5 MHz feederlink subbands (with guard bands) in the satellite to gateway band for transmission to the gateways. Both senses of circular polarization are used to minimize the need for C-band bandwidth. Likewise, on the gateway to satellite link, adjacent 16.5 MHz feederlink subbands (with guard bands) and polarization reuse are employed.

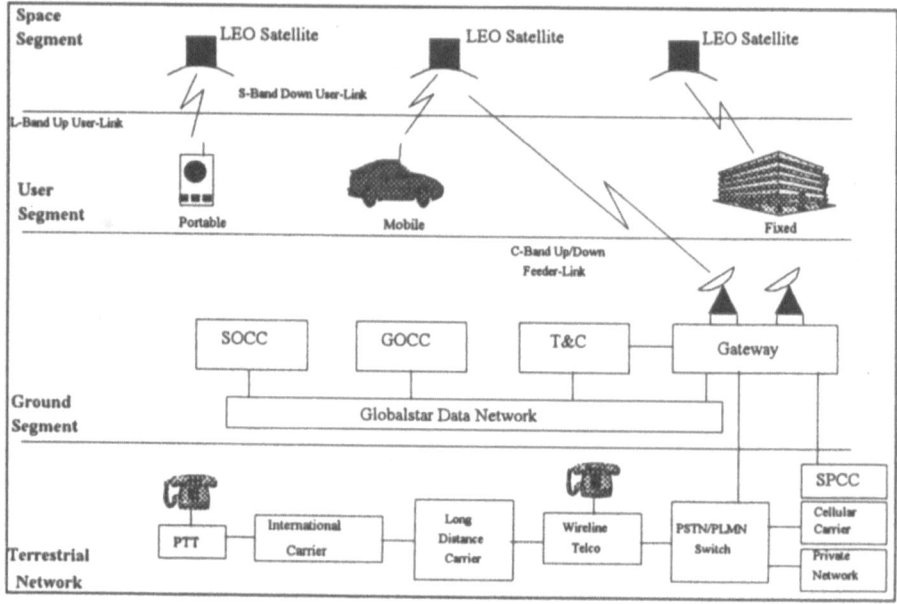

Figure 1 - Globalstar System

3 The Gateway Station

The gateway stations are the interconnection point between the Globalstar satellite constellation and the existing terrestrial telecommunications networks. They will be geographically distributed by the service providers such as to best serve their customer base. Figure 2 is a simplified block diagram of a typical gateway.

The gateways are composed of gateway earth terminals that provide the communication link between the Space Segment and the Terrestrial Segment. Each terminal consists of the RF subsystem and the CDMA subsystem. The RF subsystem includes up to four antennas to track different satellites. The antenna structure contains drive mechanism for positioning the antenna, low noise receivers and high power transmitters. The antennas are connected to a building that houses the electronics equipment. The building contains down converters and up converters to process the RF carriers, the CDMA equipment comprising all the equipment required to assemble and disassemble the CDMA waveform, PSTN/PLMN equipment that interfaces with the terrestrial telephone network, and computer equipment to operate the gateway and collect status and performance data.

Since the orbiting satellites are in continuous motion relative to the gateways, the gateway antennas follow the closest satellites among the visible ones.

Communications channels will be handed off periodically from the current satellite to the next one. This handoff is transparent to both the Globalstar and the PSTN users involved. The CDMA technnology employed also allows the intelligent choice of the "best" signal path or to combine the signals from two or more satellites to improve signal quality and availability. A single gateway has the capability to serve a limited-size geographical region (e.g. a few thousands kilometers diameter). However, more gateways may be used to serve the same region if so desired.

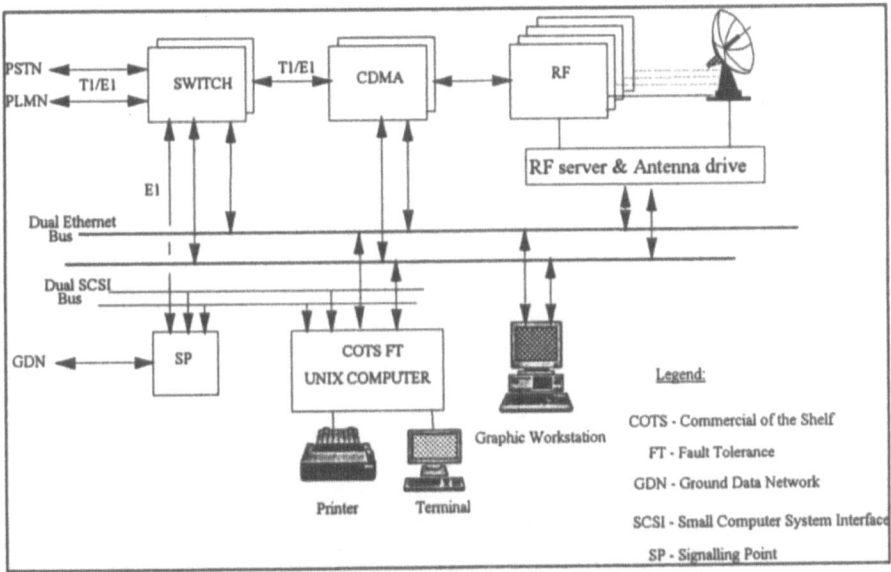

Figure 2 - Gateway Simplified Block Diagram

4 Gateway-Terrestrial Networks Interoperability Guidelines

The Globalstar system has been specifically designed as an extension of existing terrestrial telephony systems. The intent is in fact that to extend the coverage area in such a way that compatibility with existing systems be assured. In this sense, Globalstar plays a role of a complement system, rather than a competitor system, in the not (or not yet) covered areas.

Nevertheless, the Globalstar system is not precluded to operate as a stand-alone system (e.g. for rural telephony applications).

With particular reference to the services offered to Globalstar subscribers, and to the sharing of access authorization, registration and mobility management between the gateway and the terrestrial networks involved, next paragraphs include an analysis of all the possible combinations of user terminal (UT) type (i.e. fixed site Globalstar-only, Globalstar-only portable, Globalstar/GSM dual mode portable, Globalstar/CDMA/AMPS tri mode portable) and local environment (i.e. type of PLMN/ PSTN connected to the gateway), and consequent ways in which the various procedures are accomplished by cooperation among the various networks involved.

A typical reference scenario for the Globalstar sytem integrated into the existing terrestrial networks is that illustrated in Figure 3.

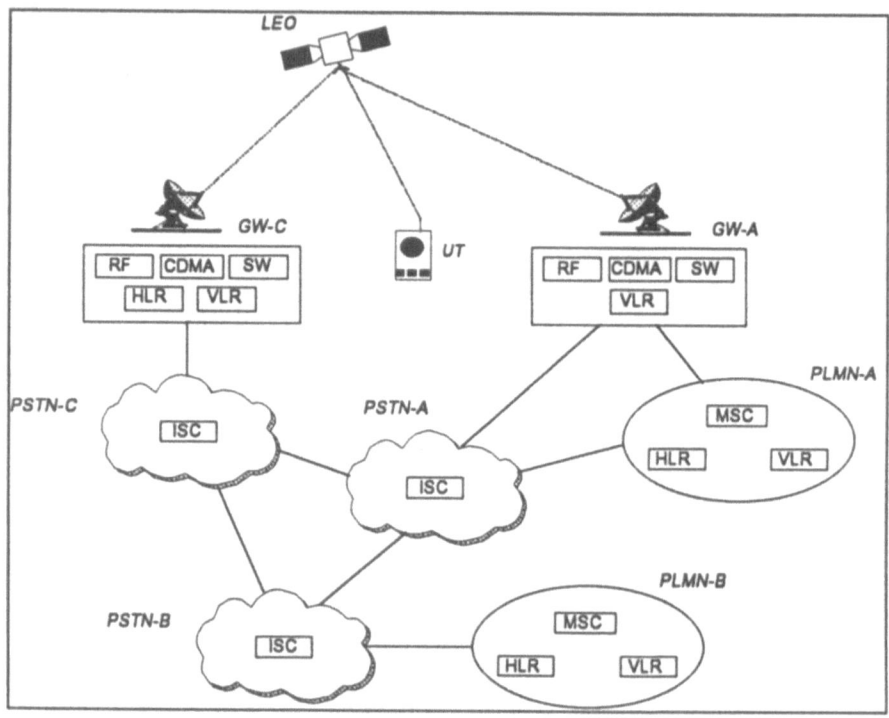

Figure 3 - Globalstar Interworking Scenario

4.1 Integration between Globalstar and PLMN

From a network and integration-to-terrestrial cellular systems point of view, one important requirement to be met is to limit as much as possible the modifications of the existing PLMNs when the Globalstar system is introduced.

The simplest way to achieve an integration between a satellite system and a cellular network should be to have the gateway as a base station (BS) by the mobile switching center (MSC) to which it is connected. In this architecture the MSC would maintain all its traditional functions versus the user terminal (connection hand-off, handling of location information, connection management).

However, the need to cope with rural telephony requirements and to manage the international roaming on a world-wide basis concur to propose an architecture where the gateway also plays the role of an MSC.

The Globalstar gateway, in order to avoid duplication of functions and equipment should be designed to be adaptive to the local environment. This means that there will be cases in which the gateway will simply emulate a BS, as well as cases in which its behaviour will include that of an MSC.

4.2   Service Types and Effect of Roaming on Services

As previously mentioned, Globalstar supports two types of services: 1) position location, and 2) telecommunication services (including voice and data communications). The first one will always be provided by the gateway to all Globalstar users connected to it. The second ones are provided by the gateway through the MSC.

A PLMN operator can further extend its coverage by signing agreements with other PLMN operators equipped with a Globalstar gateway. Accordingly, due to its inherent adaptability to the environment, the gateway would appear to provide different services to different users:

* roamers belonging to the PLMN to which the gateway is connected will receive services they have subscribed (the gateway simply acts as a BS);
* roamers of PLMNs compatible with the one to which it is connected will receive the services they have subscribed only if locally provided (the gateway still acts as a BS);
* roamers of PLMNs incompatible with the one to which it is connected and Globalstar-only user terminals will receive the services directly provided by the gateway itself (the gateway acts as an MSC).

4.3   Access Authorization and Registration

Users requiring access to the system (for call placement or reception of a call) must be authorized for transmission from the service area where they are actually located. This means that they belong to one of the following categories:

* Globalstar-only subscribers, registered in the local Home Location Register (HLR);
* subscribers of the local PLMN but also of the Globalstar services;
* subscribers of another PLMN and Globalstar services, whose operator signed an agreement with the local operator.

Registration is the process by which the user terminal notifies the gateway of its location, status, identification, terminal type.

When the gateway acts as a BS, registration is accomplished by the MSC of the PLMN to which the gateway is connected, using the Visitor Location Register (VLR), in conjunction with GSM or AMPS or CDMA HLR. The same occurs when the user belongs to another but compatible PLMN. The only difference consists in the handling of location information which is performed with the cooperation of the Intermediate Switching Center (ISC).

Registration of user terminals belonging to incompatible PLMNs is performed by the gateway itself which must include a VLR.

Registration of Globalstar-only users needs the presence of a VLR in the gateway, while the local HLR will incorporate the terminal data. In areas where there is not a PLMN, the Globalstar gateway, connected to the local PSTN, may set up a HLR. Thus, with reference to the scenario of Figure 3 the following six situations can occur:

Case 1) Globalstar/PLMN user terminal (e.g. UT-A) in its own system (e.g. PLMN-A= GSM);

Case 2) Globalstar/PLMN user terminal (e.g. UT-B) in another compatible system (e.g. PLMN-A=GSM and PLMN-B=GSM);

Case 3) Globalstar/PLMN user terminal (e.g. UT-B) in another incompatible system (e.g. PLMN-A=GSM and PLMN-B=AMPS);

Case 4) Globalstar-only user terminal (e.g. UT-G) in its subscription area where there is a PLMN;

Case 5) Globalstar-only user terminal (e.g. UT-G) in its subscription area where there is not any PLMN;

Case 6) Globalstar-only user (e.g. UT-G) out of its subscription area.

Table 2 summarizes the various network components involved for each combination of User Terminal type and PLMN to which the Globalstar gateway is connected.

Table 2 - System Components involved in the Networks Integration

	UT	GW/MSC	VLR	ISC	HLR
Case 1	UT-A	PLMN-A	PLMN-A	na	PLMN-A
Case 2	UT-B	PLMN-A	PLMN-A	PSTN-A	PLMN-B
Case 3	UT-B	GW-A	GW-A	PSTN-A	PLMN-B
Case 4	UT-G	GW-A	GW-A	na	PLMN-A
Case 5	UT-G	GW-C	GW-C	PSTN-C	GW-C
Case 6	UT-G	GW-A	GW-A	PSTN-A	GW-C

5   Gateway RF Subsystem Analysis

In order to reduce as much as possible the overall cost of the gateways, a cost effective solution for the gateway RF subsystem might be that to foresee different classes of gateway stations equipped with a different antenna diameter/HPA combination according to the maximum amount of expected traffic to be served (i.e. the maximum EIRP required).

As previously mentioned, a multiple tracking antenna configuration is to be considered for the gateways of the Globalstar system. At the middle latitudes, gateways will be equipped with four antennas: three antennas supporting traffic through three satellites and the fourth antenna available to acquire the next rising satellite. Such a logistic spreading of the gateway layout leads to the need for optimizing the apportionement of functions among the constituting gateway elements in order to minimize the total equipment count (e.g. converters) and to adopt the most cost-effective approaches (e.g. cable or fiberoptic inter-facility links, IFLs, where confirmed to be the most cost-effective solution).

5.1 RF Subsystem Configuration

According to the maximum amount of expected traffic the Globalstar system shall support within the various regions, three classes of gateway stations will exist:

1. Low Traffic Gateway (<100 ch.s);
2. Medium Traffic Gateway (<400 ch.s);
3. High Traffic Gateway (<1000 ch.s).

In order to meet the maximum EIRP requirements, different antenna diameters and different HPA sizes can been considered. For instance, by assuming antenna diameters of 4.3 and 6 meters and HPA sizes of 125, 400, and 700 W, Table 3 shows the resulting maximum amount of traffic (expressed in terms of number of channels) that each antenna diameter/HPA combination can support in accordance with the maximum EIRP required (i.e. an EIRP of 70 dBW per polarization). In this example, gateway stations equipped with a 4.3 m antenna diameter can meet traffic requirements up to 400 channels, while maximum traffic requirements can only be met by a 6 m antenna diameter and 700 W HPAs (one per polarization).

Table 3 - Number of Channels per Antenna Diameter/HPA combination

HPA Power (W)	Ant. Diam. (m)	No. of Ch.s
125	4.3	<80
125	6.0	<100
400	4.3	<150
400	6.0	<400
700	4.3	<400
700	6.0	<1000

Hence, maximum number of channels to be supported by the gateway along with the maximum EIRP requirements are two basic parameters for a proper selection of the antenna diameter/HPA configuration. However, in order to achieve a design-to-cost optimization other parameters must be taken into account.
In a multiple tracking antenna configuration, antennas must properly be spaced in order to avoid mutual structure blockage. The minimum distance between two adjacent antennas (see Annex A for an analytical determination) depends on the antenna size and the minimum operational elevation angle (10° for Globalstar). As shown in Figure A2, distances get higher as antenna diameters increase and operational elevation angles decrease, e.g. for a 3.4 m antenna diameter, the

minimum distance ranges from 20 to 60 m when the elevation angle ranges from 15° to 5°; and for a 6 m antenna diameter, the minimum distance ranges from 35 to 100 m when the elevation angle ranges from 15° to 5°. Such a separation represents the lower bound for the IFL length. In this regard, various types of IFLs can be used to connect the four antenna buildings to the control room of the gateway. These connections include cables, waveguides, and fiberoptics and to get the most appropriate IFL configuration it is needed to consider:

- as commanding requirements, the reliability, the equipment cost, and the installation cost; and
- as main constraints, the IFL length, and environmental factors (e.g. local regulations, IFL accomodation).

A preliminary analysis concerning different gateway RF subsystem configurations has been carried out based upon the following considerations:

- The gateway station consists of four antenna buildings and one control room. Three antennas are located far apart from the control room; one antenna is located on the top of the control room building.
- The adoption of fiberoptic IFLs does not imply any design constraints in the accomodation of the up/down converters which can be installed either at the antenna site or in the control room. Nevertheless, a reduction in the number of up/down converters can be obtained by installing the converters in the control room.
- The adoption of waveguide IFLs forces the up/down converters to be installed in the control room. Waveguides are not a profitable way to transmit signals at 1 GHz.
- The adoption of cable IFLs forces the up/down converters to be installed at the antenna site and implies the use of equalizers for each link. Cables are not a profitable way to transmit signals at 5 GHz.

Then, based upon the following general requirements and assumptions: a) double polarization for both transmit and receive chains, b) operational bandwidth of 200 MHz for the IF/RF Up/Down converters, c) 8 Up/Down converters to feed four antennas and two polarizations per antenna, d) redundant configurations to meet availability requirements: 2:1 for both HPA and LNA, 8:1 or 2:1 for the Up/Down converters if these are installed in the control room or at the antenna building respectively; three different options have been analysed concerning the connections between the control room and the three antenna buildings, and the different accomodations and contents of the RF subsystem components:

Option 1) Waveguide IFL and converters in a 8:1 redundant configuration accomodated in the control room;
Option 2) Fiberoptic IFL and converters in a 8:1 redundant configuration accomodated in the control room;
Option 3) Cable IFL with equalizers and converters in a fully 2:1 redundant configuration installed at the antenna site.

It should be noted that the connection between the control room and the co-located antenna was assumed by waveguide link for Options 1 and 2 and by cable link for option 3.

As an example, the block diagram of the Gateway RF subsystem relevant to the Options 2 and 3 are schematically illustrated in Figures 4 and 5.

5.2 RF Subsystem Cost Analysis

Previous sections have shown that various IFL configurations are possible, and that each IFL configuration implies a different accomodation and design of the RF gateway station. On the basis of following considerations:

- IFL length depends on the minimum separation between antennas and on the characteristics of the gateway location site;
- IFL cost depends on the type and number of links (e.g. in the present situation 4 IFLs per antenna building/control room connection are necessary), the length of each connection and the associated civil works and installation;
- fiberoptic IFL consists of a fiberoptic transmitter, a fiberoptic receiver and a custom optical fiber cable;
- cost of fiberoptic link is nearly independent on the IFL length (e.g. a 12 optical fiber cable costs less than $3 per meter);
- cost of fiberoptic link strongly depends on the Tx/Rx fiberoptic module cost (about K$20 US per module);
- cost of waveguide IFL strongly depends on the IFL length, installation and maintenance;
- cost of cable IFL strongly depends on the IFL length and installation. With regard to this, it should be noted that, depending on the IFL length, more expensive line amplifier are necessary to compensate the cable loss (an additional cost of about K$ 1.5 per cable every 150 m);

a design-to-cost tradeoff analysis was performed by evaluating the cost of the various RF subsystem configurations, also as a function of the IFL length.

Evaluations concerning the RF+IFL equipment cost, the cost per meter of IFL type and IFL installation are summarized in Table 4. Note that cost figures are given in arbitrary units, i.e. normalized with respect to the minimum overall RF+IFL equipment cost.

Table 4 - Normalized Cost of possible RF subsystem Configurations

| | RF + IFL Equipment Cost | | IFL | |
	[Control Room + 1 Ant.]	[3 Ant.s]	Length Cost/100 m	Instal. Cost/100 m
OPTION 1	0.27	0.73	0.064	0.027
OPTION 2	0.27	0.99	0.003	0.001
OPTION 3	0.26	0.78	0.015	0.022

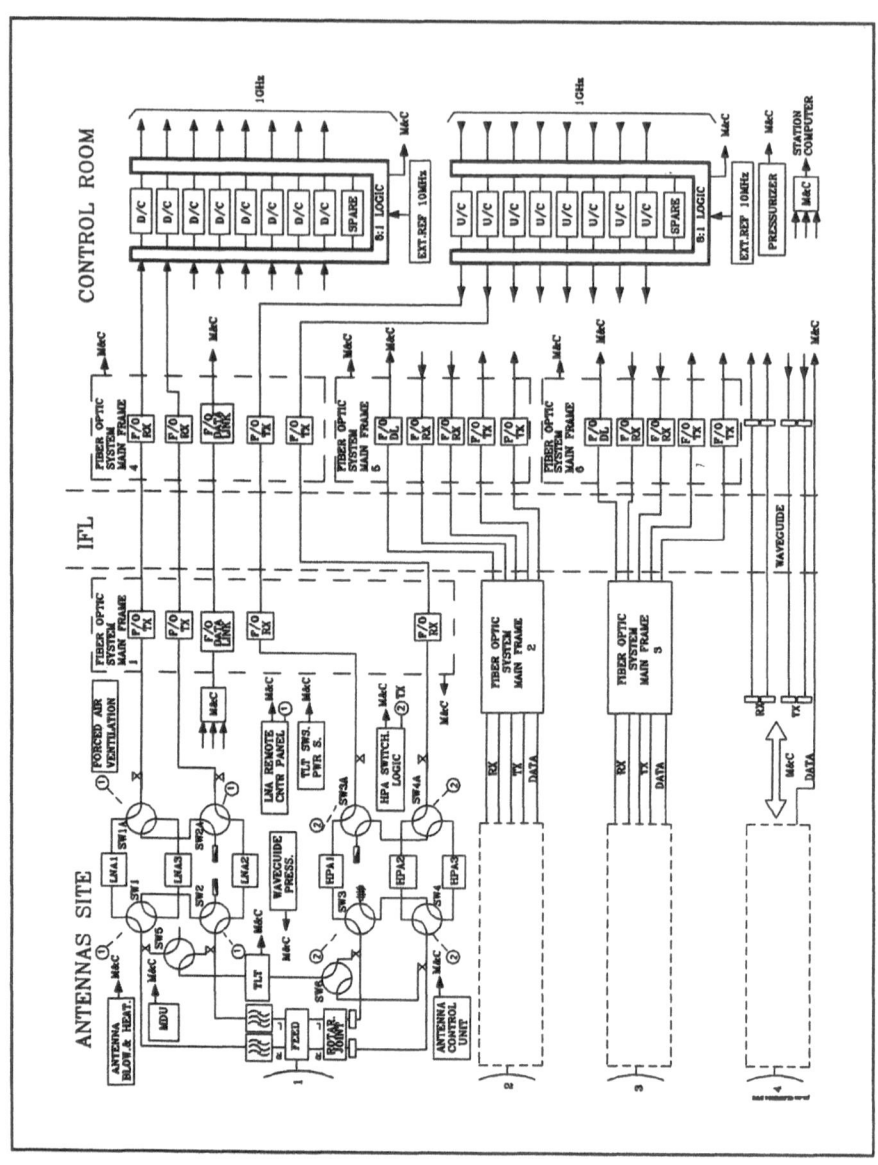

Figure 4 - Gateway RF Subsystem Block Diagram (Option 2)

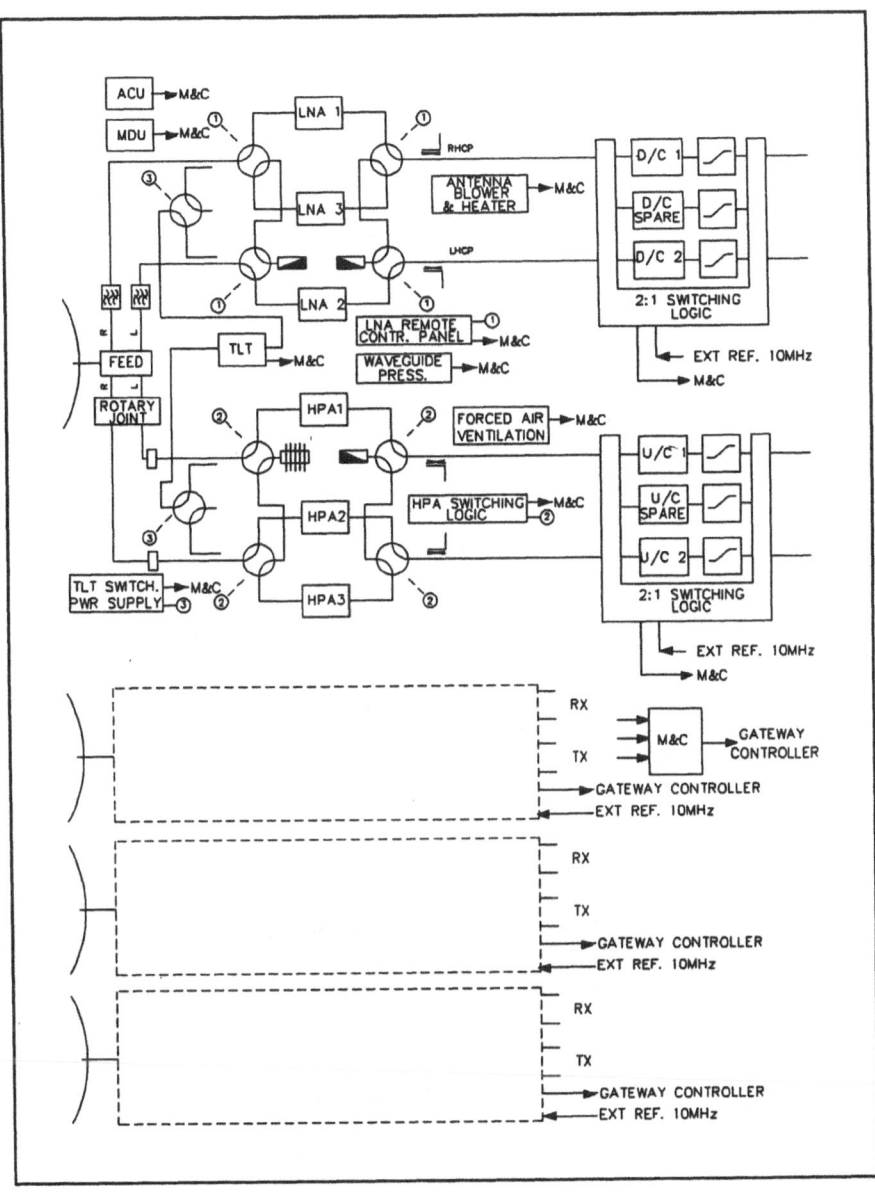

Figure 5 - Gateway RF Subsystem Block Diagram (Option 3)

For the various RF+IFL options analysed, the cost trend comparison as a function of the IFL length is shown in Figure 6, and the major characteristics, advantages, disadvantages, and final results are summarized in Table 5.

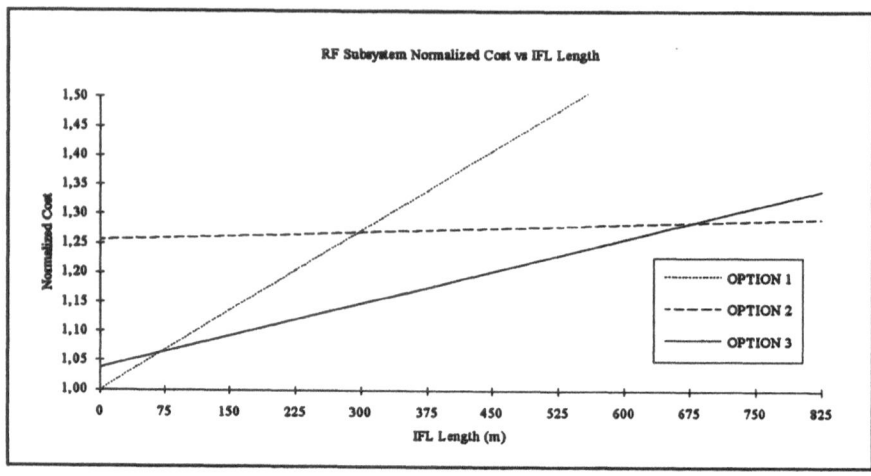

Figure 6 - Cost Trend Comparison vs IFL Length among RF Subsystems using Waveguide (Option 1), Fiberoptic (Option 2) and Cable (Option 3) IFLs.

Table 5 - Main Characteristics of Gateway RF Subsystem Configuration

	DESCRIPTION	ADVANTAGES	DISADVANTAGES	COMMENTS
OPTION 1	- Waveguide IFL - 5 GHz - Converters into CR	- RF Front-End without Rx Line Amplifiers	- High Cost IFL Instal. - High Cost IFL - Exp. Waveg. Trench - Higher Cost for Conv.s	Lowest Cost Config. for IFLs < 75 m
OPTION 2	- Fiberoptic IFL - 5 GHz - Converters into CR	- RF Front-End without Rx Line Amplifiers - Low Cost Instal. - Cheap Fiberopt. Trench - Lower Cost for Conv.s	- High Cost for Tx/Rx Fiberoptic modules	Lowest Cost Config. for IFLs > 675 m
OPTION 3	- Cable IFL - 1 GHz - Fully Redund. Config. for Converters into AB	- Low Cost IFL Instal. - Cheap Cable Trench	- Higher Conv.s Number - Equalizers required - Equalizers and Conv.s installed at Antenna site	Lowest Cost Config. for IFLs > 75 & < 675 m

6   Conclusion

The arguments discussed in this paper demonstrate that the gateway stations of a LEO MSS network is subject to delicate tradeoffs which may have a not-negligible impact on the overall system economics.

In particular, the issue of interfacing the gateway with the local terrestrial network will require careful optimizations to achieve a flexible and cost effective design when operating in conjunction with a multitude of different types of terrestrial networks (GSM, AMPS, PSTN).

The other important topic is the overall gateway layout and the relevant interconnection scheme, an issue which becomes particular significant due to the necessary displacement of several (up to four) RF terminals over quite a wide range site (typical distance between pairs of RF terminals is of the order of several tens of meters).

It is expected that optimizations today done on the basis of abstract gateway models may have to be adjusted to take into account other constraints such as those deriving from logistic and/or regulatory limitations.

Annex A - Minimum Distance between Adjacent Antennas

Given the gateway antenna diameter, D, and the minimum operational elevation angle, e, the aim of this exercise is to determine how far two antenna buildings should be separated in order to avoid mutual structural interference. Assuming the aperture field intensity uniformly distributed over the antenna dish, Equation (A1) is a closed form expression that yields the equivalent antenna diameter, D', as a function of the antenna diameter, D, the wavelength, λ, and the range, r (see Figure A1).

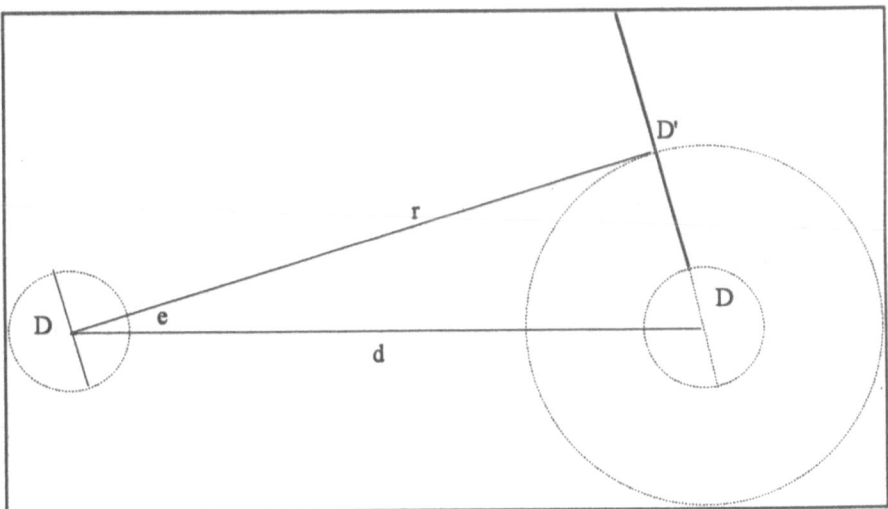

Figure A1 - Equivalent Antenna Diameter (D')

Hence, the minimum separation between antennas, d_{min}, can be derived according to the procedure here below described:

- increase the range (the initial value set equal to twice the antenna diameter) and approximately compute the corresponding equivalent antenna diameter by the following empirical formula

$$D' = \frac{1.85}{2}\left[1+exp\left(-\frac{D^2}{4r^2}\right)\right]D\sqrt{1+4\left\{\frac{1.39}{0.293}\frac{\lambda r}{\pi D^2}\left[1+exp\left(-\frac{D^2}{4r^2}\right)\right]^{-2}\right\}^2} \qquad (A1)$$

- compute the minimum elevation angle associated to the equivalent antenna diameter by

$$e_{min} = tan^{-1}\left(\frac{D/2+D'/2}{r}\right) \qquad (A2)$$

- in the case the above minimum elevation angle results less than or equal the minimum operational elevation angle (i.e. as soon as this relation holds: $e_{min} \le e_{min}^{op}$), compute the minimum separation between antennas by

$$d_{min} = \frac{D/2+D'/2}{sin(e_{min})} \qquad (A3)$$

Figure A2 shows the separation between adjacent antennas as a function of the minimum operational elevation angle obtained for various antenna diameters.

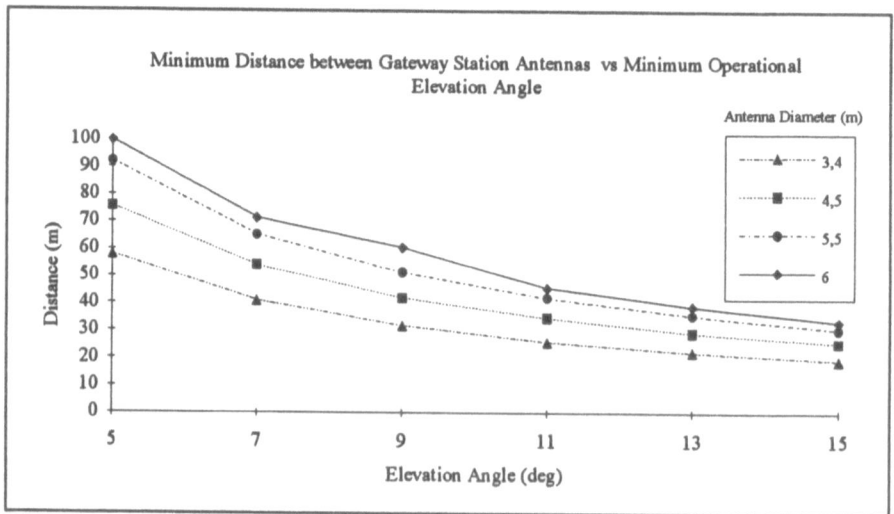

Figure A2 - Separation between Adjacent Antennas vs Minimum Operational Elev. Angle

172

References

1. L.K. Smith, et. al., "Application of Loral Cellular Systems, Corp.: Globalstar, Mobile Communications Wherever You Are", Smith, Halperin, Crowell & Moring, 1001 Pennsylvania Ave., N.W., Washington, D.C., 20004-2505, June 3, 1991.
2. P.L. Malet, et. al., "Application of Motorola Satellite Communications, Inc.: Iridium, A low Cost Orbit Mobile   Satellite System", Malet, Mamlet, Steptoe & Johnson, 1330 Connecticut Ave., N.W., Washington, D.C., \20036, December 3, 1990.
3. N.P. Leventhal, et. al., "Application of TRW Inc.: Odyssey", Leventhal, Senter & Lerman, 2000 K Street, N.W., Suite 600, Washington, D.C., 20006, May 31, 1991.
4. K.S. Gilhousen, I.M. Jacobs, R. Padovani, L.A. Weaver Jr., "Increased Capacity Using CDMA for Mobile Satellite Communications", IEEE Journal on Selected Areas in Communications, Vol. 8, No. 4, pp. 503-514, May, 1990.
5. R. Wiedeman, A. Salmasi, D. Rouffet, "Globalstar: Mobile Communications Wherever You area", AIAA International Communications Satellite Conference, March 1992.
6. M. Louie, D. Rouffet, K.S. Gilhousen, "Multiple Access Techniques and Spectrum Utilization of the Globalstar Mobile Satellite System", AIAA International Communications Satellite Conference, March 1992.
7. M. Louie, P. Monte, R. Tyner, D. Rouffet, K.S. Gilhousen, "Globalstar: Comm Payload for Global Mobile Communications", AIAA International Communications Satellite Conference, March 1992.
8. P. Monte, A. Turner, "Constellation Selection for Globalstar, A Global Mobile
9. R.J. Rusch, "Odyssey, An Optimized Personal Communications Satellite System", AIAA International Communications Satellite Conference, February-March 1994.
10. G.M. Comparetto, N.D. Hulkower, "Global Mobile Satellite Communications: A Review of Three Contenders", AIAA International Communications Satellite Conference, February-March 1994.
11. M. Louie, "National and Provincial Planning Strategies for Mobile Satellite Communications", AIAA International Communications Satellite Conference, February-March 1994.
12. P. Monte, S. Carter, "The Globalstar Air Interface Modulation and Access", AIAA International Communications Satellite Conference, February-March 1994.

Aspects of Satellite Constellation and System Connectivity Analysis

A. Böttcher [*], G.E. Corazza [^], E. Lutz [*], F. Vatalaro [^], M. Werner [*]

[*] *German Aerospace Research Establishment (DLR)*
Institute for Communications Technology
D-82230 Wessling (Germany)

[^] *University of Rome "Tor Vergata"*
Department of Electronic Engineering
Via della Ricerca Scientifica - 00133 Roma (Italy)

Abstract

Recent years have seen many efforts in the field of communication networks based on constellations of non-geostationary satellites. The paper addresses an in-depth analysis of these systems, from three different viewpoints: geometrical, transmission quality and network connectivity. The geometrical analysis yields the statistics for coverage, frequency of satellite handovers and link absence periods. The transmission quality analysis is based on a general model valid for all access techniques, which is here applied to the case of FDMA. The outage probability as a function of the specification on carrier-to-interference power ratio is evaluated for a few selected constellations, also considering some possible interference-reduction techniques (spot turn-off, intra-orbital plane frequency division, and inter-orbital plane frequency division). The approach is extended to the case of non-ideal propagation conditions, namely non-selective multipath fading and shadowing. Finally, a formal model for networks based on non-geostationary satellite constellations as well as a traffic engineering concept are introduced, both forming a basis for a detailed network connectivity analysis. Given the network topology and the traffic requirements, the main task is the assessment of capacity requirements on the different links within the network, including the radio links from the satellites to mobile users and to gateways, as well as intersatellite links and terrestrial lines. A software tool for the numerical evaluation is presented together with some representative results on worst-case and average link capacity requirements, on-board RF power figures, and propagation delays.

1. Introduction

Recently a number of non-geostationary satellite systems have been proposed for global personal communications, such as Globalstar [1], Iridium [2], Odyssey [3], etc. In the near future, they can be regarded as necessary and reasonable supplement to terrestrial digital radio networks (GSM, DCS-1800, IS-54, IS-95, etc.), which provide mobile communications services only within limited regions. In the longer term, third generation mobile telecommunications systems (UMTS, FPLMTS) with a fully integrated satellite component will globally provide seamless personal communications.

The paper addresses the analysis of non-geostationary constellation characteristics, considering both circular Low-altitude Earth Orbits (LEO) and Medium-altitude Earth Orbits (MEO). A preliminary analysis is carried out on the basis of geometrical considerations alone, leading to the evaluation of the percentage of Earth surface covered by the constellation, the statistics of the frequency of handover between satellites and of link absence periods. This analysis is preliminary to the subsequent analyses of link transmission quality and of constellation connectivity.

As far as transmission quality is concerned, it must be underlined that next generation satellite mobile systems will need an extremely high value of spectrum efficiency both in the case of competition and in that of integration with terrestrial cellular systems. Presently, the trend is evident towards multi-spot systems with increasing values for the frequency reuse factor. If the service region is covered with many relatively small spots, the satellite system virtually becomes a cellular system. Therefore, an analysis of co-channel interference plays a fundamental role in the design of these systems. Co-channel interference is a consequence of the presence of sidelobes in the on-board antenna radiation diagram, which implies non-ideal angular selectivity of the spot-beams. Given the multi-spot antenna radiation diagram, we introduce a general model for the evaluation of the carrier-to-interference power ratio, C/I, which can be applied to all access schemes. We also analyse several techniques that can be used to avoid or minimize the interference power: spot turn-off; intra-orbital plane frequency division; inter-orbital plane frequency division.

Since mobile communication channels always suffer from non ideal propagation conditions, the interference analysis must also be performed in the presence of fading. For non-geostationary satellites the elevation angle, α, changes continuously over time, leading to time-varying channel characteristics. To take this effect into account, we adopt a statistical model whose parameters are empirical functions of α [4].

In addition to the above-mentioned geometrical and transmission quality aspects, several networking issues have to be considered. The demand for telephony and its global distribution, together with upper limits for blocking probability and speech delay, essentially determine the requirements for network capacity and connectivity. The latter two mainly depend on the number of satellites and gateway earth stations and on the number of links between mobile users, satellites, and gateways, including intersatellite links (ISLs). A high degree of connectivity enhances routing alternatives and thus a good distribution of traffic flow and the flexibility of the network to cope with link or node failures. On the other hand, manufacturing and positioning of a large number of satellites and gateways means high fixed costs, and the permanent supply of large capacities causes high recurring costs. From these considerations it becomes clear that the basic analysis and design of network connectivity of a LEO/MEO-satellite system is an important task to be faced already in the initial stage of system planning.

The paper is structured as follows. Section 2 contains the definitions for the orbital characteristics and geometrical performance measures of interest. The orbital parameters of the selected constellation are listed, and the preliminary geometrical analysis is performed. Section 3 contains the general model for interference analysis, and its application to the particular case of FDMA both in unfaded and faded conditions. In Section 4, we analyse the connectivity requirements in detail. To this end, we introduce a formal model for the description of the system structure. Based on this, we follow a traffic engineering approach to investigate different constellations with regard to link capacity requirements. Finally, numerical results of computer-based investigations are presented exemplarily for some constellations.

2. Geometrical Constellation Analysis

A preliminary but insightful analysis of satellite mobile systems based on different orbital constellations can be carried out on the basis of geometrical considerations alone. According to the satellite altitude, H, circular orbits can be classified as: Low-altitude Earth Orbits (500 < H < 2000 km), Medium-altitude Earth Orbits (5000 < H < 20000 km), the Geostationary Orbit (GEO) (H = 35800 km). The choice of H has fundamental consequences for all system characteristics, and some immediate considerations are possible:

- the lower is H the larger the constellation size, N_s. Fig. 2.1 shows a lower bound on the number of satellites needed to provide global Earth coverage (polar regions excluded) as a function of H and with the minimum elevation angle, α_m, as a parameter. The bound is simply obtained by dividing the service area by the coverage area of a single satellite.
- the lower is H the smaller is the free-space attenuation. However, for a fixed extent of the satellite coverage area, the smaller is H the less uniform is the free-space attenuation going from center to edge-of-coverage;
- the lower is H the smaller is propagation delay. In particular, the reduced propagation delay for LEO and MEO allows a double hop via satellite for mobile to mobile connections in accordance with CCITT Recommendation G.114;
- given a satellite antenna aperture, the lower is H the smaller is the spot footprint (cell). Smaller cells lead to more extensive frequency reuse, hence to larger system capacity;
- for a given cell area, the lower is H the longer is the slant path in the atmosphere at the edge of coverage, leading to larger excess attenuation and thus partially reducing the benefit of smaller free-space attenuation;
- the lower is H the higher the satellite velocity, which increases the handover and Doppler rates.

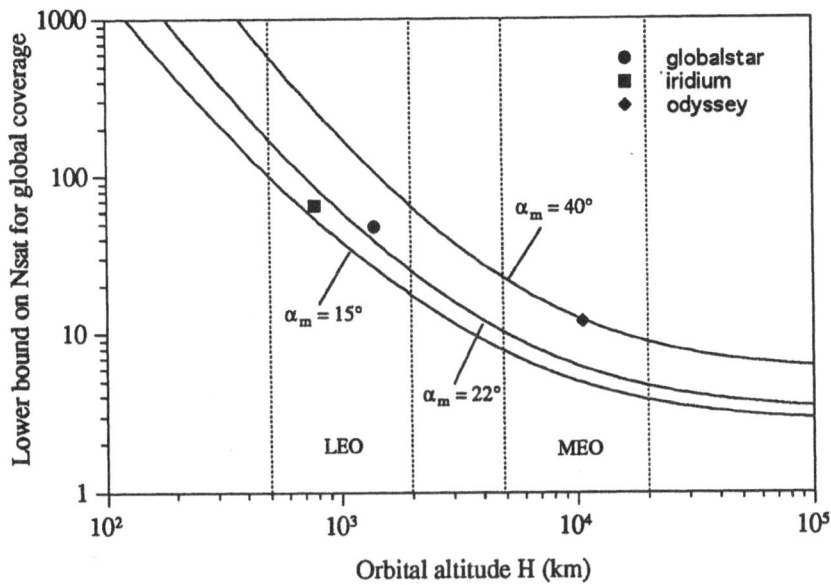

Fig. 2.1 - Lower bound on constellation size to achieve global coverage and characteristic points of some proposed systems

A more in-depth analysis of the constellation dynamics should evaluate the coverage areas, handover times and link absence times. This multi-satellite system geometrical analysis has been performed by means of a purposely developed software package [5]. The package contains a kernel which simulates the periodic constellation motion around the Earth, whose surface is substituted with a spherical grid of points. All the relevant events are observed for every instant on each grid point; the results are first averaged over the complete grid and, subsequently, time averaged over the constellation period. Assuming stationarity, we exchange time with probability to obtain a statistical characterization.

In this part of the study, the analysis is performed on three constellations, all designed to provide global Earth coverage:

- LEO1 (Iridium-like [2]): $N_s = 66$ satellites located at $H = 780$ km on 6 quasi-polar orbital planes (11 satellites per orbit), with inclination over the equatorial plane i = 86°, and $\alpha_m = 15°$;
- LEO2 (Globalstar-like [1]): $N_s = 48$ located at $H = 1389$ km on 8 orbital planes (6 satellites per orbit), with i = 52°, and $\alpha_m = 15°$;
- MEO (Odyssey-like [3]): $N_s = 12$ located at $H = 10600$ km on 3 orbital planes (4 satellites per orbit), with i = 55°, and $\alpha_m = 22°$.

A. Coverage

The instantaneous coverage achieved by a constellation, $\mathscr{C}(t; \alpha_m)$, can be defined as the weighted fractional area of the Earth surface S covered with an elevation angle greater than α_m [6]:

$$(1) \qquad \mathscr{C}(t; \alpha_m) \overset{\Delta}{=} \frac{\displaystyle\int_S \Phi(t; \alpha_m, P) w(P) dP}{\displaystyle\int_S w(P) dP} \quad ,$$

where P is a point over S; $\Phi(t; \alpha_m, P) = 1$ if P is covered at time t at least with angle α_m, and $\Phi(t; \alpha_m, P) = 0$ otherwise; the weight function w(P), $0 \le w(P) \le 1$, is tailored to the system service area and, possibly, to the expected traffic demand. In all subsequent analyses, we will assume w(P) = 0 above +70° lat. and below -70° lat., and w(P) = 1 elsewhere. The results of the coverage analysis for the above mentioned systems are reported in Fig. 2.2, which shows that coverage rapidly decreases as we increase the elevation angle. For relatively high values of α_m coverage is worse for LEO constellations, which can therefore be expected to be more affected by link blockage in the presence of obstacles.

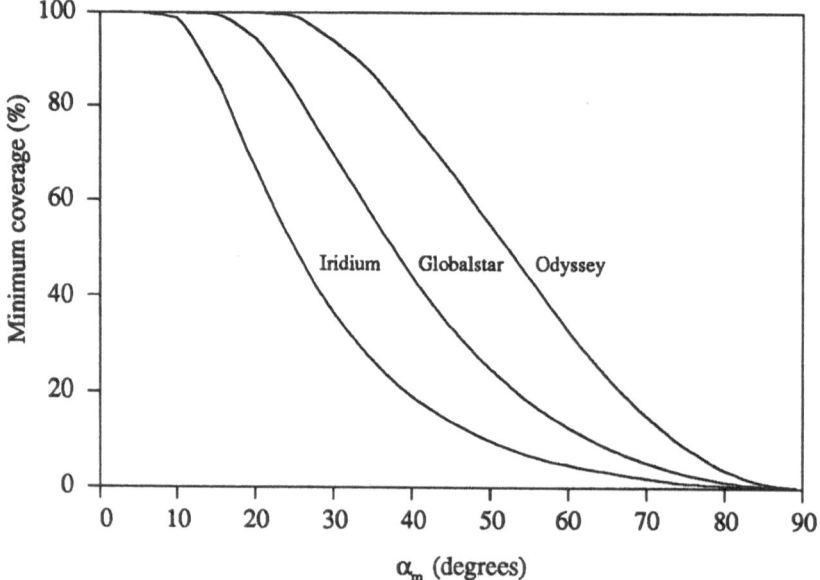

Fig. 2.2 - Minimum coverage achieved by different constellations on the globe (excluding polar regions) as a function of the minimum elevation angle

B. Handover

Several types of handover events are possible in a non-geostationary satellite system:
- *satellite handover*: a new satellite is chosen (according to some strategy) to provide the link between the Mobile Terminal (MT) and the Fixed Earth Station (FES);
- *spot handover*: a new beam in the satellite multi-spot antenna is necessary to provide the MT-FES link;
- *FES handover*: a new FES is needed by the serving satellite to provide the link from the MT to the terrestrial network;
- *FES and satellite handover*: a new satellite is selected, and this satellite is also served by a different FES.

To analyse spot handovers the multi-spot antenna configuration must be specified. To analyse handovers involving FES, the geographical coordinates of all gateways must be known. For simplicity, in the following we restrict our attention to the analysis of satellite handovers.

Different strategies can be adopted for handing over the communication from one satellite to another. A first possibility is to try to instantaneously maximize the elevation angle by handing over the communication every time a new satellite rises over the horizon higher than the serving satellite. At the opposite extreme, one could try to minimize the handover rate, by maintaining the link through a satellite as long as it guarantees a minimum elevation angle [7]. In practice, the adopted strategy will likely be a compromise between these two contrasting approaches.

Some results adopting the minimization of handover rate approach are shown in Table 2.1 for the selected constellations. The handover analysis yields the first statistical moments of the time interval between handovers, T_{HO}, with results weighted through the function w(P). Generally speaking, the time between handovers is smaller for constellations with lower H, due to the higher speed of the satellites.

Constellation	Avg (T_{HO}) (s)	Min (T_{HO}) (s)	Max (T_{HO}) (s)	std. dev. (s)
LEO1	277.7	10.3	523.7	102.9
LEO2	485.1	21.3	894.1	173.1
MEO	3690.3	34.4	7017.5	1824.3

Tab. 2.1 - Time between handovers statistics

C. Link absence

While searching for all satellites in visibility from a certain Earth location at a time instant t, it is possible that none of the satellites be above the minimum required elevation angle. In this case the link is considered absent, and the observation of the absence time intervals around the Earth (possibly weighted through w(P)) leads to an estimate of the service availability in clear-sky for a given constellation, which is a component of the overall system availability. The link absence analysis yields the first statistical moments of the link absence periods, T_{AB}. Table 2.2 contains the results of the link absence analysis for the selected constellations

Constellation	Avg (T_{AB}) (s)	Min (T_{AB}) (s)	Max (T_{AB}) (s)	std. dev. (s)
LEO1	97.7	3.4	1198.0	101.7
LEO2	76.3	7.1	220.0	60.4
MEO	210.9	34.4	550.4	141.9

Tab. 2.2 - Link absence time statistics
(LEO1: $\alpha_m = 15°$, LEO2 : $\alpha_m = 15°$, MEO: $\alpha_m = 22°$)

3. Interference Analysis

The software simulation package can be exploited to perform interference analyses, the results of which are a fundamental part in the link budget of these global systems with large served traffic.

As before, N_s indicates the number of satellites in the constellation, while N_c is the number of cells per satellite. A specified cell in the area of service is identified through the couple (j, k), j = 1, ..., N_c, k = 1, ..., N_s. Each cell is covered by the main lobe of a spot-beam whose gain, $G_j(\theta)$, is related to the normalized far-field radiation pattern, $F_j(\theta)$, by:

(2) $G_j(\theta) = G_{Mj} F_j^2(\theta)$,

where G_{Mj} is the maximum gain of the j-th spot. The spot-beams may have different radiation patterns to compensate for the different angles of incidence and free space losses. The radiation patterns have been modeled through suitable masks [5] enveloping the maxima of the generic tapered-aperture antenna radiation pattern. Also, for the case $N_c = 49$, an empirical model has been derived to synthesise the effective aperture diameters, d_{aj}, which give good multispot antenna footprints for a wide range of orbital altitudes and minimum elevation angles. This model is useful for a fair comparison of systems with different constellation characteristics. As an example, in Fig. 3.1 the resulting antenna footprint is plotted for the MEO case.

Mobile users in the jth cell in the area of coverage of the kth satellite are numbered from 1 to N_u, so that any user or its communication channel may be identified through a unique set of indices (i, j, k), i = 1, ..., N_u , j = 1, ..., N_c , k = 1, ..., N_s . Let the wanted mobile be assigned the set (m, n, p). The angles of interest in the model description are shown in Fig. 3.2. θ_{ijk} is the angle under which user (i, j, k) is seen from the serving satellite with respect to the boresight of the main lobe covering the cell containing the wanted mobile, while α_{ijk} is the elevation angle under which user (i, j, k) sees the serving satellite. The same definitions apply for the wanted mobile replacing (i, j, k) with (m, n, p). Note that, in general, these angles are time-varying.

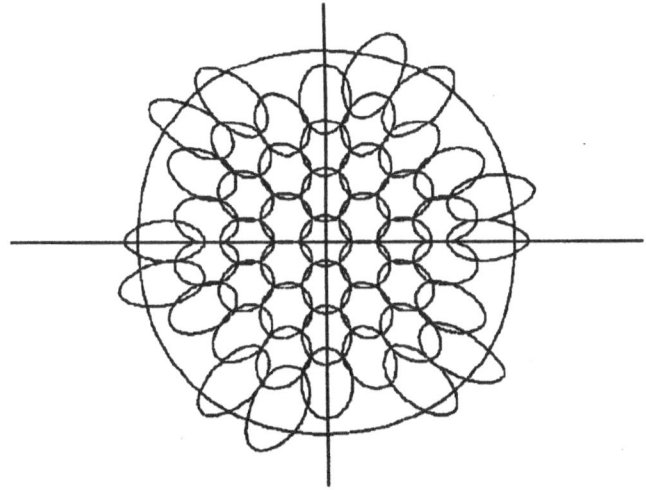

Fig. 3.1 - Multi-spot antenna footprint

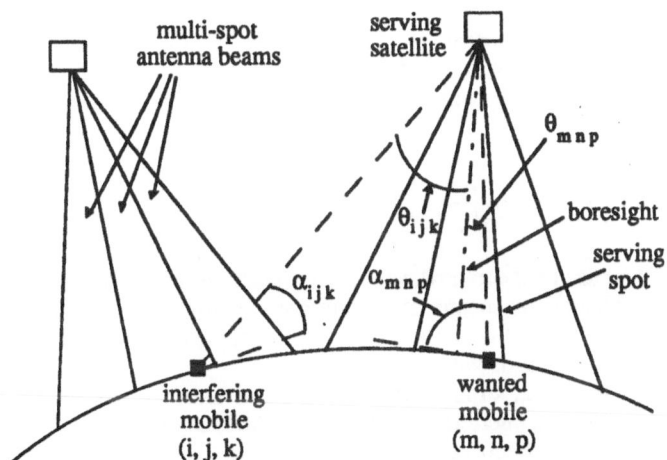

Fig. 3.2 - Angles of interest for wanted and interfering mobiles

Let us consider the case of a return link, from the wanted mobile to the Fixed Earth Station (FES). Given a certain position of the satellite constellation at a certain instant, denote as V the set of interfering mobiles for which $\alpha_{ijk} \geq \alpha_0$ (e.g. $\alpha_0 = 5°$). The signal to interference power ratio in the up-link can be written as [8]:

$$(3) \qquad \frac{C}{I} = \frac{\beta_{mnp}\, \varepsilon_{mnp}(\alpha_{mnp})\, G_n(\theta_{mnp})}{\displaystyle\sum_{\substack{k=1 \\ (i,j,k)\,\in\,V}}^{N_s} \sum_{j=1}^{N_c} \sum_{i=1}^{N_u} \gamma_{ijk}\, \mu_{ijk}\, \beta_{ijk}\, \varepsilon_{ijk}(\alpha_{ijk})\, G_n(\theta_{ijk})} \quad ,$$

where
- β_{ijk}^{-1} is the attenuation in the path from user (i, j, k) to the serving satellite;
- $\varepsilon_{ijk}(\alpha)$ is the product of transmitted power and antenna gain in the direction α for mobile terminal (i, j, k);
- γ_{ijk} is the orthogonality factor: $\gamma_{ijk} = 0$ means that channels (i, j, k) and (m, n, p) are perfectly orthogonal (not interfering), while for $0 < \gamma_{ijk} \leq 1$ these channels are interfering;
- μ_{ijk} is the activity factor: the mean period of time in which channel (i, j, k) is active.

The above specified model can be used to evaluate the signal to interference power ratio in the case of FDMA, TDMA, or CDMA depending on the value assumed by γ_{ijk} [8].

Given a multiple access technique and a digital modulation method, a specification for the minimum carrier to interference power ratio, $(C/I)_{TH}$, can be identified. We define the outage probability, P_{out}, as the probability of failing to achieve the specified C/I:

$$(4) \qquad P_{out} = \text{Prob}\left\{ \frac{C}{I} \leq \left(\frac{C}{I}\right)_{TH} \right\} .$$

As for the case of the geometrical performance measures, the instantaneous C/I has been evaluated by means of a time-domain simulation of the constellation motion over the discretized Earth, and the statistical characterization has been obtained exchanging time with probability.

Generally speaking, strong interference is not unlikely due to the intersecting orbital planes. Several interference reduction methods can be devised:
A) *Spot turn-off.* According to this technique, whenever two spots overlap exceedingly, one of the two spots is turned-off. In our approach, the measure for overlapping is based on the distance on the Earth surface between the intersections of the spots boresight.
B) *Intra-orbital plane frequency division.* According to this technique, satellites on the same orbital plane are assigned different frequency subsets up to a specified modulo R. For example, in the case of R = 2 there are two subsets, each one reused every other satellite. It is evident that R should be preferably chosen to be a divisor of the total number of satellites on an orbital plane.

C) *Inter-orbital plane frequency division.* According to this technique, the available frequency spectrum is subdivided into as many subsets as the number of orbital planes, so that satellites on different orbital planes are non-interfering.

D) *Spot turn-off and intra-orbital frequency division.* This technique follows simply from the simultaneous use of techniques A and B.

All these techniques bring some reduction in the reuse of the available frequency spectrum. Therefore, the improvement in outage probability must be gauged against the reduction in spectrum efficiency.

3.1 Analysis in the Absence of Fading

Assume FDMA as the multiple access scheme and $N_c = 49$, with a 7-cell frequency reuse pattern. Figs. 3.3 (a), (b), (c) contain the results of the analysis for the LEO1, LEO2 and MEO systems, respectively, and for the different techniques, while Tab. 3.1 shows the number of times the frequency spectrum is reused correspondingly. In all cases technique C brings the greatest reduction in outage probability, as it produces an almost static interference situation as in the GEO case. On the other hand, it is also the one that uses the spectrum less efficiently: nevertheless, if the number of orbital planes is small (as in the MEO case) the efficiency reduction could be tolerable. Technique B is generally inferior with respect to technique A, at least for the interesting range of outage probabilities (below 10^{-1}). Technique A is also the one that ensures the maximum spectrum efficiency. Technique D obviously yields lower outage probabilities than A or B, with a small reduction in frequency reuse with respect to B.

Constella-tion	No tech.	A (average)	B (R=2)	C	D (R=2) (average)
LEO1	462	430.7	231	77	224.1
LEO2	336	224.8	168	42	146.6
MEO	84	50.6	42	28	34.8
GEO	24	-	-	-	-

Tab. 3.1 - Frequency reuse for different interference reduction techniques

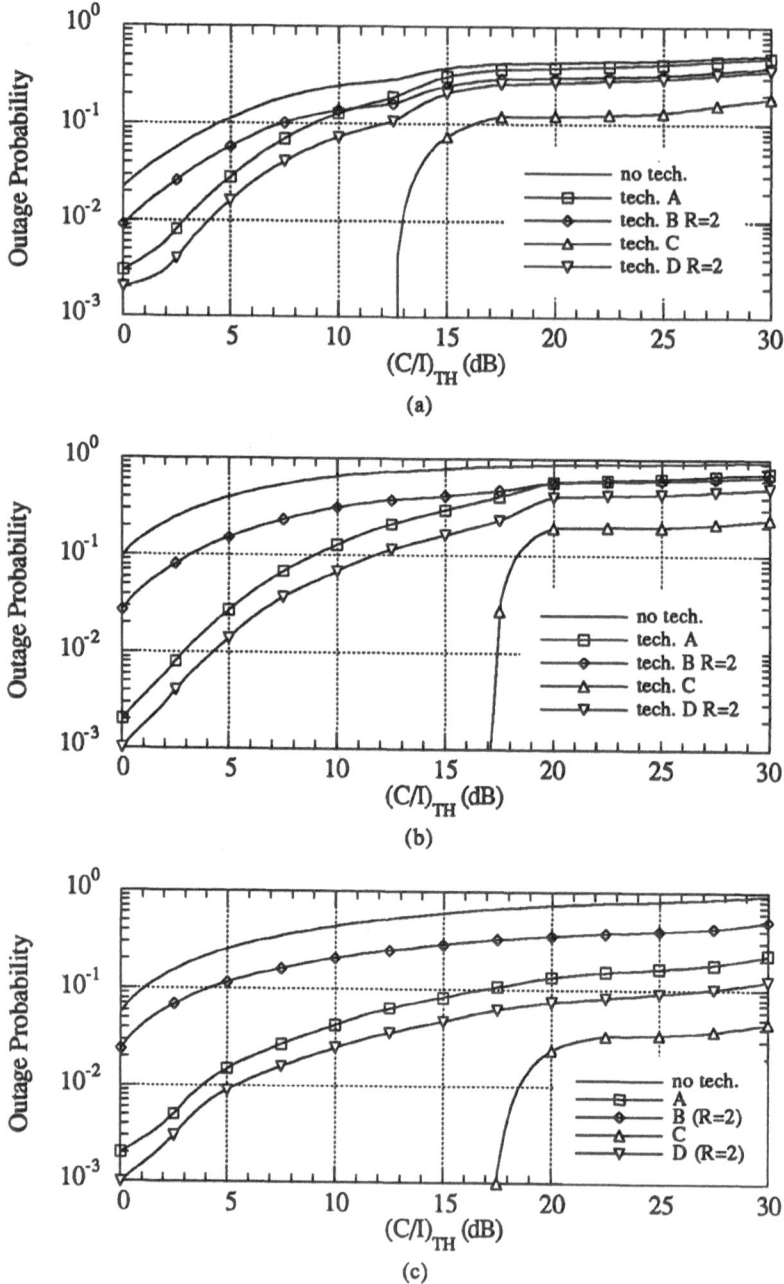

Fig. 3.3 - Outage probability for different interference reduction techniques and for the selected constellations: (a) Iridium, (b) Globalstar, (c) Odyssey

3.2 Analysis in the Presence of Fading

Assuming that every user experiences non-selective fading, (3) can be easily modified to obtain the carrier to interference ratio in the presence of fading, C_f / I_f, as follows:

$$(5) \qquad \frac{C_f}{I_f} = \frac{r_{mnp}^2 \ C}{\displaystyle\sum_{(i,j,k) \in W} r_{ijk}^2 \ I_{ijk}} \quad ,$$

where W is the subset of V containing the users reusing the same frequency channel as (m,n,p).

In (5) C, $\{I_{ijk}\}$ are the deterministic useful and interfering signals power in the absence of fading, while each fading envelope, r, is described by a Rice-lognormal probability density function (p.d.f.) which takes into account both multipath fading and shadowing [9]:

$$(6) \qquad p_r(r) = \int_0^\infty p(r|S) \ p_S(S) \ dS \ ,$$

where

$$(7) \qquad p(r|S) = 2(K+1)\frac{r}{S^2} e^{-(K+1)\frac{r^2}{S^2} - K} \ I_0\left(2\frac{r}{S}\sqrt{K(K+1)}\right) \quad (r \geq 0),$$

and

$$(8) \qquad p_S(S) = \frac{1}{\sqrt{2\pi} \ h \ \sigma \ S} e^{-\frac{1}{2}\left(\frac{\ln S - h \ \mu}{h \ \sigma}\right)^2} \quad (S \geq 0) \ .$$

In (7), I_0 is the zero order modified Bessel function of the first kind, and K is the so called Rice factor. In (8), h = (ln10)/20. The model parameters K, μ, σ can be expressed as empirical functions of the elevation angle α, which for non-geostationary systems is continuously varying [9]. The coefficients of the empirical function have been derived for different environments (urban, tree-shadowed, suburban, open) [4], allowing to model the propagation channel in a vast variety of situations, as necessary for systems with global coverage.

Here, we present only some results of an approximate evaluation of the system outage probability, under the simplifying assumption that all users experience Rice fading with equal K [5]. Figs. 3.4 (a) and (b) report outage probability for LEO2 and MEO constellations, respectively, for three different values of the Rice factor K (0 dB, 5 dB, 10 dB). The analysis has been carried out for European coverage and adopting interference reduction technique C (inter-orbital plane frequency division).

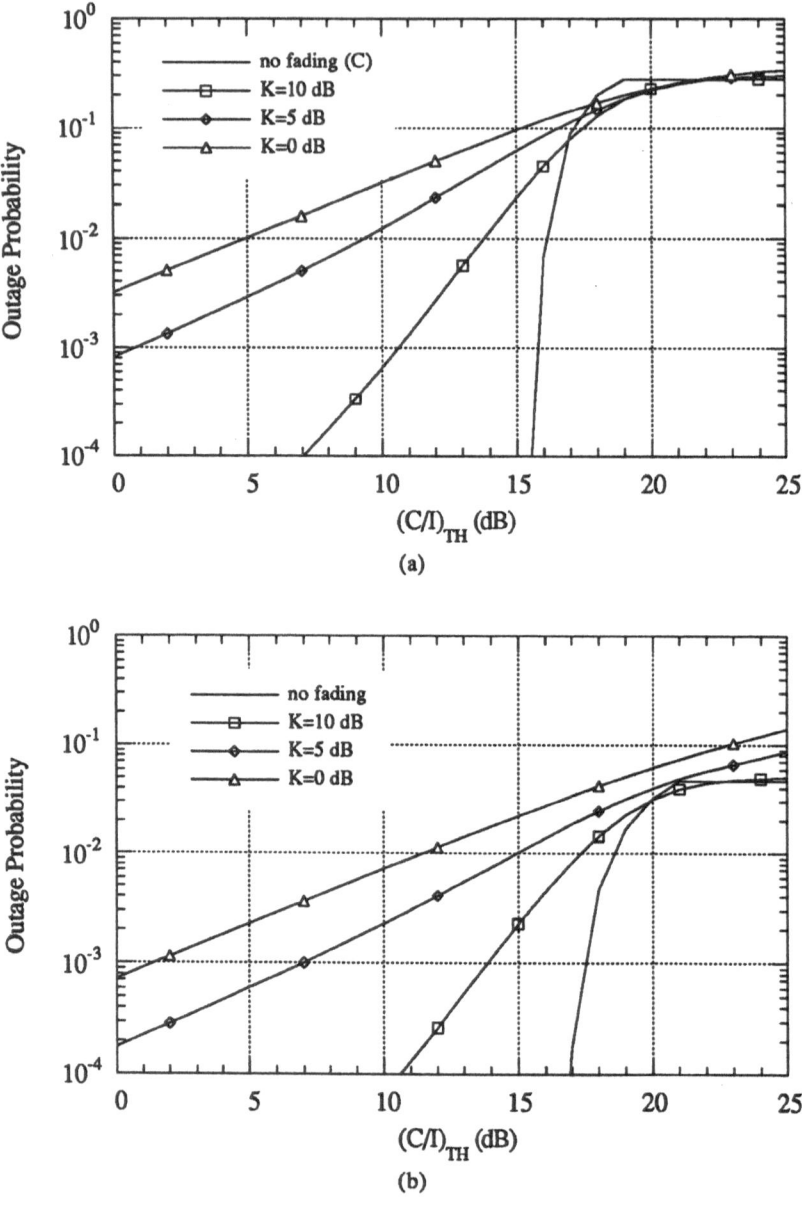

Fig. 3.4 - Outage probability in the presence of fading for different values of the Rice factor K. (a) Globalstar with technique C, (b) Odyssey with technique C

4. System Connectivity Analysis

From the general viewpoint of system analysis and design a network connectivity analysis becomes necessary. We understand *connectivity* as the "degree of possible/provided network connections" rather than in a restricted graph theoretical sense. Thus, connectivity does not only depend on the time-varying geometry of system components and on the presence of links, but also on the capacity provided on the various links within the network.

To investigate both topology and capacity requirements we introduce a traffic engineering concept and a capacity evaluation procedure based on an elaborate network model. Numerical results are derived from computer-based investigations incorporating this approach.

Specific topology details (primarily w.r.t. ISLs) and numerical results will be given examplarily for two constellations, namely Iridium as a typical LEO system, and LEONET as a representative MEO concept developed in the framework of an ESA study[1] [10]. LEONET, as the four constellations introduced in Section 2, was designed to provide global Earth coverage: N_s = 15 satellites located at H = 6370 km on 3 orbital planes (5 satellites per orbit), with i = 54°, and α_m = 20°.

4.1 Network Topology

As an input to the detailed network model, first the main elements of typical LEO/MEO topologies are discussed. The network nodes of a LEO/MEO system are:

*   *Communications satellites* - satellites in LEO or MEO (polar or inclined) with transmission and (optionally, with OBP) switching functions.
*   *Gateway stations (GWs)* - terrestrial stations interfacing the space segment with existing terrestrial public networks. Besides that, their main tasks are switching and network management.
*   *User terminals* - terminals of mobile and fixed users, representing the sources and destinations of traffic.

The network nodes are connected by several kinds of links:

*   *Mobile User Links (MULs)* - links between satellites and those mobile users within their footprint, who communicate via the satellite.
*   *Gateway Links (GWLs)* - links between satellites and gateways in the coverage area of the satellite.
*   *Links through the Public Switched Telephone Network / Public Data Networks (PSTN/PDNs)* - totality of existing telephone and data networks providing the possibility for all fixed users to communicate with mobile LEO/MEO system users via gateway stations.
*   *Intersatellite Links (ISLs)* - direct connections between satellites; their use in a LEO/MEO satellite system is optional. They may be used for the routing of long-distance traffic and/or the exchange of system specific signalling and network management information.

---

[1] The name LEONET is historically derived from the title of this study.

4.2 ISL Topology

The connectivity of a LEO/MEO-satellite network substantially depends on the presence of intersatellite links (ISLs). Therefore, as network connectivity is the central aspect of our considerations, we include the presence of an ISL subnetwork as a possible option in our investigations.

From the topology point of view, two types of ISLs can be identified, namely intra-plane ISLs connecting satellites within the same orbit plane and inter-plane ISLs connecting satellites in adjacent orbit planes. Two satellites on different orbit planes "see" each other under time-varying pointing angles. Therefore, inter-plane ISLs generally require antenna steering, whereas intra-plane ISLs can be maintained with fixed antennas. Moreover, inter-plane ISLs may not be permanently maintained, because as the satellites follow their orbits, their distance may vary within a large range and the earth may interrupt the line of sight. In this case, the inter-plane ISL would have to be switched off and on again, requiring the formidable task of pointing, acquisition, and tracking (PAT). Due to the latter fact, we have restricted our investigations on ISL topologies where either all ISLs can permanently be maintained (for LEONET) or the switching of ISLs is possible with very modest pointing angle requirements (for Iridium).

Fig. 4.1 shows the according topology of satellites and ISLs for the LEONET constellation. Each satellite must be equipped with four bidirectional ISL ports (ISL antenna, transmitter, and receiver); two of the four ISL antennas must be steerable. Fig. 4.2 gives an impression of the time-varying inter-plane pointing angles in azimuth and in elevation, related to the flight direction of a satellite. Quite a large range of pointing angles is necessary, causing rather challenging PAT requirements.

For the Iridium constellation, intra-plane ISLs and three equatorial rings of inter-plane ISLs are considered, as shown in Fig. 4.3. It is assumed that for each time those inter-plane rings are maintained which are closest to the equator. For the north-bound satellites new inter-plane ISLs south of the equator are switched on when the oldest inter-plane ISLs north of the equator are switched off. The corresponding procedure is followed for the south-bound satellites. For the boundaries between north-bound and south-bound satellites we assume appropriate inter-plane ISLs at any moment of time. Assuming a vertical two-sided 2-dB-beamwidth of 5° and a 1:2 beam ellipticity (corresponding to approximately 36 dBi gain at 23 GHz), as suggested in [11], it is possible to keep the vertical antenna pointing fixed between latitudes of approximately 60° south or north, respectively, and to perform a limited horizontal pointing with a remaining steering range of 31°.

In the following we assume that ISLs are used for Iridium; for LEONET we consider ISLs as a system option.

For the routing of long-distance traffic, there exist two major alternatives. In a system providing ISLs this traffic may be transported as far as possible through the space segment. In connection with a terrestrial backbone it is possible to significantly reduce the number of world-wide necessary GW stations. If the system does not provide ISLs, then the whole long-distance traffic has to be transported through public lines. In this case it is necessary for global connectivity that every satellite has connection to at least one GW station at any instant of time. A reasonable realization of a LEO/MEO-satellite based communications system could advantageously combine the positive features of both alternatives.

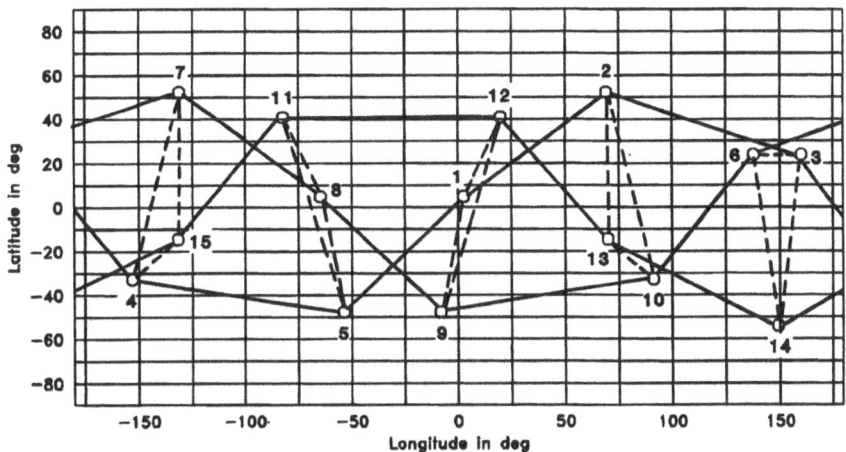

Fig. 4.1. Topology of satellites and ISLs for LEONET at t = 240 sec.

o Subsatellite points   —— intra-plane ISLs   - - - - - inter-plane ISLs

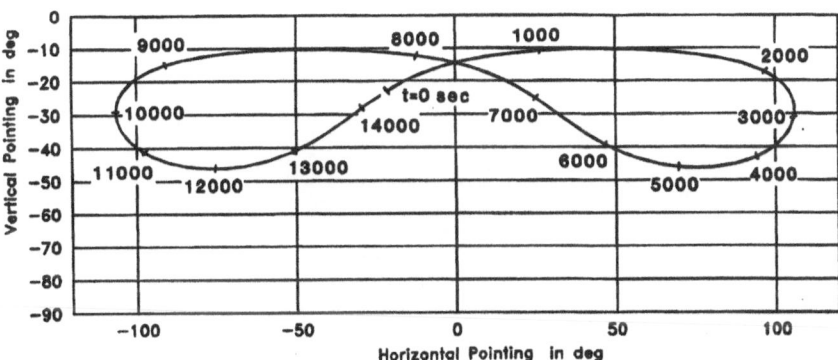

Fig. 4.2. Time-varying inter-plane pointing angles for LEONET. Pointing from satellite 1 to satellite 12.

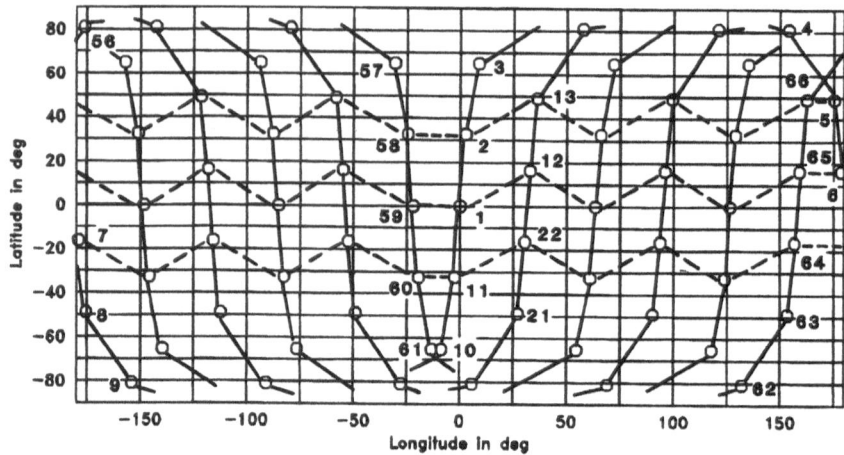

Fig. 4.3. Topology of satellites and ISLs for Iridium at t = 0 sec.

o Subsatellite points ———— intra-plane ISLs - - - - - inter-plane ISLs

4.3 Network Model

As a base for the following capacity evaluation procedure we present a graph-theoretical model describing the network configuration at a certain instant of time [12]. Time variance of network topology and traffic are taken into account on the computer calculation level and will be discussed later. The model consists of the following components:

1. n satellites $s_1,\ldots,s_n$.
2. n traffic sources/destinations $m_1,\ldots,m_n$; m_i represents all mobile users communicating via satellite s_i.
3. n network nodes $f_1,\ldots,f_n$ in the footprints of the corresponding satellites. The whole amount of PSTN traffic "belonging" to the footprint of s_i is assumed to be concentrated in node f_i. Furthermore, the f_i are assumed to be completely connected with each other via the PSTN. The presence of at least one GW in the footprint of s_i results in a direct connection between f_i and s_i via a GWL; therefore, such a GW can be regarded as physically representing the fictitious node f_i.

The whole traffic generated at an instant of time can be described by a traffic matrix T of dimension $2n \cdot 2n$:

$$T \cong
\begin{array}{c|c|c}
 & m_1,\ldots\ldots\ldots,m_n & f_1,\ldots\ldots\ldots,f_n \\
\hline
\begin{array}{c} m_1 \\ \cdot \\ \cdot \\ \cdot \\ m_n \end{array} & t_{m_i m_j} & t_{m_i f_j} \\
\hline
\begin{array}{c} f_1 \\ \cdot \\ \cdot \\ \cdot \\ f_n \end{array} & t_{f_i m_j} & \begin{array}{c} 0 \\ \text{(no traffic to} \\ \text{be considered)} \end{array}
\end{array}$$

In this matrix the elements t_{ij} describe the amount of traffic between nodes i and j in Erlang (Erl). Traffic between two fixed users f_i and f_j is assumed as pure PSTN traffic and therefore not included in the analysis. In case only voice communication over full duplex channels is considered, the traffic matrix T becomes symmetric.

The totality of connections within the network is described by a symmetric connectivity matrix C of dimension $3n \cdot 3n$, where "1" denotes an existing connection between adjacent nodes:

$$C \cong
\begin{array}{c|c|c|c}
 & m_1,\ldots,m_n & f_1,\ldots,f_n & s_1,\ldots,s_n \\
\hline
\begin{matrix} m_1 \\ \cdot \\ \cdot \\ \cdot \\ m_n \end{matrix} & 0 & 0 & E \\
\hline
\begin{matrix} f_1 \\ \cdot \\ \cdot \\ \cdot \\ f_n \end{matrix} & 0 & \begin{matrix} 1 \\ (PSTN) \end{matrix} & c_{f_i s_j} \\
\hline
\begin{matrix} s_1 \\ \cdot \\ \cdot \\ \cdot \\ s_n \end{matrix} & E & c_{s_i f_j} & \begin{matrix} c_{s_i s_j} \\ (ISLs) \end{matrix} \\
\end{array}$$

Here $E = (e_{ij})$ denotes the unit matrix, i.e. $e_{ij} = 1$ for $i = j$ and $e_{ij} = 0$ for $i \neq j$. The $c_{s_i s_j}$ describe connections with ISLs, and the $c_{s_i f_j}$ and $c_{f_i s_j}$ mark a connection between a satellite and a GWstation in its footprint, i.e.

(9)
$$c_{s_i f_j} = \begin{cases} 1 & \text{for } i = j \text{ and a GW present in the footprint} \\ 0 & \text{for } i = j \text{ and no GW present in the footprint} \\ 0 & \text{for } i \neq j \end{cases}$$

4.4 Traffic Engineering Concept and Capacity Evaluation Procedure

Any LEO/MEO satellite system is in principle highly dynamic due to the time-varying topology; furthermore, according to dynamic user activity, also the demand for link capacities permanently varies. In order to take this two-fold time variance into account, investigations are made for several successive instants of time, in which the system is assumed as static. By choosing the time interval small enough, it is possible to gain reliable worst-case and average results for required link capacities and for delay values. The generalized flow chart of the C computer program developed for this capacity evaluation in Fig. 4.4 illustrates the evaluation procedure for a certain instant of time.

Starting point is a satellite constellation, characterized through orbit height (respectively orbit period), number and inclination of orbits, and number and phasing of satellites in these orbits. With a given minimum elevation angle for the connection to the satellites the coverage areas can now be calculated.

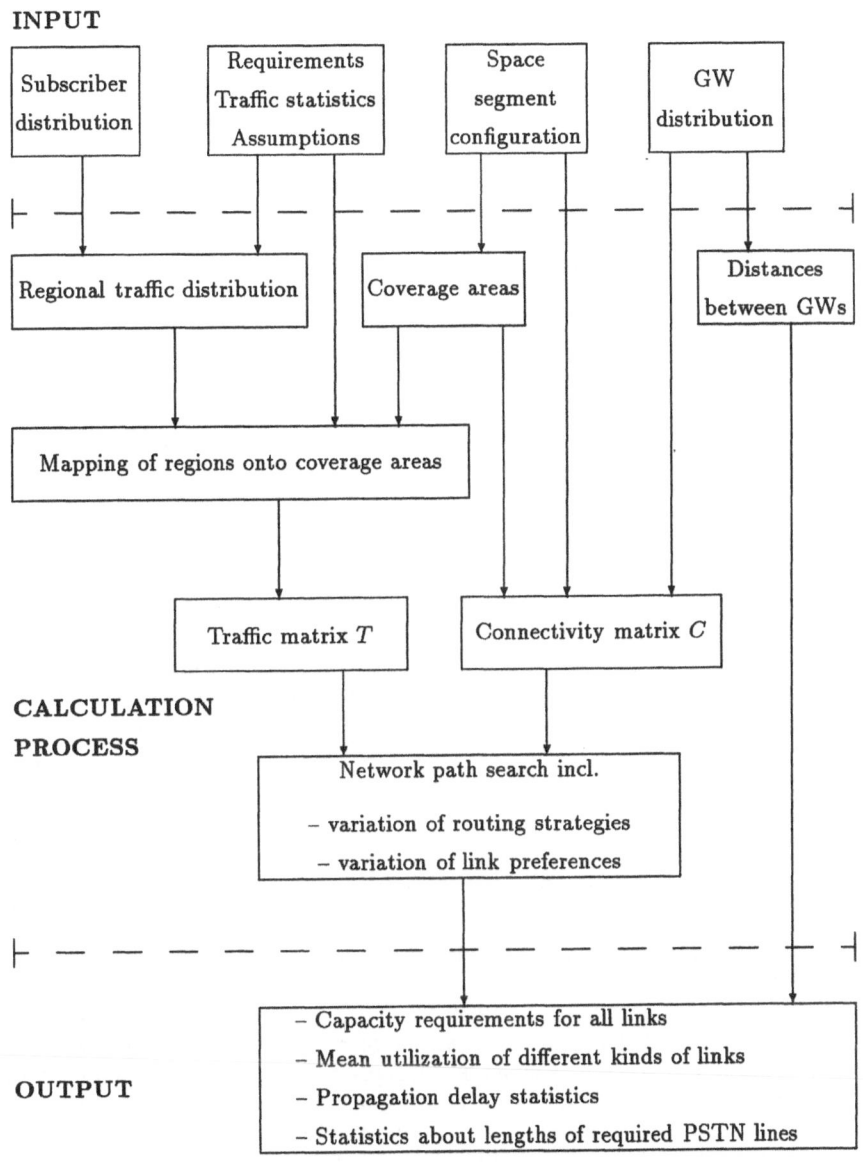

INPUT

| Subscriber distribution | Requirements Traffic statistics Assumptions | Space segment configuration | GW distribution |

Regional traffic distribution

Coverage areas

Distances between GWs

Mapping of regions onto coverage areas

Traffic matrix T

Connectivity matrix C

CALCULATION PROCESS

Network path search incl.

– variation of routing strategies

– variation of link preferences

OUTPUT

– Capacity requirements for all links
– Mean utilization of different kinds of links
– Propagation delay statistics
– Statistics about lengths of required PSTN lines

Fig. 4.4. Schematic description of network connectivity analysis.

With respect to communications traffic, the following assumptions have been made:

- Voice service is provided with a blocking probability less than 5 % for 1 million subscribers distributed within six regions according to Table 4.1. These (land mass) regions are visualized in Fig. 4.5
- All users are permanently generating traffic with 5 mErl, no matter which time zone they are actually in. This "permanent world-wide busy hour" is legitimated by the objective to gain worst-case traffic requirements.
- The traffic flow between the six regions is according to the regional traffic flow matrix in Table 4.2.
- Mobile-to-fixed and mobile-to-mobile connections are distinguished assuming 10 % of the system users to be mobile, which yields a ratio of mobile-to-fixed traffic : mobile-to-mobile traffic = 18:1.

For further network analysis the regional traffic is allocated to the different satellites according to their percentage of coverage of the respective land mass. This mapping of land mass regions onto coverage areas is indicated by the hatchings in Fig. 4.6. For coverage determination, $\alpha_m = 20°$ is assumed in this case. The result is a traffic matrix T for the specific configuration. Together with a given GW distribution, also the connectivity matrix C can be given.

In order to perform the transport of generated traffic, paths from source i to destination j with $t_{ij} > 0$ have to be determined, considering the underlying conditions (blocking probability, delay, etc.). From connectivity matrix C it can be seen that no direct connections ($c_{ij} = 0$) exist between source and destination nodes (mobile and fixed users). The basic criterion for path search is a generalized cost function, and "cheapest" paths are selected for the traffic transport; this is performed with the Dijkstra shortest path algorithm [13]. On the basis of different cost functions the following strategies for the routing of long-distance traffic are implemented:

1. ISLs are preferred (high pseudo-costs attached to PSTN links, low ones to ISLs).
2. PSTN links are preferred (high pseudo-costs attached to ISLs, low ones to PSTN links).
3. Advanced Routing Strategy (ARS): If not every satellite has connection to at least one GW station at all times (e.g. in the case of non-optimally positioned GW sites), then a certain amount of ISL capacity is necessary to mantain global connectivity. This capacity is in the whole network used as far as possible and only the traffic exceeding this amount is routed through PSTN. In our investigations, numerical values for this capacity were extracted from simulations.

As output the algorithm yields the required (worst-case) link capacities in Erlang for the different connections (MUL, GWL, ISL, PSTN). Furthermore, the average utilization is calculated for every link type, so that the calculation of average capacity requirements is possible, too.

region	percentage	absolute #
North America	25%	250 000
Europe	25%	250 000
Asia	20%	200 000
South America	10%	100 000
Africa	10%	100 000
Australia/NZ	10%	100 000

Table 4.1. Regional distribution of subscribers.

	N.America	Europe	Asia	S.America	Africa	Australia
N.America	1030 Erl	48 Erl	48 Erl	36 Erl	24 Erl	24 Erl
Europe	48 Erl	1030 Erl	48 Erl	36 Erl	36 Erl	12 Erl
Asia	48 Erl	48 Erl	825 Erl	10 Erl	20 Erl	39 Erl
S.America	36 Erl	36 Erl	10 Erl	436 Erl	10 Erl	5 Erl
Africa	24 Erl	36 Erl	19 Erl	10 Erl	436 Erl	5 Erl
Australia	24 Erl	12 Erl	39 Erl	5 Erl	5 Erl	441 Erl

Table 4.2. Regional traffic flow in Erlang.

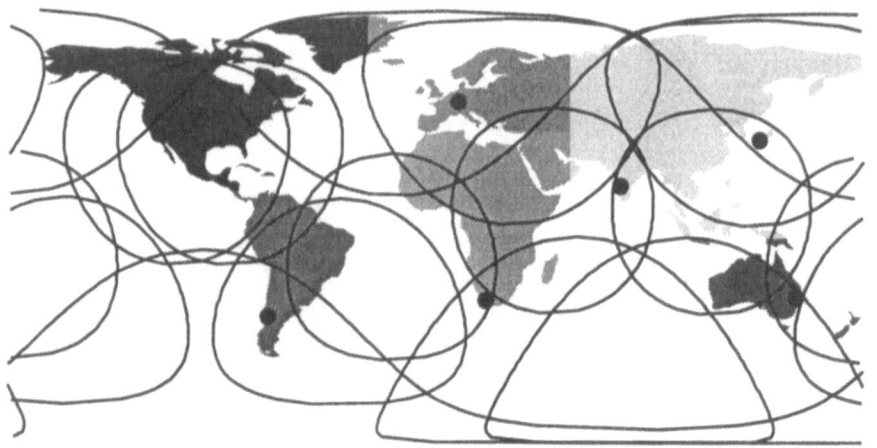

Fig. 4.5. Traffic regions on earth; footprints and example GW distribution for constellation LEONET.

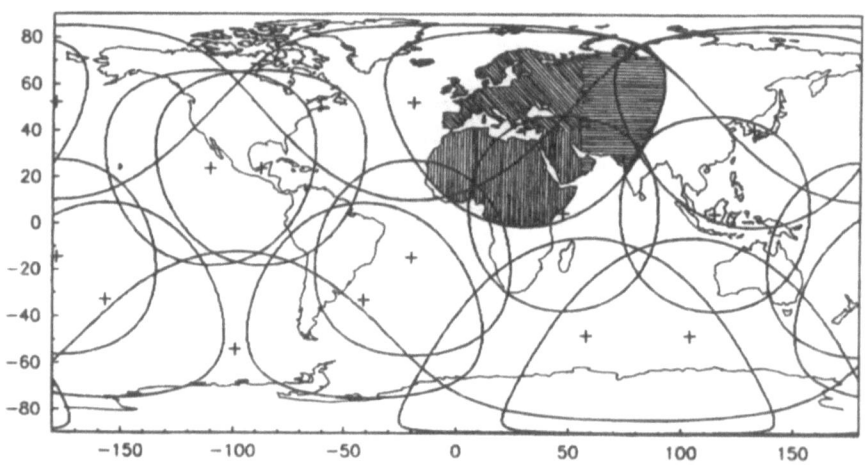

Fig. 4.6. Land mass coverage and traffic mapping for constellation LEONET.

For network dimensioning the maximum number of required channels on any link is of importance. Therefore, channel numbers are calculated according to the Erlang B formula [14]

$$(10) \qquad B = \frac{\dfrac{A^C}{C!}}{\sum_{k=0}^{C} \dfrac{A^k}{k!}}$$

where B is the blocking probability, C the number of channels, and A the traffic in Erlang. In addition, from all point-to-point-connections the maximum propagation delay is extracted and an average value for the whole network is calculated.

4.5 Numerical Results of Computer-Based Analysis

With the developed software tool a variety of investigations has been performed for a number of different system constellations. In order to evaluate the long distance traffic requirements, all constellations were investigated for the different routing strategies and for several GW distributions, with "reasonable" locations for the GW stations (e.g. preferably in most industrialized regions, on dry land). Figures 4.7 and 4.8 exemplarily show worst-case full duplex channel requirements per ISL and in the PSTN, respectively, for the LEONET constellation with 4 ISLs per satellite as presented above.

A look on both Figures shows the following:

- With the "extreme" routing strategies, "prefer ISLs" and "prefer PSTN", worst case channel requirements can monotonously be decreased with an increasing number of GWs either for PSTN or for ISL links, respectively. The requirements for the preferred traffic transport resource remains high.
- The advanced routing strategy (ARS) is capable of significantly reducing the worst case channel requirements for both long-distance traffic resources at the same time.
- With 7 GW stations reasonably distributed over dry land, the LEONET system is able to guarantee global connectivity without use of ISLs; alternatively, the whole long-distance traffic within the system could exclusively be routed via ISLs. The corresponding GW distribution is depicted in Fig. 4.5.
- A number of 6 or 7 GW stations seems to be a clear lower limit for a reasonable LEONET system realization; considering some other relevant aspects as well (e.g. regulatory and political issues, multiple visibility between satellites and GWs, robustness w.r.t. failure or regional traffic overload), one may regard a number of 10 -- 20 GW stations as good choice for LEONET.

However, the 7 GW constellation is well suited to show the characteristics of the LEONET system. Numerical results for a constellation without and with 4 ISLs per satellite, respectively, are given in Table 4.3, in comparison with the corresponding figures for the Iridium system, which is essentially based on the presence of an extensive ISL infrastructure.

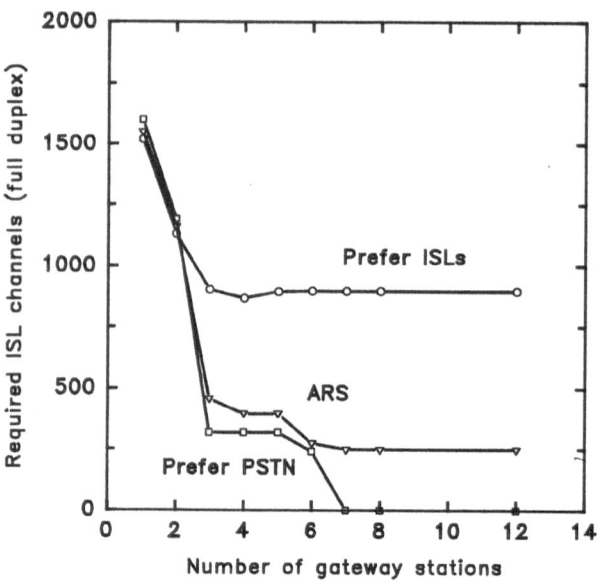

Fig. 4.7. Required full duplex channels per ISL. ARS = Advanced routing strategy. *Constellation:* LEONET with 4 ISLs per satellite.

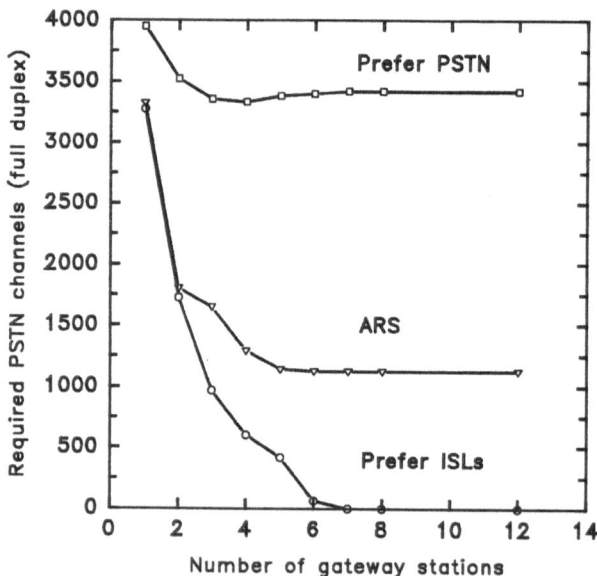

Fig. 4.8. Required full duplex PSTN lines for long-distance traffic in the whole network. ARS = Advanced routing strategy. *Constellation:* LEONET with 4 ISLs per satellite.

Constellation		Iridium 55 GWs	Iridium 55 GWs	LEONET 7 GWs	LEONET 7 GWs
ISLs	per satellite	2/4	2/4	4	0
	intra-orbit	2	2	2	0
	inter-orbit	0/2	0/2	2	0
Routing strategy		Pref. ISLs	ARS	ARS	Pref.PSTN
Elevation parameters:					
Min. elevation angle (mob. user)		8.2°	8.2°	20°	20°
Min. elevation angle (GWs)		5°	5°	5°	5°
Results:					
Required channels	per MUL	488	488	1047	1047
(5 mErl, worst-case)	per ISL	902	82	248	0
	per GWL	455	480	1018	1047
Long distance	≤ 7500 km	59	2573	676	1851
PSTN lines	≤10 000 km	14	192	304	1118
(5 mErl)	≤12 500 km	7	105	78	271
	≤15 000 km	6	70	35	81
	≤ 20 000 km	6	35	31	100
Peak power	MUL (1.6 GHz)	81 W	81 W	393 W	393 W
per satellite	ISL (23 GHz)	7 W	0.7 W	11.5 W	0
(5 mErl)	GWL (4 GHz)	3.9 W	4.2 W	8.3 W	8.5 W
Average power	MUL (1.6 GHz)	13 W	13 W	124 W	124 W
per satellite	ISL (23 GHz)	2 W	0.1 W	5.5 W	0
(5 mErl)	GWL (4 GHz)	0.6 W	0.7 W	2.7 W	2.7 W
Max. propagation delay		171 ms	171 ms	265 ms	198 ms
Aver. propagation delay		46 ms	32 ms	95 ms	75 ms

Table 4.3. Numerical results for different constellations.

The considered number of 55 Iridium GWs is due to the smaller footprints and was found as a good representative constellation during the investigations. Besides the higher number of Iridium GW stations, two other characteristic differences can be seen from the figures:

- The scope for traffic flow optimization by means of adapted routing strategies is quite limited for Iridium in comparison to LEONET. The preference of ISLs for long distance traffic routing is the best choice. The ARS approach is not suitable for Iridium.
- The worst-case channel requirements for MUL (per link!), GWL (per link!), and PSTN are significantly lower for Iridium.

Based on the worst-case channel figures for MUL (downlink), ISL, and GWL, link budget calculations were performed to determine peak RF power requirements on board the satellite, which is of basic importance for satellite design. The average power values were calculated by considering the mean utilization of the respective links. The figures given in Table 4.3 show that the power consumption on board the satellite is mainly determined by the MUL. Consequently, the overall power requirements are significantly lower for Iridium satellites.

In the last two lines of Table 4.3 figures for maximum and average propagation delay are denoted, showing a better performance for Iridium due to the lower orbit altitude. Considering a typical packet length of 20 ms and current switching technology, the sum of additional delays through codec, assembling/disassembling processes and switching/buffering operations should be in the range of 100 ms for a typical point-to-point connection, so that the CCITT limit of 400 ms for voice transmission can be guaranteed within either system.

5. Conclusions

The paper addressed the analysis of mobile satellite systems based on non-geostationary constellations. These systems are foreseen as key elements in the future Universal Personal Communication system.

In the first part, a purely geometrical analysis has been carried out, based on a time domain simulation of the constellation motion around the Earth. This allows the evaluation of performance functionals such as coverage, handover and link absence times. An interesting conclusion that can be drawn is that LEO systems cover most of the Earth with low elevation angles (below 30°), and can therefore be expected to frequently suffer from link obstructions.

In the second part, transmission quality is analysed from the point of view of carrier-to-interference power ratio. A general model for the evaluation of C/I has been applied here to the case of FDMA, also considering a few possible interference reduction techniques: spot turn-off, intra- and inter-orbit frequency division. The analysis has been performed both with and without the presence of non-selective fading. The numerical results show relatively high outage probabilities due to C/I, implying that link budgets for these systems cannot be adequately calculated without a precise estimate of interference power.

The third part of the paper was devoted to the discussion of networking aspects. A formal model for mobile communications networks based on LEO/MEO satellites and a traffic engineering concept to handle their characteristics were introduced and a method for the analysis of network connectivity requirements was proposed. This evaluation procedure can efficiently be used in the initial process of planning and dimensioning LEO/MEO networks. For numerical analysis, the application in form of a specially developed software tool was shown exemplarily for a comparison between two proposed systems, Iridium and LEONET.

Numerical results of the network connectivity investigation show that the system constellation basically influences the link capacity requirements. Thus the presented results provide important input information for the design of network components and for comprehensive system cost calculations [15].

References

1. R.A. Wiedeman, A.J. Viterbi, "The Globalstar mobile satellite system for worldwide personal communications," in Proc. 3rd Int. Mobile Sat. Conf., IMSC '93, Pasadena, pp. 291-296, June 1993.
2. J.E. Hatlelid, L. Casey, "The Iridium system personal communications anytime, anyplace," in Proc. 3rd Int. Mobile Sat. Conf., IMSC '93, Pasadena, pp. 285-290, June 1993.
3. C.J. Spitzer, "Odyssey personal communications satellite system," in Proc. 3rd Int. Mobile Sat. Conf., IMSC '93, Pasadena, pp. 297-302, June 1993.
4. G.E.Corazza, A.Jahn, E.Lutz, F.Vatalaro, "Channel Characterization for Mobile Satellite Communications", Proc. EMPS '94.
5. F.Vatalaro, G.E.Corazza, C.Caini, C.Ferrarelli, "Analysis of LEO, MEO and GEO Global Mobile Satellite Systems in the Presence of Interference and Fading", to be published on IEEE Journ. Selec. Areas in Comm. .
6. G.E.Corazza, F.Vatalaro, "Comparison of Low and Medium Orbit Systems for Future Satellite Personal Communications" IEEE Pac. Rim Conf. on Comm., Computer and Signal Proc., IEEE 93CH32-88, 1993, pp. 678-681.
7. A.Böttcher, M.Werner, "Strategies for Handover Control in Low Earth Orbit Satellite Systems", Proc. IEEE Veh. Tech. Conf., Stockholm, 1994. pp. 1616-1620.
8. G.E.Corazza, F.Vatalaro, "Interference Analysis in Satellite Cellular Systems", Proc. IEEE Int. Symp. on Personal, Indoor and Mobile Radio Comm., Boston, October 19-21, pp. 377-381, 1992.
9. G.E.Corazza, F.Vatalaro, "A Statistical Model for Land Mobile Satellite Channels and Its Application to Non-Geostationary Orbit Systems", IEEE Trans. on Vehicular Technology, Vol. 43, N.3, August 1994.
10. A. Böttcher et al., "Networking requirements for user-oriented low earth orbit satellite systems," Final Report, ESA Contract No.: 9732/91/NL/RE, 1992.
11. R. Del Ricco, "LEO system study for a definition of MINISTAR (Iridium-like)," Final Report, ESA Contr. 8564/89/NL/DG CCN N. 1, Dec. 1991.
12. F. Harary, "Graph Theory", Addison-Wesley, 1969.
13. C.H. Papadimitriou and K. Steiglitz, Combinatorial Optimization: Algorithms and Complexity. Englewood Cliffs, NJ: Prentice-Hall, 1982.
14. L. Kleinrock, "Queueing Systems", Vol.1. New York: J. Wiley, 1975.
15. A. Dutta und D.V. Rama, "An optimization model of communications satellite planning," IEEE Trans. Commun., Vol. COM-40, No. 9, pp. 1463-1473, 1992.

Networking and Signalling Aspects of a Satellite Personal Communications Network

C. Cullen, A. Sammut, R. Tafazolli, B.G.Evans.
Centre for Satellite Engineering Research,
University of Surrey, Guildford, GU2 5XH, England.

Abstract

A possible network architecture for the European Space Agency's MAGSS-14 (Medium Altitude Global Satellite System) constellation is proposed and described. Network connectivity statistics are examined and a physical implementation of the ground segment control network, using 11 globally positioned gateway stations, is described. This example helps clarify the complex nature of S-PCN system connectivity and control. Using this example as the baseline architecture, an examination is made of GSM LAPDm signalling procedures over satellite. A slightly modified version of this protocol is used for these satellite network simulations. Mobility management signalling is initially examined, with a trade-off being made between location update and paging signalling allowing an optimum user location uncertainty distance to be found. User originated and terminated call setup signalling is also simulated. The results provide some useful insights into the use of modified LAPDm over satellite.

1 Introduction

This paper concerns MEO constellations, to be used for the provision of satellite PCN (S-PCN) services. Many aspects are also relevant to LEO constellations although these are not specifically discussed here. There are two sections. Section 1 sets up the later work by describing the constellation used in this study, explaining the network architecture chosen for this constellation and suggests a possible layout for the controlling ground segment. The MAGSS-14 constellation [1] - a 14 satellite medium altitude constellation similar in style to 'Odyssey'[2] and to the constellations currently being considered for Inmarsat-P - is used as a baseline. Integration with terrestrial mobile communication networks is not directly considered. In section 2 of this paper, use of a modified air-interface LAPDm protocol is considered. Handheld terminals providing voice (4.8 kbps) and data services are assumed as the baseline user terminal.

2 A MEO Network Architecture

2.1 The MAGSS-14 constellation [1]

The MAGSS-14 satellite constellation (M-14) is a constellation proposed by the European Space Agency and used by them as a research baseline. It consists of 14 satellites, excluding spares, orbiting at an altitude of 10,350 km. This gives an orbit period of approximately 6 hours which results in a ground track 'resonance', or repetition, every $23^h 56^m$. The orbit plane is inclined at an angle of 56° to the equatorial plane, providing enhanced satellite visibility to a user terminal (UT) between 30° and 60° latitude. The satellite altitude, together with the specified minimum elevation angle to a user terminal of 28.5° , gives a coverage area beneath each satellite with a radius of approximately 4,500 km. The minimum elevation angle considered available to fixed earth stations is 5 degrees. Figure 1 illustrates a 'snapshot' of an ideal M-14 user coverage, together with the ground track of one of the satellites for the resonance period. The complete global coverage provided by the constellation should be noted, as well as the large areas of coverage overlap, which provide multiple satellite diversity. It can also be seen that the beginning and end of the ground track meet in the same position, illustrating the orbit resonance. The distortion of the circular coverage areas with increasing latitude is due to the earth projection used.

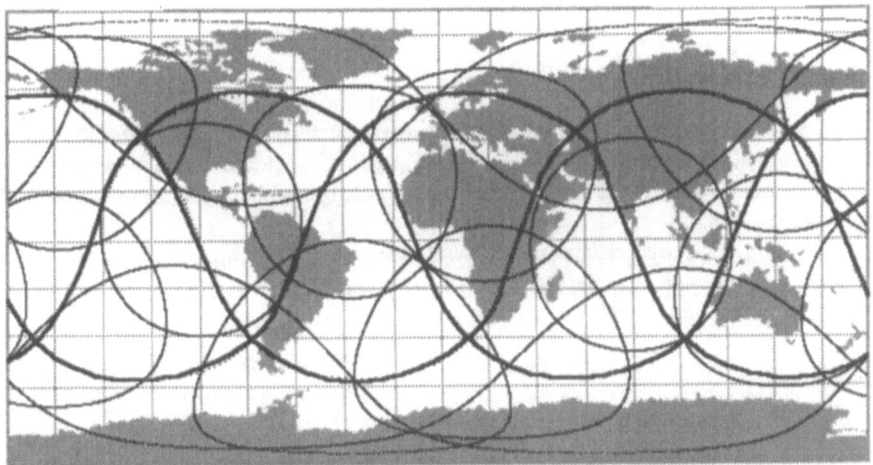

Figure 1 - M-14 constellation coverage and a satellite ground track

The satellite coverage to user terminals is divided into 37 spot-beam cells. The spot-beams provide higher gain to UTs compared with a single global beam, while simultaneously enabling a more efficient utilisation of the available radio spectrum. Communications to and from a user terminal is always conducted through one of these spot-beams, usually through the satellite with the highest elevation angle available to the user, although in circumstances when shadowing occurs (or for route optimisation) a satellite with lower elevation angle, if available, may be used. Figure 2 illustrates the shape of the coverage area formed by the use of these spot-beams. The resulting coverage zone shapes are seen to diverge from the ideal circular coverage due to the distortion effects of the curved earth. The distorted shape of the satellite coverage areas has important implications for overlap between satellites and the diversity offered to users. It also effects constellation frequency planning. The coverage area for 5 degrees minimum elevation angle, envisaged for earth station coverage, is also shown in figure 2 (outer circle), it is currently envisaged that this coverage will be a circular single beam coverage.

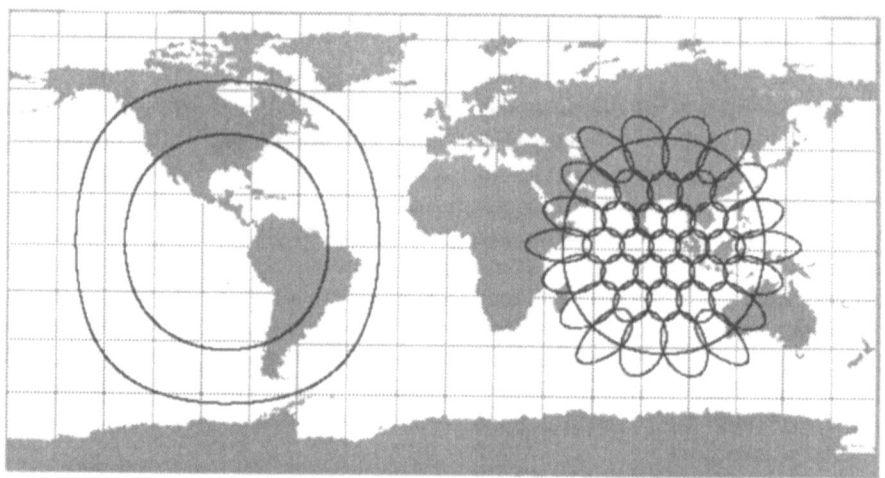

Figure 2 - M-14 satellite coverage

2.2 M-14 network architecture[3]

There are clearly many differences between terrestrial and space-based mobile communication systems Many terrestrial network features cannot be directly mapped onto the M-14 S-PCN system being considered here. The most significant differences are:

> 1) *Connectivities between network elements are continuously changing (user terminal to satellite coverage zone, user terminal to satellite, earth station to satellite).*

> 2) *There is no physical carry-over of the terrestrial concept of an administrative area for S-PCN network management entities (due to continuously changing satellite coverage areas).*

The requirements of the S-PCN system architecture are that it should provide a suitable and flexible hierarchy for the control of a global network - both from a user and an operator point of view. It should not introduce excess levels of signalling and delay and should use S-PCN resources efficiently. On the regulatory side, it should allow a competitive structure for system operation. These aspects are considered below in choosing an appropriate S-PCN architecture. The use of fully regenerative On-Board Processing (OBP) or Inter-Satellite Links (ISLs) is not considered in M-14. Also, integration with terrestrial networks, although considered in [3], is not referred to here.

2.2.1 Spotbeam level control

From the network management point of view, each spotbeam can be considered as the equivalent of a terrestrial cell. Each spotbeam therefore requires at least one set of control channels. A spotbeam with two or more sets of control channels would allow an important aim of the ETSI directive to be immediately implemented - that of ensuring a competitive structure. Service providers - using the same or different earth stations but with their own signalling channels - can all have direct access to S-PCN users within whichever spot beams they choose.

But there is a problem here. For an S-PCN system, as with all satellite systems, the satellite power budget is a critical factor. To maximise the possible number of subscribers, a maximum amount of satellite transmit power should be allocated to traffic channels. Signalling channels, because they require high link margins to guarantee high availability, use relatively high levels of satellite power[1]. The use of two or more sets of signalling channels within each spotbeam results in poor use of satellite resources.

=> Each spotbeam in the S-PCN system will have only one set of signalling channels associated with it.

2.2.2 Satellite level control

The next point concerns the choice of earth station(s) to control these spotbeams. An important thing to remember is that each satellite is a finite unit of resources (channels / bandwidth / power). Another aspect to be considered in S-PCN is the large variation in satellite traffic demand as a satellite orbits the earth and large variations in adjacent satellite coverage overlap. These factors result in each satellite being required to coordinate its resources in a highly dynamic manner. This must be done with due consideration to adjacent spotbeams and satellites, which ultimately requires inter-satellite coordination. Using just one earth station per satellite localizes all intra satellite coordination and reduces signalling overheads. Inter satellite coordination is always necessary and is again most efficiently performed at a single centre for each satellite.

=> Each satellite in the S-PCN system will, at any single time, be controlled by only one gateway station (GS) which is responsible for the signalling channels in each of the satellites coverage zones.

2.2.3 Constellation control

The choice of the global control architecture now arises - distributed or centralised. The choice depends on which architecture is able to manage network resources in the most efficient manner. A centralised architecture requires only one global control centre. This centre controls the resources of each satellite in the S-PCN system. It would have terrestrial links to earth stations providing it with continuous (although indirect) access to each satellite in the constellation. In a distributed architecture, control of a satellites resources is being continuously passed around the globe, between gateway stations which are within the satellite's coverage region. Each satellite is therefore controlled by a gateway station within its range. As the satellite passes out of range, control of that satellite is handed over to a different having improved connectivity with that satellite[2]. Each gateway station has an associated visitor location register (VLR) for the space segment. The presence of a home location register (HLR) depends on the level of integration with the terrestrial mobile networks. Clearly, a distributed architecture is preferable from a redundancy point of view.

---

[1] Any signalling channel requires a high propagation margin so that UTs can transmit and receive signalling information over a wide range of location types. Propagation margins of 15 to 20 dB may be required. Providing such large margins over satellite channels results in reduced channel bit rates.

[2] From M-14 satellite dynamics, the average connectivity window between a satellite and an earth station is 100 minutes. This results in about 200 earth station to earth station handovers per M-14 per day.

To help decide between the use of a centralised or a distributed architecture a user-terminating call is considered. It is assumed that the call is directed towards the user's HLR and then on to the S-VLR. In a centralised architecture, there is only one S-VLR. The use of this single global register results in unnecessarily long terrestrial signalling lengths for most calls. Basically, locating the register at only one global location makes it local for that region but remote for most regions. Measuring signalling cost as a function proportional to both bit requirement and distance, and considering also the large delays that can be introduced with such a setup, a centralised architecture can be seen as inefficient.

In a distributed architecture, each satellite is controlled by a gateway station which has direct connectivity with the satellite. With an integrated architecture, each of these gateway stations has an associated S-VLR and can therefore take proper call control decisions. For the user-terminating call, the call is therefore switched to the appropriate register which is 'local' to the users location. In this case, while the signalling bit requirement should not change significantly, both the distance and delay are reduced. It seems more appropriate to distribute network intelligence and control around the S-PCN network. A distributed network architecture is favoured on this examination.

The same conclusion is reached with traffic channel requests. In order to obtain a channel allocation in either network architecture, the relevant controlling centre needs to be interrogated. With a centralised architecture, large signalling lengths are typical. With a distributed architecture, traffic channel allocation can be made with reduced signalling overheads. Yet there is a cost to keeping satellite control information local to each satellite. Since M-14 satellites orbit the earth every six hours, so also must the satellite control information. This control information must be passed between the gateway station of each satellite. This adds to the terrestrial signalling cost but the low delay for user-oriented signalling is not affected. For network monitoring purposes, centralized management centre(s) can be assumed.

=> For the M-14 S-PCN system considered, a globally distributed terrestrial control network is favoured.

2.3 Competitive structure

To introduce a competitive structure a two layered approach is proposed. Layer one involves a set of earth stations which are involved in administering satellite traffic and control channels. These are the gateway stations referred to above. Each satellite in the M-14 constellation is controlled, at all times, by one of these gateway stations.

A layer two Traffic Station (TS), does not get involved in immediate S-PCN control aspects. Traffic stations are linked with local gateway stations using a terrestrial mesh network. All initial signalling, required to set up any call, either user-originated or user-destined, is performed by the gateway station which is currently controlling the satellite and hence is in charge of the signalling links with the user terminal. When a call request arrives at a gateway station, this information is sent to relevant traffic stations. These traffic stations then bid for the call, with the cheapest offer being accepted by the gateway station. It is envisaged, provided some basic rules are adhered to, that a gateway station can also act as a traffic station and so make a bid for appropriate calls. The choice depends on the optimal route to the call destination in terms of terrestrial distance, satellite loading and ultimately the cost of the overall link. Factors such as handover possibility (which may result in a new call routing requirement) also need to be taken into account. The final routing decision is made by the gateway station.

The number of traffic stations might greatly outnumber the gateway stations. Their existence will reflect the needs of the local market. They therefore provide the network with a competitive structure independent of the network operator. Traffic stations also offer improved space segment to ground segment traffic channel connectivity as well as improved inter ground segment connectivity, both of which provide the network with greater routing flexibility. Reference [4] provides further detail on S-PCN route optimization within such a network architecture. Figure 3 shows the proposed S-PCN architecture[3].

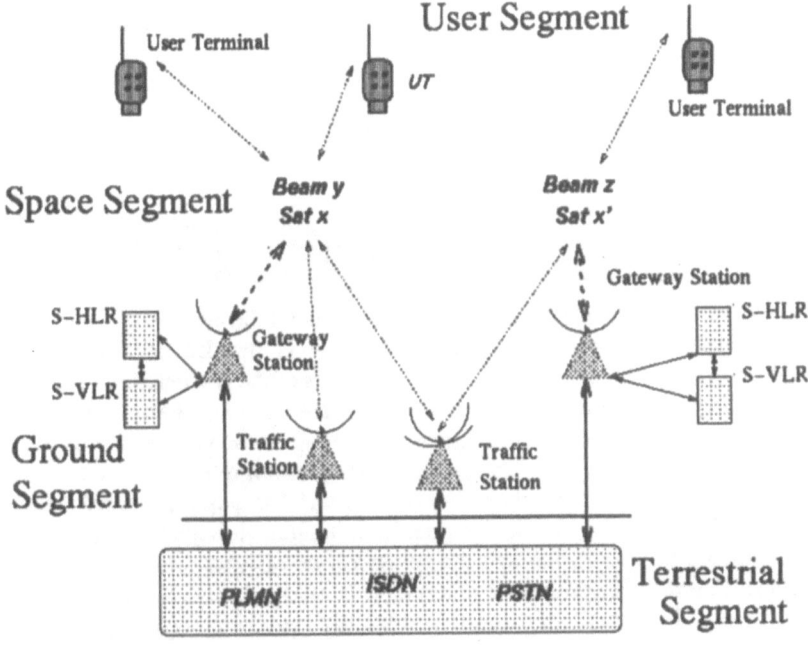

Figure 3 - Proposed S-PCN architecture

2.4 S-PCN ground segment layout

Considering the above network architecture structure, the network operator needs only to concern itself with providing an appropriate number of gateway stations. Because gateway stations can also act as traffic stations, such a structure would automatically create a functional network. Gateway stations are controlled by the S-PCN network operators. They should be located in suitable positions to ensure that satellites can always be controlled. When a satellite reaches the end of it's period of connectivity (or visibility) with a gateway station it must already have connectivity

---

[3] Due to the dynamically varying network connectivities which result in S-PCN systems, the idea of a S-VLR always being able to directly contact a UT does not hold. This is an important difference between a GSM VLR and an S-PCN S-VLR.

with a new gateway station. An overlap in connectivity ensures a seamless handover of control from one gateway to the next. A possible world-wide network of 11 gateway stations is shown in figure 4 - the coverage areas shown are the 5° earth station visibility contours. This set of gateway stations enables full control of all M-14 satellites as well as providing 100% backup connectivity ensuring full gateway station redundancy. This can be used, for example, during heavy rain fades or in the event of a catastrophic failure of a single gateway station.

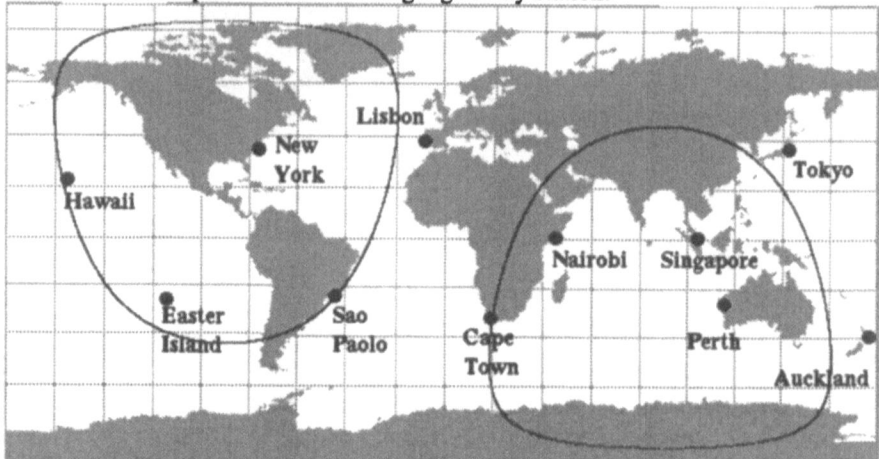

Figure 4 - Example gateway station layout

The control of satellites by gateway stations is decided by the constantly changing physical relationship between the positions of the two entities. As satellites progress about their orbits and the earth spins, the set of gateway earth stations visible to each satellite slowly changes, resulting in a pattern which repeats approximately every 24 hours. This repetition is the result of the previously referred to orbit resonance and is clearly a useful property. The gateway station which is designated to control any particular satellite is chosen through a combination of elevation angle to the satellite and the suitability of each gateway station to carry traffic [5]. Control of each satellite at any time is chosen to ensure that the gateway station being used has the most appropriate terrestrial telecommunications infrastructure. In this network two of the gateway earth stations are used only as backup to other gateways - those located in Nairobi and Easter Island. Figure 5 illustrates the control of one of the M-14 satellites by the ground station network. Figure 6 indicates the satellite(s) being controlled by one of the gateway stations in the network. Both figures cover the full resonant period. Darker shading is used to indicate satellite control. Note that the first rows in figures 5 and 6 are similar as they refer to the same satellite / gateway station pair.

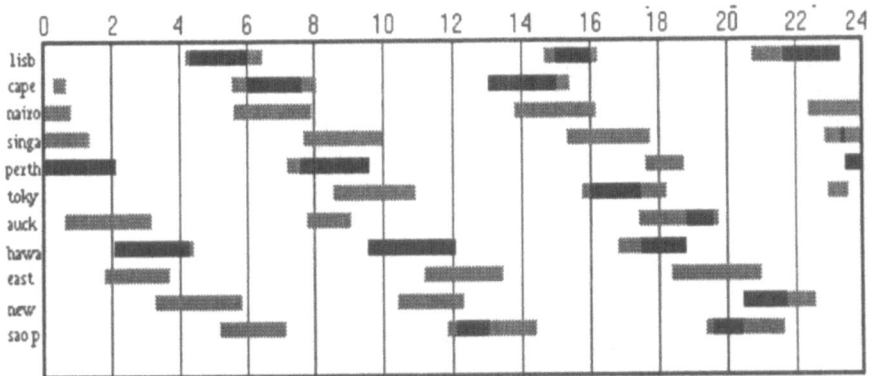

Figure 5 - 24-hour control and connectivity of a satellite

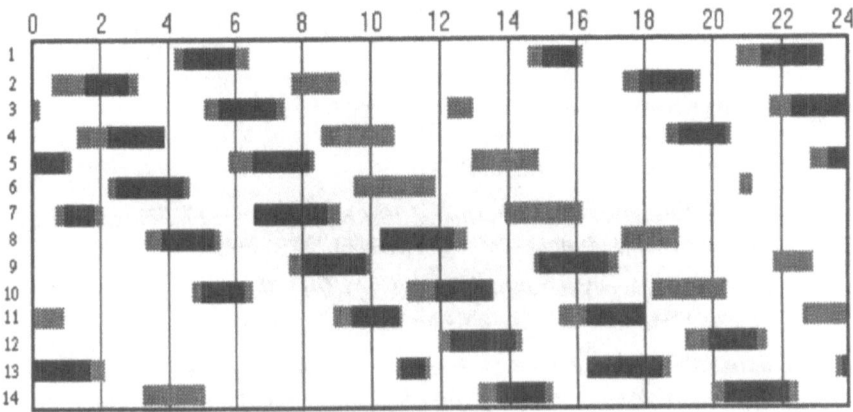

Figure 6 - Control periods of M-14 satellites by Lisbon gateway station

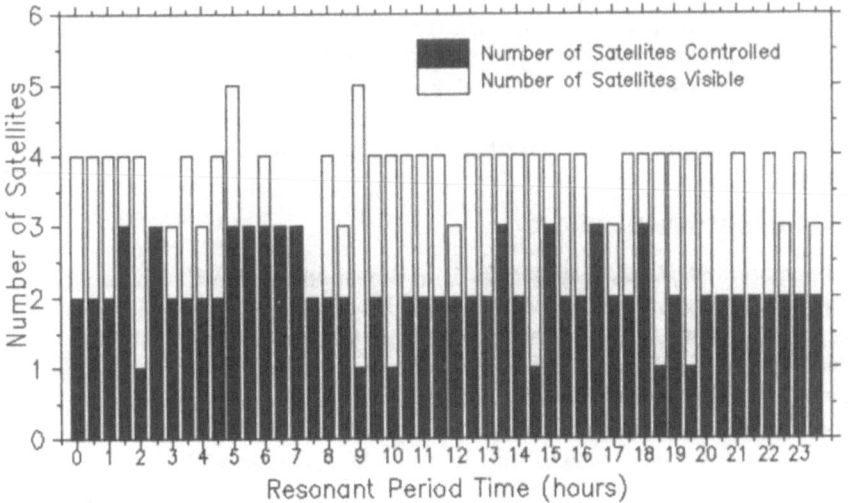

Figure 7 - Satellite control and visibility for Lisbon gateway station.

Note that a gateway station may control more than one satellite simultaneously. Figure 7, above, shows the total number of satellites visible to the Lisbon gateway station together with the number of satellites under it's control in normal circumstances. If Lisbon were to be used as a backup to a failed gateway station, the number of satellites under its control would increase, the gateway therefore must have the potential for controlling up to 5 satellites. As before, darker shading indicates satellite control.

3 S-PCN Signalling

This section concentrates on aspects of Link Access Protocol D modified (LAPDm) signalling [6] but applied to satellites. Three signalling procedures are examined - location registration/update, user originated call setup and user destined call setup. The signalling procedures described occur independently in each spotbeam of each satellite in the S-PCN. For first generation S-PCN systems, the use of GSM based protocols is desireable. The next generation (S-UMTS) is more likely to be based on BAP and INAP standards which are currently under development. The control channels used by the satellite and which are relevant here are:

1) *Satellite Paging CHannel (S-PCH): forward link only, used by a gateway station to page user terminals.*

2) *Satellite Random Access CHannel (S-RACH): return link only, used by user terminalss to request a dedicated control channel.*

3) *Satellite Access Grant CHannel (S-AGCH): forward link only, used by a gateway station to allocate a dedicated control channel.*

4) *Satellite Stand-alone Dedicated Control CHannel S-SDCCH): bidirectional channel, used for dedicated signalling between a user terminal and a gateway station.*

In the following two main subsections, these channels are used to perform the different signalling procedures mentioned above. Within each spotbeam there are also other signalling channels which are not mentioned above. The main one of these is the broadcast control channel from which the identities of the paging, random access and access grant channels are obtained. For the messages outlined below, reference [7] provides further detail on the elements which make up these messages.

3.1 Mobility management

The purpose of mobility management is to provide the network with an efficient way of contacting active user terminals. When a terminal is turned on it must first register itself with the network. To do this it provides its current 'location' or 'position'. In [3], different mobility management approaches are described in greater detail. As a user terminal travels during its active time and depending on the distance it travels, it may be required to make further location updates. Finally, when a user terminal is switched off it should deregister itself from the network. This eliminates the possibility of the network paging a user terminal which is no longer responsive.

For dual mode terminals, mobility management instability might occur if a user terminal passes through islands of terrestrial coverage - the signalling requirement might increase rapidly. In order to avoid such a potential instability, a certain amount of internetworking between the different location registers is proposed. This would ensure that a user terminal can deregister from and reregister with the space segment through the terrestrial network. In this way, excessive space segment mobility management related signalling is avoided. Interworking between the terrestrial and space systems, based on the GSM MAP protocol, is required in order to provide this. Mobility management functions are divided into three separate procedures, user terminal (de)registration, user terminal update and user terminal paging. In S-PCN, these procedures are performed in the following situations:

1) *Location (de)registration - when the user terminal is (de)activated.*

2) *Location (de)registration - when, for the first time in its current active period, the user terminal leaves (returns to) a terrestrial service area and enters (leaves) a satellite only service area.*

3) *Location update (external) - when the user terminal moves between two different environments (satellite/terrestrial) of integrated networks. The terrestrial network ensures correct location register call switching.*

4) *Location update (internal) - when the user terminal has moved more than a specified distance (location uncertainty radius, S_c) from its last registered network location, while still remaining within satellite service.*

5) *Paging is performed by gateway stations for UT-terminated contacts.*

3.1.1 Location registration / updating procedure

This section describes an S-PCN modified LAPDm location updating procedure on the radio interface. It consists of specific combinations of elementary messages[4]. Location registration / updating is always initiated by the UT. In the following discussion new LAPDm bit estimations are made for S-PCN oriented applications. Figure 8 indicates the signalling sequence over the air interface. Note that only the gateway station is involved in this signalling.

Channel request. This message is sent on the random access channel from the user terminal to the gateway station. The S-PCN estimation for this message is tabulated below.

Message	octets (uncoded)	octets (coded)	final tx. length
Chan req	1	2	5

---

[4] Before transmission, half rate coding is applied on all these network layer messages. A further three octets is added for channel synchronisation.

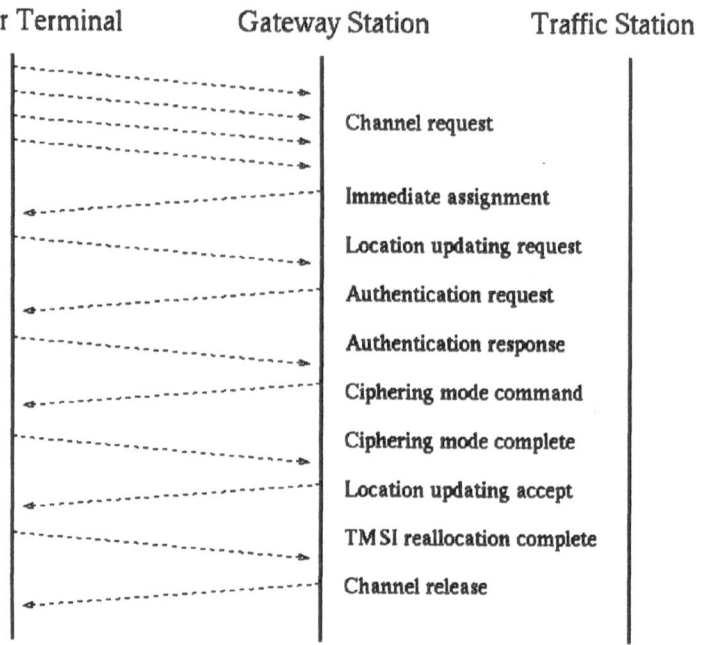

Figure 8 - Location update signalling sequence

Immediate Assignment. This message is sent on the access grant channel by the gateway station to the user terminal in idle mode to change the channel configuration to a dedicated one. The S-PCN estimation for this message is tabulated below.

Message	octets (uncoded)	octets (coded)	final tx. length
Imm assign	6	12	15

Location updating request. This message is sent by the user terminal to the gateway station on the newly allocated dedicated control channel either to request update of its location file or to request IMSI/s attach. The S-PCN estimation for this message is tabulated below.

Message	octets (uncoded)	octets (coded)	final tx. length
Loc upd req	10	20	23

Authentication request. This message is sent by the gateway station to initiate authentication of the user terminal identity. The S-PCN estimation for this message is tabulated below.

Message	octets (uncoded)	octets (coded)	final tx. length
Auth req	8	16	19

Authentication response. This message is sent by the user terminal to deliver a calculated response to the gateway station. The S-PCN estimation for this message is tabulated below.

Message	octets (uncoded)	octets (coded)	final tx. length
Auth res	6	12	15

Ciphering mode command. This message is sent on the dedicated control channel from the gateway station to the user terminal to indicate that the gateway station has started deciphering and that enciphering and deciphering should be started in the user terminal, or to indicate that ciphering will not be performed. The S-PCN estimation for this message is tabulated below.

Message	octets (uncoded)	octets (coded)	final tx. length
Ciph mod cmd	3	6	9

Ciphering mode complete. This message is sent on the dedicated control channel from the UT to the GS to indicate that enciphering and deciphering has been started in the UT. The S-PCN estimation for this message is tabulated below.

Message	octets (uncoded)	octets (coded)	final tx. length
Ciph mod com	2	4	7

Location updating accept. This message is sent by the GS to the UT to indicate that updating or IMSI attach in the network has been completed. The S-PCN estimation for this message is tabulated below.

Message	octets (uncoded)	octets (coded)	final tx. length
Loc upd acc	9	18	21

TMSI reallocation complete. This message is sent by the UT to the GS to indicate that reallocation of the new TMSI has taken place. The S-PCN estimation for this message is tabulated below.

Message	octets (uncoded)	octets (coded)	final tx. length
TMSI real com	2	4	7

Channel release. This message is sent on the dedicated control channel from the GS to the UT to initiate deactivation of the dedicated channel used. The S-PCN estimation for this message is tabulated below.

Message	octets (uncoded)	octets (coded)	final tx. length
Chan rel	3	6	9

3.1.2 Simulation results

The above LAPDm signalling sequence was simulated using the simulation tool BONeS for the M-14 S-PCN system described. A 28.5% efficient random access channel (slotted aloha) with uniform backoff of between 1 and 10 slots was assumed. Two channel rates are used. This is done to highlight the effect of the signalling channel rate on system performance. The rates used - 240 bps and 1200 bps - result in margins of 13 dB and 6 dB, respectively, over a 4.8 kbps voice channel. Since user cooperation cannot be assumed for location update related signalling, the larger signalling margin is likely to be required - see footnote 1. Processing delays at both the user terminal and the gateway station of between 400 and 800 ms were included. The results, showing the delay distribution for 98% of cases, are shown in figure 9. Note the differences in scale between the two graphs.

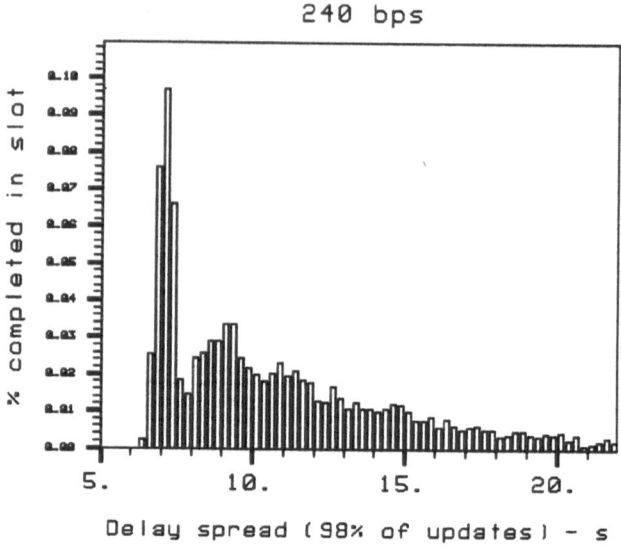

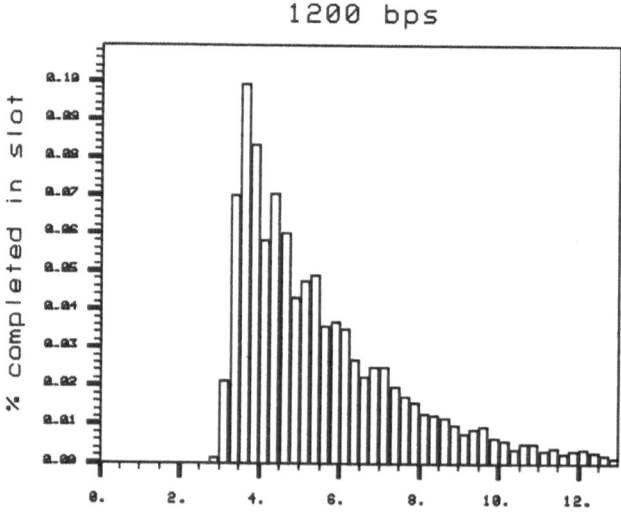

Figure 9 - Location update signalling delay distributions

These results show the significance of the different signalling channel rates. For the 240 bps channel (which uses the same amount of power as a voice channel) the average signalling duration is 9 seconds - which is equivalent from a satellite power point of view to use of a traffic channel except that from the network operators point of view there is no revenue. It is clearly in the interest of the network operator to minimise the signalling overheads resulting from mobility management. In [3], a requirement of 1.00 location signallings per coverage zone per second was estimated. Using the 240 bps signalling channel rate, a random access requirement of 40 bits and the above slotted aloha channel, it can be estimated that sufficient throughput is provided - 1.43 successful accesses per second - with a slot duration of 48 bits (200 ms). But there is not enough further capacity to allow for call setup related accesses. This leads to the requirement of two separate random access channels - one for location update signalling and the other for call setup signalling.

3.1.3 Mobility management signalling trade-off

Based on the above signalling bit estimates and an assumption for user mobility characteristics, a trade-off between location update related signalling and network paging signalling is now performed. The assumption is that the probability of a user location update - P(u) - decreases exponentially as the location uncertainty radius (S_c) increases and that each user performs location registration / deregistration daily[5].

Based on the mobility management techniques described in [3] and the above revised LAPDm message lengths, location update related signalling was calculated for S_c between 100 and 1000 km. Using M-14 paging statistics [8] as well as the bit estimation for the paging channel (21 octets - see later) the paging bit requirement was also calculated. One user terminating call per user per day was assumed. These results are then combined - see figure 10. From this there is clearly an uncertainty radius of minimum signalling per mobile user per day.

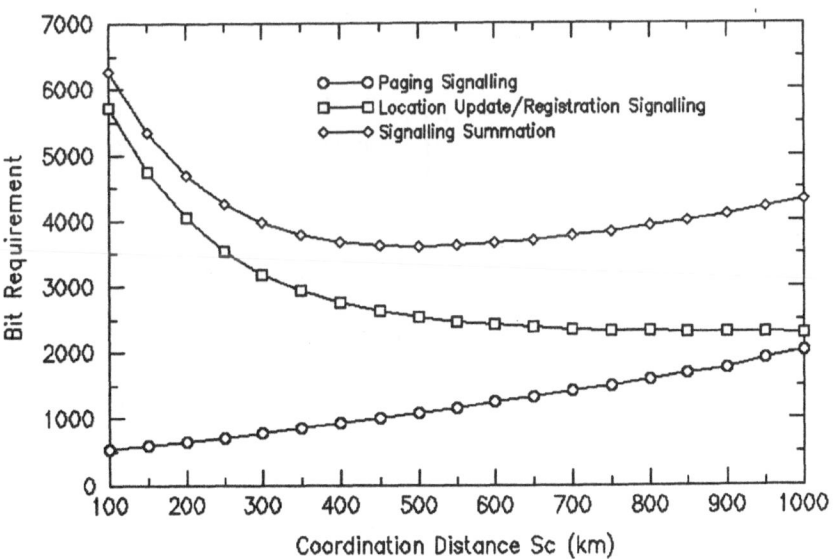

Figure 10 - Location update vs paging trade-off

---

[5] The following equation is used: $P(u) = 2 + 3\exp^{-(Sc-100)/150}$

The key point to be read from this graph is the optimum radius for the minimisation of location update/registration and paging signalling. This can easily be read from the graph and is seen to be about 500 km. However, the assumptions concerning user mobility and user terminated calls do have an important influence on the final result. So also does the satellite coverage zone geometry as this effects the number of zones that need to be paged. For example, with the smaller zones associated with LEO constellations, the paging bit requirement is likely to be much higher. What is important to remember is the significance of the P(u) distribution chosen. This distribution assumed 4 registration/updates per day for a location radius of 100 km. As the radius increased to 1000 km, P(u) drops off exponentially towards a value of 2 registration/update per day. Changes in the shape of this assumed distribution would influence the radius of minimum signalling.

3.2 S-PCN Call Control

3.2.1 Mobile originating call establishment - late assignment

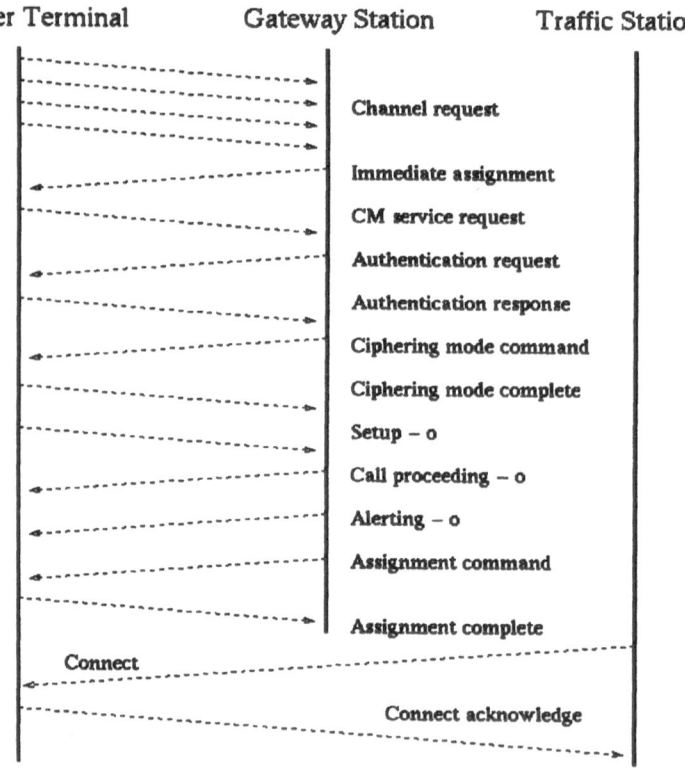

Figure 11 - User originated call signalling sequence

Figure 11 above indicates the LAPDm signalling sequence, with late assignment, over the air interface. Late assignment means that the traffic channel is not allocated until the called user has already answered the call. It results in a more efficient use of system resources as no traffic channels are allocated while preliminary call setup is taking place (while the called user's phone is ringing).

CM service request. This message is sent by the UT to the GS to request a service for the connection management sublayer. The message is constructed as follows:

Message	octets (uncoded)	octets (coded)	final tx length
CM serv req	7	14	17

Setup. This message is sent from either the UT or the GS, to initiate call establishment. After the network receives this message it can begin route optimisation. The message is constructed as follows:

Message	octets (uncoded)	octets (coded)	final tx length
Setup	46	92	95

Call proceeding. This message is sent by the GS to the calling UT to indicate that the requested call establishment information has been received. The message is constructed as follows:

Message	octets (uncoded)	octets (coded)	final tx length
Call proc	3	6	9

Alerting. This message is sent by the GS to the calling UT or by the called UT to the GS, to indicate that called user alerting has been initiated (phone begins to ring). The message is constructed as follows:

Message	octets (uncoded)	octets (coded)	final tx length
Alerting	11	22	25

Assignment Command. This message is sent by the GS to the UT to change the channel configuration. The message is constructed as follows:

Message	octets (uncoded)	octets (coded)	final tx length
Assign cmd	18	36	39

Assignment Complete. This message is sent from the UT to the network to indicate that the UT has established the main link successfully. The message is constructed as follows:

Message	octets (uncoded)	octets (coded)	final tx length
Ass com	3	6	9

Connect. This message is sent by the called UT to the network or by the network to the calling UT to indicate call accept by the called user. The message is constructed as follows:

Message	octets (uncoded)	octets (coded)	final tx length
Connect	11	22	25

Connect acknowledge. This message is sent by the network to the called UT or by the calling UT to the network to indicate that the UT has been awarded the call. The message is constructed as follows:

Message	octets (uncoded)	octets (coded)	final tx length
Conn ack	2	4	7

3.2.2 Simulation results

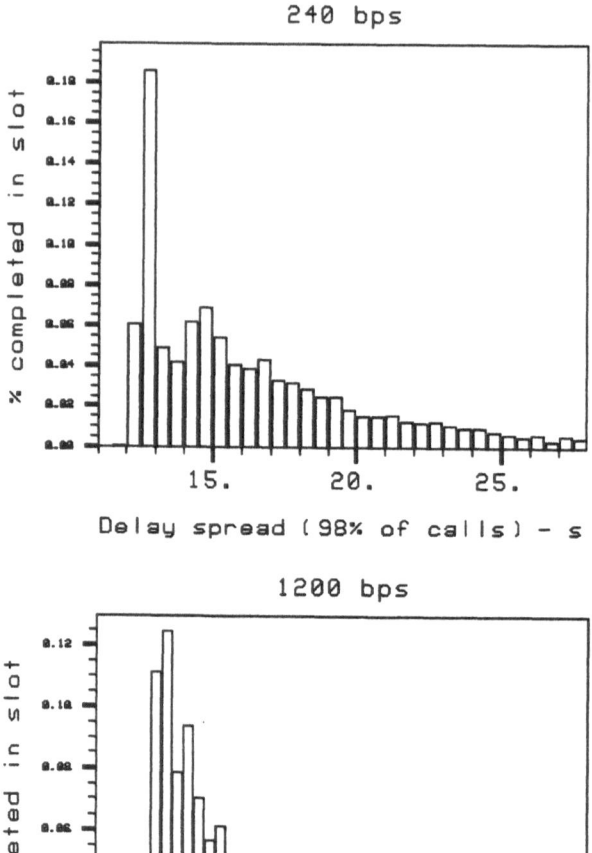

Figure 12 - Mobile originated call setup delay distributions

Figure 12 above shows the results of the BONeS simulations for UT originated call setup. Similar assumptions as before are made. The final two messages were not included as their execution depends on the called user answering their phone. Differences in scale between the two graphs should be noted. Since, for mobile originated calls, user cooperation can be assumed, lower signalling channel margins might be used and so the higher channel rate (1.2 kbps) can be assumed. This implies the use of different signalling channel rates depending on the level of user co-operation. Variable signalling channels rates, depending on the received signal quality, are suggested as a highly useful property for future S-PCN systems.

For the 1200 bps channel, the average call setup duration is 8 seconds. As with location update/registration signalling, this 'dead time' is an important cost factor which needs serious consideration by the network operator. In [3], a requirement of 0.55 random accesses per coverage zone per second for mobile originated calls, was estimated. This level of throughput can be provided even at 240 bps.

3.2.3 Mobile terminating call establishment - late assignment

This section describes a LAPDm mobile terminating call setup procedure with late assignment. The three elementary messages not already described are broken into their constituent parts and estimates are made for S-PCN. Figure 13 indicates the signalling sequence over the air interface.

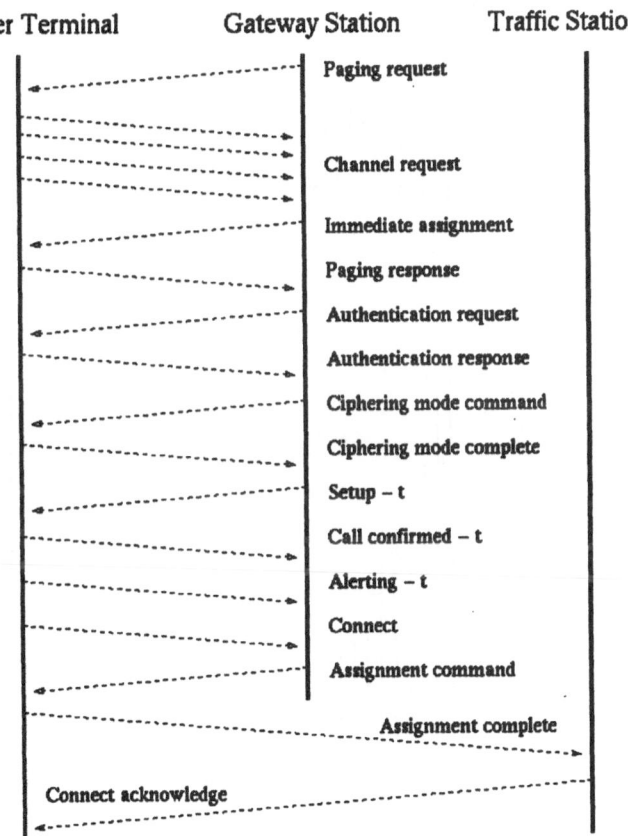

Figure 13 - User terminating call signalling sequence

Paging request type 1. This message is sent on the paging channel by the GS to a UT to trigger a channel access attempt. The message is constructed as follows:

Message	octets (uncoded)	octets (coded)	final tx length
Call proc	19	18	21

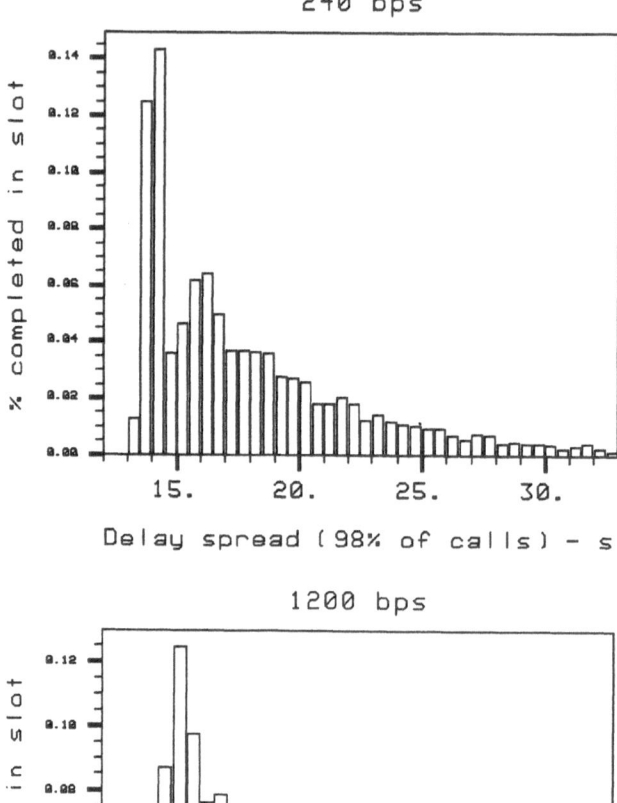

Figure 14 - Mobile terminated call setup delay distributions

Paging response. This message is sent by the UT to the GS in response to the paging request message after establishment of the main signalling link. The message is constructed as follows:

Message	octets (uncoded)	octets (coded)	final tx length
Pag res	6	12	15

Call confirmed. This message is sent by the called UT to the GS to confirm an incoming call request. The message is constructed as follows:

Message	octets (uncoded)	octets (coded)	final tx length
Call proc	3	6	9

3.2.4 Simulation results

The above LAPDm signalling sequence was also simulated. The results are shown in figure 14 on the previous page. Since, for mobile terminating calls, user cooperation cannot be assumed, the lower signalling channel rate is most likely to be used here. However, depending on the signal received quality and if variable signalling channel rates were available, higher rates might be used if suitable. What is very clear is the very large delay for the 240 bps channel rate. The average delay is about 16 seconds but for some users this can be more than 30 seconds. As with the previous simulations, the final two messages were not simulated. The effect of this signalling time on system capacity and costs needs careful consideration.

.In [3], a requirement of 0.36 random accesses per coverage zone per second for mobile terminated calls was estimated. This level of throughput can be provided even with the lower rate random access channel of 240 bps. A total of 0.91 random accesses per coverage zone per second is the final estimate in [4]. With a 28.5% efficient access channel and a slot duration of 48 bits, this is within the capabilities of the 240 bps channel. The paging requirement for this depends on the location radius used - see section 4.1 or [8].

4 Conclusions

In this paper an overview was made of key network aspects for future S-PCN systems. This initially concerned a description of the MEO constellation used followed by a proposed network architecture. In this globally distributed architecture, as each satellite orbits the earth, it is controlled by a gateway station in its current visibility region. A possible implementation of these gateway stations providing complete system redundancy is then described. Constellation resonance is seen here to be a useful property for network control.

S-PCN air interface signalling aspects were then examined. LAPDm location update and call setup procedures were simulated over likely channels. A trade-off between location registration / update signalling and paging signalling based on an assumed probability of update distribution, resulted in an optimum location radius being found. Signalling channel rates were seen to have an important impact. Due to the long delays resulting from low channel bit rates, a variable rate signalling interface, which depends on the channel quality, is proposed. The high amount of system capacity and time spent on signalling is an important factor which needs to be considered by the network operator.

Two random access channels were proposed - one for location update signalling and the other for call set-up signalling. The advantage here is that each can be dimensioned seperately and stability ismore assured. With only one random access channel per coverage zone and considering the lowest channel rate of 240 kbps, stability would be a real concern. The modified LAPDm messages, the different channel rates as well as the very different propagation distances, result in the requirement for different timers within the S-PCN protocols. Another aspect requiring consideration is mobile to mobile calls. An average setup duration of about (6 + 16) 22 seconds, not including ringing time, could be required here. Finally, attention is drawn to the large differences between the service connection delays estimasted above (comparable to Inmarsat-M delays) and those which can be found in current mobile terrestrial communication systems.

References

1. J. Benedicto, J. Fortuny, P. Rastrilla, 'MAGSS-14: a medium altitude global satellite system for personal communications at L-band', ESA Journal 16, 1992
2. J. Rusch, P. Cress, M. Horstein, R. Huang, E. Wiswell, 'Odyssey, a constellation for personal communications', 14th International communication satellite systems conference, March 1992, Washington, DC, USA
3. C. Cullen, 'Network and signalling aspects of satellite personal communication networks (S-PCNs), Estec Working Paper (EWP) 1736, September 1993
4. C. Cullen, R. Tafazolli, B.G. Evans, 'Networking and Signalling Aspects of a Satellite Personal Communications Network (S-PCN)', 15th AIAA conference, Washington DC, USA, February 1994.
5. A. Sammut, C. Cullen, R. Tafazolli, B.G.Evans, 'Ground Segment control network for a MAGSS-14 based satellite personal communications network (S-PCN)', COST 227 TD(94) 01, ESTEC, January, 1994
6. ETSI/PT GSM Recommendations, March 1991
7. C. Cullen 'LAPDm modifications for S-PCN', Temporary internal document, University of Surrey
8. A. Sammut, C. Cullen, R. Tafazolli, B.G.Evans, 'Mobility management related signalling for a MAGSS-14 based satellite personal communications network (S-PCN)', COST 227 TD (94) 22, National Technical University of Athens, May, 1994

Part IV

Channel and radiofrequency aspects

Channel Characterization for
Mobile Satellite Communications

Giovanni E. Corazza[1], Axel Jahn[2], Erich Lutz[2], and Francesco Vatalaro[1]

[1] Dipartimento di Ingegneria Elettronica, Università di Roma Tor Vergata, Via della Ricerca Scientifica, 00133 Roma, Italy
[2] German Aerospace Research Establishment (DLR), Institute for Communications Technology, D-82230 Wessling, Germany

Summary. This paper concentrates on channel modelling for mobile personal satellite systems. A single-environment channel model is introduced suitable to quasi-stationary flat-flat channels. The model is compared with results taken from the literature. Also a two-state generative channel model with parameters fitted from measurements is presented. Finally, the structure of a wideband channel model is proposed. In order to derive its parameters, DLR currently performs a measurement campaign for channel effects related to non-geostationary satellite systems. The programme comprises narrowband and wideband propagation measurements as well as interference measurements.

1. Introduction

Several system alternatives are being considered for provision in the near future of mobile and personal satellite (MPS) services. Many proposals adopt non-geostationary satellite constellations, so that the channel characteristics are not stationary, even for a fixed user terminal. As well-known, two classes of non-geostationary (non-GEO) circular orbit constellations are competing: the Low-altitude Earth Orbits (LEO), having an orbit heigth H typically between 500 and 2000 km and the Medium-altitude Earth Orbits (MEO), with $5000 < H < 12000$ km. Furthermore, multiple access techniques under consideration range from narrow-band (e.g. SCPC), to wide-band (e.g. CDMA) solutions. Finally, due to the requirement of being virtually global, a MPS system should provide service in a wide variety of environmental conditions, ranging from open areas to urban areas, though with some operational limitations in the latter case.

The above-mentioned wide variety of system characteristics imposes to MPS channel models to be accurately tailored, since the more accurate is the channel modelling, the tighter can be the link budget margin with important impacts on cost reduction, system capacity increase and technological feasibility.

This paper concentrates on channel modelling for MPS systems and is divided in four parts. First, in Section 2 it introduces a channel model which is suitable to quasi-stationary MPS channels, i.e. channels characterized by not too varying environmental conditions, such as those encountered in hand-held

personal communications, under the assumption of time and frequency flat-
ness (flat-flat fading channel). This channel model is suitable for performance
evaluations when SCPC or narrowband TDMA access schemes are assumed
without any limitation on the type of propagation environment. The channel
model is based on a combination of Rice and lognormal statistics. Section
2 provides first and second order probabilistic modelling, model validation
vis-a-vis some measurement results collected in the literature and, finally,
evaluation of error probability under some sample conditions.

A generative model for frequency flat land mobile satellite channels is presented in Section 3. This model provides an approach for characterizing the
MPS channel under wider and abrupt variations of the environmental conditions, such as those encountered in vehicular communications. Its parameters
have been fitted from measurements using a MARECS satellite.

Wideband channel models for satellite systems suffer from a lack of propagation experiments. A bulk of data currently used as an indicator of wideband satellite channels is actually drawn from terrestrial trials. In Section
4, a tap model is envisaged the parameters of which have to be determined
from wideband channel measurements.

In cooperation with Inmarsat and ESA, the German Aerospace Research
Establishment (DLR) is currently performing measurement campaigns for
mobile-to-satellite channel effects related to non-geostationary satellite systems. The programme comprises narrowband propagation measurements,
wideband propagation measurements in a bandwidth of 10...30 MHz, as well
as interference measurements for the satellite up- and down-link. In Section
5, these experiments are described and first results are presented.

2. Channel model for single-environment flat-flat fading channels

2.1 First order probabilistic model

We assume the following probability density function (p.d.f.) of the received
signal envelope, r:

$$p_r(r) = \int_0^\infty p(r|S)p_S(S)\,\mathrm{d}S, \tag{2.1}$$

where $p(r|S)$ is a Rice p.d.f. conditioned on the shadowing variable, S (i.e.,
the square root of the short-term average power):

$$p(r|S) = 2(K+1)\frac{r}{S^2}e^{-(K+1)\frac{r^2}{S^2}-K}I_0(2\frac{r}{S}\sqrt{K(K+1)}) \quad (r \geq 0) \tag{2.2}$$

$I_0(x)$ being the zero order modified Bessel function of the first kind, and K
the so called *Rice factor* (i.e. the ratio between the direct plus specular signal
power to the diffuse multipath power). The shadowing, S, is lognormal with
p.d.f.:

$$p_S(S) = \frac{1}{\sqrt{2\pi}h\sigma S} e^{-\frac{1}{2}(\frac{\ln S - h\mu}{h\sigma})^2} \quad (S \geq 0), \tag{2.3}$$

where $h = (\ln 10)/20$, μ and σ^2 are the mean and the variance of the short-term average power expressed in dB, respectively (σ is often referred to as the *dB spread*).

An important property of the assumed Rice-lognormal (RLN) channel model is that the received signal envelope, r, can be rigorously factored as the product of two independent processes, i.e. $r = SR$, where S is lognormal with p.d.f. provided by (2.3), and R is a Rice process with Rice factor K and parameter $\sigma_R^2 = 1/2(K+1)$, i.e. with probability density function:

$$p_R(R) = \frac{R}{\sigma_R^2} e^{-\frac{R^2}{2\sigma_R^2} - K} I_0(\frac{R}{\sigma_R}\sqrt{2K}) \quad (R \geq 0). \tag{2.4}$$

Due to the independence between R and S we can show that:

$$\begin{aligned}
p_r(r) &= \int_0^\infty \frac{p_S(S)}{S} p_r(\frac{r}{S}) \, dS \tag{2.5} \\
&= \mathcal{E}_S\left\{\frac{1}{S}p_R(\frac{r}{S})\right\},
\end{aligned}$$

where $\mathcal{E}_X\{\}$ stands for the average with respect to X.

The cumulative distribution function (c.d.f.) of the received envelope is:

$$\begin{aligned}
P_r(r) &= \text{Prob}(x < r) \tag{2.6} \\
&= 1 - \mathcal{E}_S\left\{Q(\sqrt{2K}, 2\frac{r}{S}\sqrt{K(K+1)})\right\},
\end{aligned}$$

where Q is the Marcum's Q-function [1].

From the moments of the lognormal process [2] and those of the Rice process [1], the n-th order moment of r can be derived as:

$$\begin{aligned}
\mathcal{E}_r\{r^n\} &= \mathcal{E}_R\{R^n\}\mathcal{E}_S\{S^n\} \tag{2.7} \\
&= (K+1)^{-\frac{n}{2}}e^{-K}\Gamma(1+\frac{n}{2})e^{nh\mu}e^{\frac{1}{2}n^2h^2\sigma^2} \, {}_1F_1(1+\frac{n}{2}, 1; K),
\end{aligned}$$

where Γ is the gamma function and $_1F_1$ is the confluent hypergeometric function [3]. Depending on the combination of K, μ, σ all the relevant non-selective models in mobile communications can be derived as limiting cases of the RLN channel model:

- when $K = 0$, (1) provides a Rayleigh-lognormal p.d.f. for $p_r(r)$ and the channel is the so-called *Suzuki-channel* [4];
- when $K \to \infty$, $p_R(R)$ tends to a Dirac-pulse located at $R = 1$ and $p_r(r)$ tends to $p_S(r)$, i.e. the channel is lognormal;
- in the limit for $\sigma \to 0$, $p_S(S)$ tends to $\delta(S - e^{h\mu})$, a Dirac-pulse located at its mean value: therefore $p_r(r) \to p(r|e^{h\mu})$ and the channel is Rice; when also $K = 0$ the channel is Rayleigh.

2.2 Average number of level crossings and average duration of fades

Following Rice, the average number of upward crossings of a fixed signal level X, $N(X)$, can be put in terms of $p(r, r')$, the joint p.d.f. of the envelope, r, and its derivative, r', i.e. [5]:

$$N(X) = \int_0^\infty r'\, p(X, r')\, \mathrm{d}r' \ . \tag{2.8}$$

Analytical derivation of $p(r, r')$ for the RLN behaviour is a difficult task, in general. However, it can be shown rigorously that r and r' are independent in some limiting cases, e.g. if r is a Rice process [6], and if r is a lognormal process [7], in both cases r' being normally distributed with zero mean and variance $\sigma_{r'}^2$. Therefore, we heuristically extend to the RLN process, at least approximately, the condition $p(r, r') = p_r(r)p_{r'}(r')$, so that:

$$N(X) = p_r(X) \int_0^\infty r'p_{r'}(r')\, \mathrm{d}r' = \frac{\sigma_{r'}}{\sqrt{2\pi}}p_r(X) \ , \tag{2.9}$$

where we also retain the gaussian assumption for the derivative of the signal envelope.

Since we can think of r' as the output of a filter with transfer function $H(f) = j2\pi f$ having r at its input, we have:

$$\sigma_{r'}^2 = 4\pi^2 \int_{-\infty}^\infty f^2 W_r(f)\, \mathrm{d}f \ , \tag{2.10}$$

where $W_r(f)$ denotes the power spectrum of the envelope. Let us now introduce the *Doppler effective bandwidth*, β:

$$\beta^2 = \frac{\int_{-\infty}^\infty f^2 W_r(f)\, \mathrm{d}f}{\int_{-\infty}^\infty W_r(f)\, \mathrm{d}f} = \frac{\sigma_{r'}^2}{4\pi^2\sigma_r^2} \ , \tag{2.11}$$

so that finally we have:

$$N(X) = \sqrt{2\pi}\beta\sigma_r p_r(X) \ , \tag{2.12}$$

where the variance of the envelope $\sigma_r^2 = \mathcal{E}(r^2) - \mathcal{E}^2(r)$, evaluated through (2.8), is a function of K, μ, σ, and $p_r(X)$ is given by (2.5).

Under the same assumptions we also have the following expression for the average fade duration, $T_F(X)$ [5]:

$$T_F(X) = \frac{1}{N(X)}P_r(X) = \frac{1}{\sqrt{2\pi}\beta\sigma_r}\frac{P_r(X)}{p_r(X)} \ , \tag{2.13}$$

where $P_r(X)$ is the probability that the envelope $r < X$, given by (2.6). Figs. 2.1 and 2.2 respectively show the normalized average number of level crossings, $N(X)/\beta$, and the normalized average fade duration, $\beta T_F(X)$, as a function of the normalized level X/σ_r (dB) for some values of K, μ, σ.

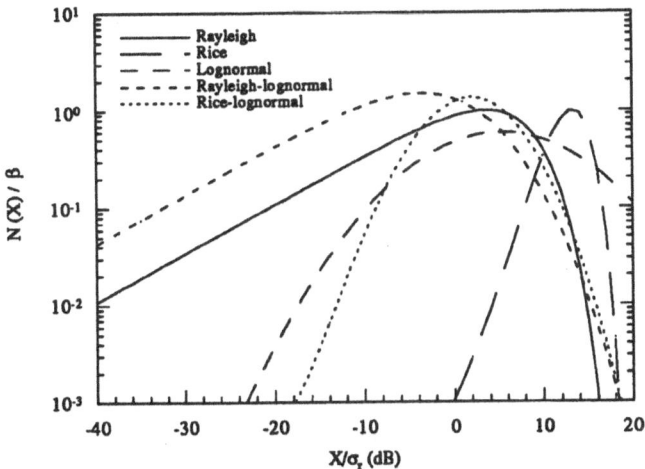

Fig. 2.1. Normalized average number of level crossings, $N(X)/\beta$, as a function of the normalized envelope level X/σ_r (dB) for some values of K, μ, σ: (a) Rayleigh, (b) Rice ($K = 10$), (c) lognormal ($\mu = 0$ dB, $\sigma = 4$ dB), (d) Rayleigh-lognormal ($\mu = 0$ dB, $\sigma = 4$ dB) , (e) Rice-lognormal ($K = 10$, $\mu = 0$ dB, $\sigma = 4$ dB).

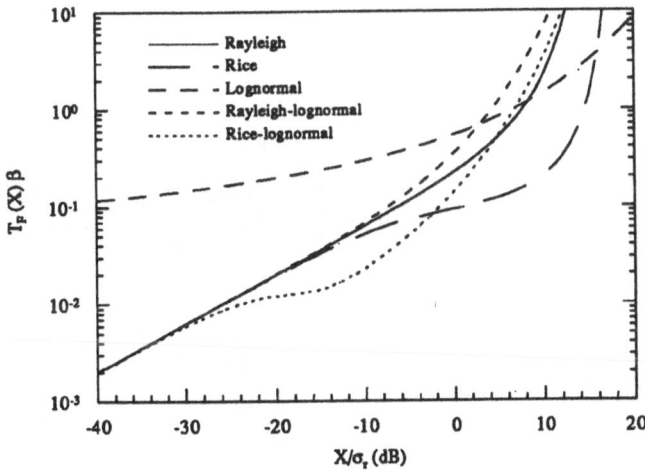

Fig. 2.2. Normalized average fade duration, as a function of the normalized envelope level for some values of K, μ, σ: a) Rayleigh, (b) Rice ($K = 10$), (c) lognormal ($\mu = 0$ dB, $\sigma = 4$ dB), (d) Rayleigh-lognormal ($\mu = 0$ dB, $\sigma = 4$ dB), (e) Rice-lognormal ($K = 10$, $\mu = 0$ dB, $\sigma = 4$ dB).

2.3 Rice-lognormal model validation and empirical formulas

The RLN channel model was validated with respect to measurement data available in the literature, always achieving very good fit. As an example, in [8][1] we reported comparisons with data measurements presented by Loo [10].

The channel model has been also fitted with results of ESA measurement campaigns over a wide range of elevation angles, α, of signal arrival [11]. This last comparison is very significant for non-GEO systems in which the elevation angle under which the serving satellite is seen by the mobile terminal is continuously varying in a wide range ($\alpha \approx 10° \ldots 90°$). Fig. 2.3 shows the c.d.f. of the RLN model for several values of α compared with ESA measurement results taken in different propagation environments ranging from a rural tree-shadowed area (Fig. 2.3 (a)), an urban area (Fig. 2.3 (b)), a suburban area (Fig. 2.3 (c)), and, finally, an open area (Fig. 2.3 (d)). Fitting was performed by determining the optimum triplet (K, μ, σ) for each considered environment, by means of the following polynomial empirical model:

$$
\begin{aligned}
K(\alpha) &= K_0 + K_1\alpha + K_2\alpha^2 + K_3\alpha^3 \\
\mu(\alpha) &= \mu_0 + \mu_1\alpha + \mu_2\alpha^2 + \mu_3\alpha^3 \\
\sigma(\alpha) &= \sigma_0 + \sigma_1\alpha + \sigma_2\alpha^2 + \sigma_3\alpha^3
\end{aligned}
\tag{2.14}
$$

where optimum coefficients K_i, μ_i, and σ_i $(i = 0, 1, 2, 3)$, are provided in Table 1 for the considered environmental conditions. The above polynomial empirical formulas are also assumed to interpolate model parameters in the range $20° < \alpha < 90°$.

	Rural tree-shadowed area	Urban area	Suburban area	Open area
K_0	2.731	1.750	-13.600	26.430
K_1	$-1.074 * 10^{-1}$	$6.700 * 10^{-2}$	$9.650 * 10^{-1}$	-2.644
K_2	$2.774 * 10^{-3}$	0.0	$-1.663 * 10^{-2}$	$8.337 * 10^{-2}$
K_3	0.0	0.0	$1.187 * 10^{-4}$	$-4.111 * 10^{-4}$
μ_0	-20.25	-52.12	-1.998	3.978
μ_1	$9.919 * 10^{-1}$	2.758	$-9.919 * 10^{-3}$	$-1.742 * 10^{-1}$
μ_2	$-1.684 * 10^{-2}$	$-4.777 * 10^{-2}$	$1.520 * 10^{-3}$	$2.647 * 10^{-3}$
μ_3	$9.502 * 10^{-5}$	$2.714 * 10^{-4}$	$-1.266 * 10^{-5}$	$-1.367 * 10^{-5}$
σ_0	4.500	7.800	8.000	0.0
σ_1	$-5.000 * 10^{-2}$	$-3.542 * 10^{-1}$	$-3.741 * 10^{-1}$	0.0
σ_2	0.0	$6.500 * 10^{-3}$	$6.125 * 10^{-3}$	0.0
σ_3	0.0	$-3.958 * 10^{-5}$	$-3.333 * 10^{-5}$	0.0

Table 2.1. Empirical formulas coefficients for different environments.

---

[1] Note that the parameter μ in [8] is here μ/h.

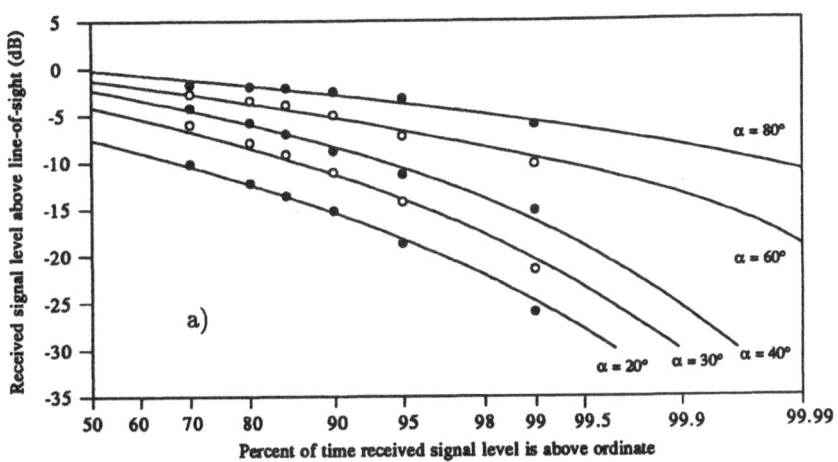

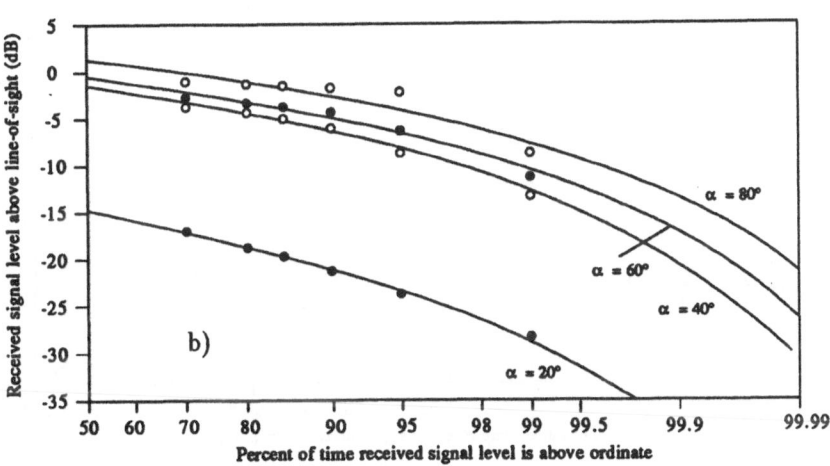

Fig. 2.3. Cumulative distribution function of the Rice-lognormal model for several values of the elevation angle compared with ESA measurement results taken in different propagation environments: (a) rural tree-shadowed area, (b) urban area.

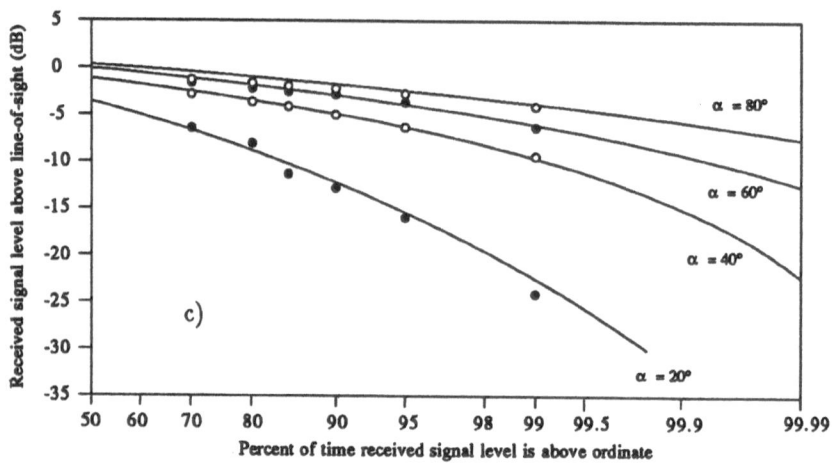

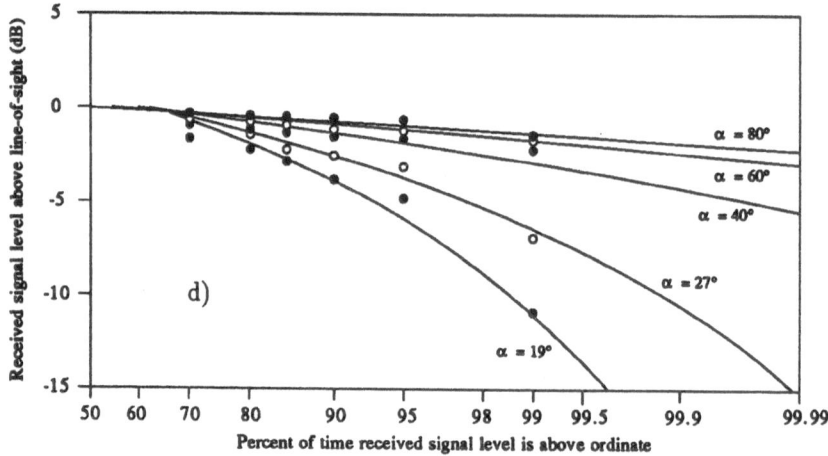

Fig. 2.3. Cumulative distribution function of the Rice-lognormal model for several values of the elevation angle compared with ESA measurement results taken in different propagation environments: (c) suburban area, (d) open area.

2.4 Probability of error in the flat-flat channel

The symbol error probability for transmission in a flat-flat fading channel can be written as:

$$P_e = \int_0^\infty P(e|r) p_r(r) \, dr \; , \tag{2.15}$$

where $P(e|r)$ is the symbol error probability conditioned on a certain value of r and $p_r(r)$ is the received signal envelope p.d.f.. For given modulation and coding $P(e|r)$ depends on the kind of disturbance acting within the channel (noise, interference, etc.).

By substituting (2.5) into (2.15) and interchanging the order of integration we have:

$$P_e = \int_0^\infty p_S(S) \left\{ \int_0^\infty P(e|RS) p_R(R) \, dR \right\} dS \; , \tag{2.16}$$

where $R = r/S$. The inner integral represents the average error probability in the presence of Rice fading only (i.e. for a given value of S):

$$\mathcal{E}_R\{P(e|RS)\} = \int_0^\infty P(e|RS) p_R(R) \, dR \doteq f(S) \; , \tag{2.17}$$

while the outer integral in (2.16) is the average of $f(S)$ in the presence of lognormal shadowing. In conclusion, we have:

$$P_e = \mathcal{E}_S\{f(S)\} = \mathcal{E}_S\{\mathcal{E}_R\{P(e|r)\}\} \; . \tag{2.18}$$

The error probability provided by (2.18) depends on the model parameters K, μ, σ which in a given site are constant for GEO or slow-varying for non-GEO personal communication applications. As an example Fig. 2.4 shows the probability of symbol error in a rural tree-shadowed environment, $P_e(\alpha)$, as a function of ϱ_b, i.e. the (bit energy)/(noise spectral density) ratio, for different values of α in case of BPSK and DPSK modulations.

The actual average error probability which is experienced by a user terminal located in a given site, $\bar{P}_e$, is evaluated by weigthing $P_e(\alpha)$ through $p_\alpha(\alpha)$, the p.d.f. of the elevation angle:

$$\bar{P}_e = \mathcal{E}_\alpha\{P_e(\alpha)\} = \int_0^{\pi/2} p_\alpha(\alpha) \, P_e(\alpha) \, d\alpha \; . \tag{2.19}$$

In a given site the elevation angle p.d.f., $p_\alpha(\alpha)$, depends on the characteristics of the selected orbit constellation parameters and can be estimated through appropriate software simulation [8, 9]. Fig. 2.5 shows $\bar{P}_e$ in case of BPSK and DPSK modulations, as experienced in Rome, Italy, for the 66-satellites Iridium polar constellation [12] for different environmental conditions.

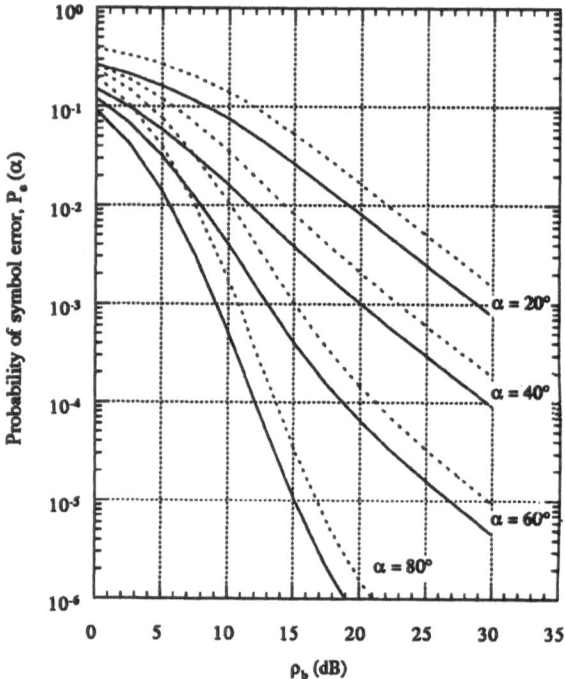

Fig. 2.4. Probability of symbol error in a rural tree-shadowed environment as a function of the (bit energy)/(noise spectral density) ratio for different values of the elevation angle for BPSK (——) and DPSK (- - -) modulations.

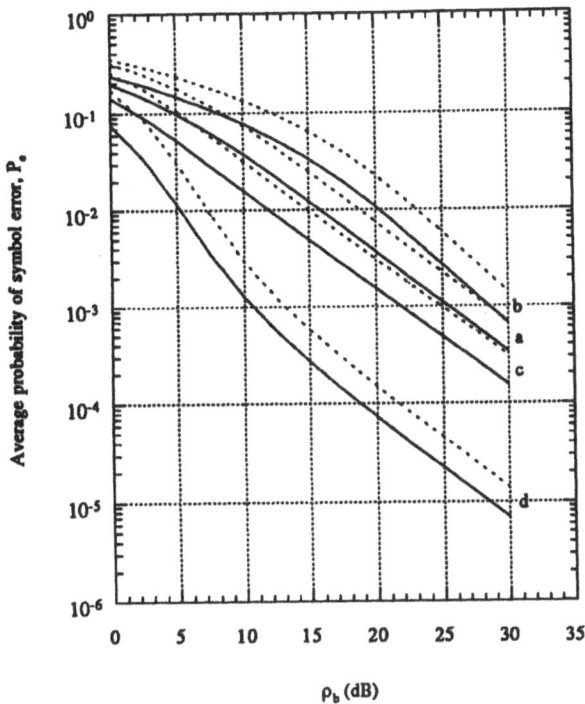

Fig. 2.5. Average probability of error for BPSK (——) and DPSK (- - -) modulations, as experienced in Rome, Italy, for the 66-satellites Iridium polar constellation for different environmental conditions: (a) rural tree-shadowed , (b) urban, (c) suburban, (d) open area.

3. A multi-environment generative model for frequency flat mobile satellite channels

The RLN model is suitable for describing a single-environment and is "quasi-stationary" in the sense that a given environment can be characterized by a set of twelve coefficients (K_i, μ_i, σ_i), $i = 0, \ldots, 3$ (see (2.14)). Quasi-stationarity of channel characteristics is due in general to the slow variation of the elevation angle under which the satellite is seen.

When the mobile terminal moves in different environments, the quasi-stationary assumption is no more valid, and the empirical model coefficients must change. Though statistical channel characteristics can significantly vary over a large area, this can always be discretized into M types of sub-areas each having constant environmental properties (i.e., with a fixed set of (K_i, μ_i, σ_i), $i = 0, \ldots 3$). A nonstationary channel model can thus be represented by M quasi-stationary RLN models.

The nonstationary channel is modelled by a Markov chain, having a finite number of states, s_i, $i = 1, \ldots M$, each state describing a type of sub-area [13]. A Markov chain is specified in terms of its state probability vector $\mathcal{P}$, having elements $\pi_i = \text{Prob}(\text{state} = s_i)$, and the transition probability matrix:

$$\Pi = \begin{bmatrix} \pi_{11} & \cdots & \pi_{1M} \\ \cdots & \cdots & \cdots \\ \pi_{M1} & \cdots & \pi_{MM} \end{bmatrix}, \tag{3.1}$$

where $\pi_{ij} = \text{Prob}(\text{next state} = s_j \mid \text{present state} = s_i)$, with conditions:

$$\sum_{j=1}^{M} \pi_{ij} = 1; \qquad \pi_j = \sum_{i=1}^{M} \pi_{ij} \pi_i. \tag{3.2}$$

In the following, a channel model characterized by $M = 2$ will be introduced.

In [14] a channel model was derived, distinguishing an unshadowed Rice state (the *good state*) and a shadowed Rayleigh-lognormal state (the *bad state*). By a proper choice of the state transmission probabilities and the parameters characterizing the fading statistics related to both states, respectively, this model can be used to describe the land mobile satellite channel in various macroscopic environments such as rural, residential, and city areas.

The channel model investigated in [14] is generative in the sense that it can be used to produce a sample of the complex fading process $\underline{r}(t)$, e.g. for simulation purposes. This model is based on channel measurements performed with a geostationary satellite under various elevation angles and with a constantly driving van; therefore, it is more applicable to land mobile satellite communications, rather than personal communications.

The channel model shown in Fig. 3.1 assumes that the transmitted signal $\underline{s}(t)$, for which $|\underline{s}(t)|^2 = 1$ is assumed, is multiplied with the fading process $\underline{r}(t)$. The concept of multiplicative fading can be applied when the bandwidth

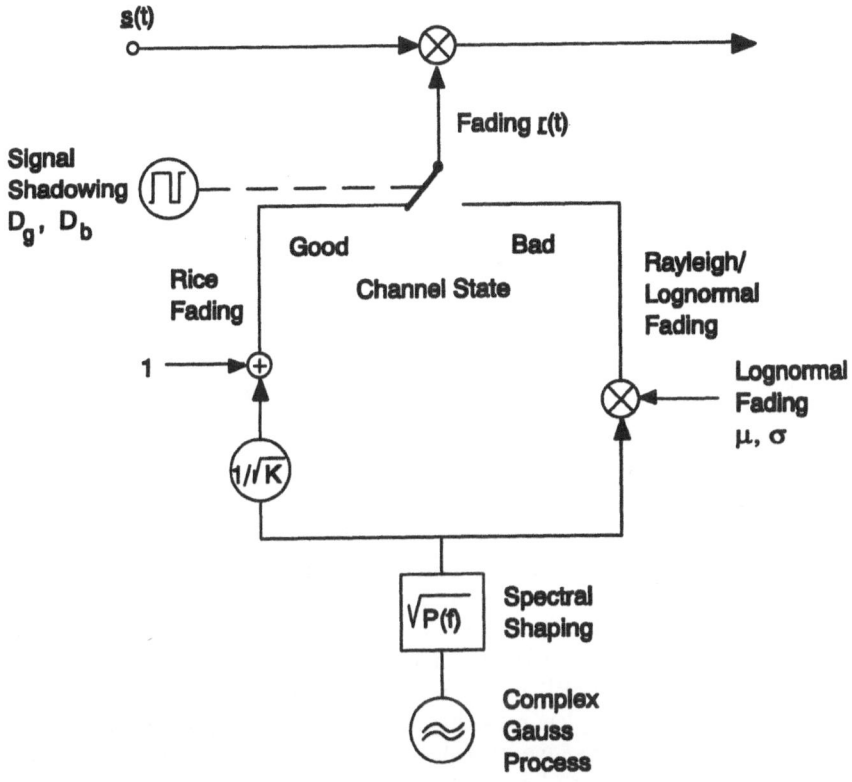

Fig. 3.1. Generative model of the land mobile satellite channel

of the transmitted signal is small compared to the coherence bandwidth of the fading.

According to Fig. 3.1 the fading process $\underline{r}(t)$ is "switched" between Rician fading, representing unshadowed areas with high received signal power (*good* channel state) and Rayleigh/lognormal fading, representing shadowed areas with low received signal power (*bad* channel state).

In the good channel state, the amplitude $r(t)$ of the fading process obeys a Rician probability density $p_{Rice}(r)$ according to (2.2), with mean received total power $1 + 1/K$. When shadowing is present, it is assumed that no direct signal path exists and that the multipath fading has a Rayleigh characteristic $p_{Rayl.}(r|S_0)$ with short-term mean received power S_0, [14].

The slow shadowing process results in a time varying short-term mean received power S_0 for which a lognormal distribution $p_{LN}(S_0)$ corresponding to (2.3) is assumed.

In order to get the resulting probability density function of the received signal envelope, the time-share of shadowing, A, is defined, and the resulting probability density function becomes

$$p_r(r) = (1 - A) * p_{Rice}(r) + A * \int_0^\infty p_{Rayl.}(r|S_0)p_{LN}(S_0)\,\mathrm{d}S_0 \qquad (3.3)$$

where the integral expression results from the theorem of total probability. $p_r(r)$ is independent of vehicle velocity v, as long as it is assumed constant.

In order to generate the fading process, the received multipath signal is modeled as a complex Gauss process having Rayleigh distributed amplitude and uniformly distributed phase. This process may be generated by complex addition of two independent stationary Gaussian processes. The spectral properties of the complex Gauss process are defined by an appropriate filter. In order to model the fading in city environments, a filter may be chosen that approximates [19, eq. (20)]. For rural environment, the filter should rather have a slowly decreasing low-pass characteristic [14]. As a compromise, a flat low-pass filter with a bandwidth equal to the maximum doppler frequency may be used for all types of environments.

Based on the filtered Gauss process, the Rician fading is produced by attenuating the Gauss process to power of $1/K$ and adding a value of unity to represent the direct satellite signal component. The Rayleigh/lognormal fading is generated by multiplying the Rayleigh process with a slow lognormal shadowing process. In order to approximate the dynamic behavior of the lognormal shadowing, the short-term mean received signal power S_0 may be chosen according to its lognormal probability density independently for each shadowing interval and kept constant during that interval.

The characteristics of the switching process between shadowed and unshadowed sections can be approximated by a Markov model, cf. Fig.3.2. For a given speed $v[m/s]$ and bit rate $R[b/s]$ the transition probabilities π_{gb} and π_{bg} can be related to the bit duration. According to the Markov model, the mean duration of a period of good (bad) channel state is given by

238

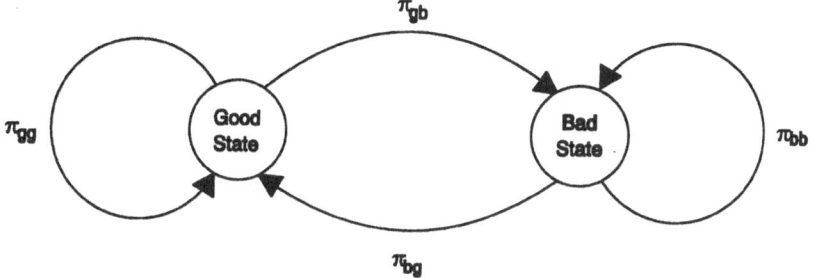

Fig. 3.2. Markov model for the channel state.

$$D_g[bits] \quad = \quad \frac{1}{\pi_{gb}} = \frac{R}{v} D_g[m] \tag{3.4}$$

$$D_b[bits] \quad = \quad \frac{1}{\pi_{bg}} = \frac{R}{v} D_b[m] \ .$$

The time-share of shadowing, A, is related to the durations D_g and D_b by

$$A = \frac{D_b}{D_g + D_b}. \tag{3.5}$$

In [14] it is shown that the Markov model for the switching process fits very well to the shadowing behaviour experienced during measurement runs. Overall, the channel model is determined by five independent parameters K, μ, σ, D_g, and D_b. For different satellite elevations, different types of environments, and different antennas, these parameters have been determined from the statistics of the recordings by a least square curve-fitting procedure and are compiled in Table 3.1 for city and highway environments.

The generative channel model, Fig. 3.1, reproduces the probability density function of the received signal power als well as the dynamic behavior of the fading and shadowing process for the land mobile satellite communication channel.

4. Channel model for wideband satellite channels

For system planning and design the knowledge of the frequency-selective channel is of crucial influence. The development and verification of wideband models for mobile satellite channels is at the very beginning and therefore only a preliminary approach can be presented. In our work, we will derive a channel model which is based on a tap model, cf. Fig. 4.1. This model is a generative model able to produce subsequent impulse responses. The model

Satellite Elevation	Environment	Antenna	A	$10 \log K$	μ	σ	D_g	D_b
13°	Highway	C3	0.24	10.2 dB	−8.9 dB	5.1 dB	90 m	29 m
	City	C3	0.89	3.9 dB	−11.5 dB	2.0 dB	9 m	70 m
18°	City	C3	0.80	6.4 dB	−11.8 dB	4.0 dB	8 m	32 m
	City	D5	0.80	5.5 dB	−10.0 dB	3.7 dB	8 m	33 m
21°	New City	D5	0.57	10.6 dB	−12.3 dB	5.0 dB	45 m	60 m
	Highway	D5	0.03	16.6 dB	−7.1 dB	5.5 dB	524 m	15 m
	Highway	S6	0.03	18.1 dB	−7.9 dB	4.8 dB	514 m	17 m
24°	Old City	C3	0.66	6.0 dB	−10.8 dB	2.8 dB	27 m	52 m
	Old City	D5	0.78	9.3 dB	−12.2 dB	4.4 dB	21 m	76 m
	Old City	S6	0.79	11.9 dB	−12.9 dB	5.0 dB	24 m	88 m
	Highway	C3	0.25	11.9 dB	−7.7 dB	6.0 dB	188 m	62 m
	Highway	S6	0.19	17.4 dB	−8.1 dB	4.2 dB	700 m	160 m
34°	City	C3	0.58	6.0 dB	−10.6 dB	2.6 dB	24 m	33 m
	City	M2	0.72	10.0 dB	−11.9 dB	4.9 dB	21 m	55 m
	City	S6	0.60	9.5 dB	−12.2 dB	2.0 dB	20 m	31 m
	Highway	C3	0.008	11.7 dB	−8.8 dB	3.8 dB	1500 m	12 m
	Highway	S6	0.007	16.7 dB	−13.4 dB	5.3 dB	1500 m	11 m
43°	City	C3	0.54	5.5 dB	−13.6 dB	3.8 dB	42 m	49 m
	City	M2	0.65	11.0 dB	−15.4 dB	5.4 dB	16 m	29 m
	City	S6	0.56	6.5 dB	−15.6 dB	3.8 dB	51 m	65 m
	Highway	C3	0.002	14.8 dB	−12.0 dB	2.9 dB	8300 m	17 m
	Highway	M2	0.002	17.3 dB	−13.8 dB	2.0 dB	8300 m	17 m

Table 3.1. Parameters of the generative channel model. For a definition of antenna types, cf. [14]

parameters will be determined by fitting them with statistics of measured echo-delay-profiles.

Let us denote the channel impulse response $h(\tau, t)$, where τ is the delay of an echo. The physical background of the model is a number of reflectors causing variable echoes with different delays. The received impulse response is then a superposition of impulses. The complex impulse response can be

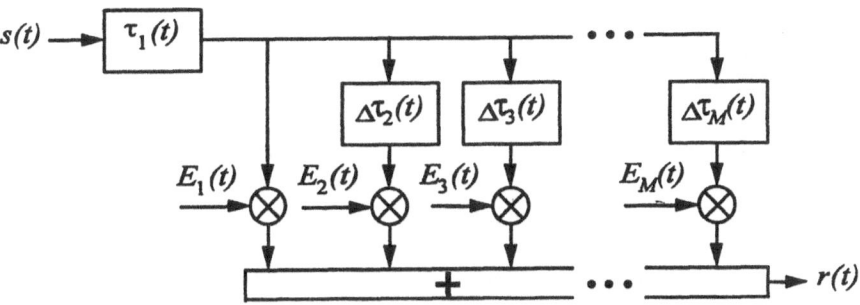

Fig. 4.1. Structure of the wideband tap model

expressed as a sum of $k = 1 \ldots M$ echoes $E_k(t)$ having delays $\tau_1(t)$ and $\tau_k(t) = \tau_1(t) + \Delta\tau_k(t)$, $k = 2 \ldots M$, respectively:

$$h(\tau, t) = \sum_{k=1}^{M} E_k(t) \cdot \delta(\tau - \tau_k(t)) \,. \tag{4.1}$$

Similarly to the approach of the terrestrial CODIT/RACE model, [18], each echo

$$E_k(t) = a_k(t) \cdot e^{j\phi_k(t)} \tag{4.2}$$

can be considered to consist of subpaths including a main echo and a number of secondary echoes causing short-term fluctuations. The statistics of the time-varying amplitudes $a_k(t)$ and phases $\phi_k(t)$ have to be investigated on the basis of the results of wideband measurements. From preliminary considerations, it can be expected that the amplitude of the direct path, $a_1(t)$, can be modeled by a Rice process, whereas $a_k(t)$, $k = 2 \ldots M$ have Rayleigh characteristics. The phases $\phi_k(t)$ can be assumed to be uniformly distributed in $[0, 2\pi)$.

5. DLR measurements for mobile satellite services

Narrowband measurements are still of great value in extending the databases of propagation measurements relevant to handheld mobile satellite services.

Also the time variance of the channel parameters due to the high velocity of the satellites and their time-varying elevation and azimuth angle has not been considered so far. Very little is known about wideband propagation experiments for satellite systems.

Besides the propagation effects, interference by other systems has to be determined for a proper system design. On the satellite up-link mainly the influence of terrestrial Fixed Service (FS) links may cause interference whereas man-made noise with sources nearby the mobile user dominates on the down-link. Depending on the frequency band, microwave ovens and sodium-vapour lamps show strong spectral emission. Areas nearby FS towers might also be contaminated by sidelobes of the FS transmitters or wide beams of incoming FS links.

In cooperation with Inmarsat and ESA, the German Aerospace Research Establishment (DLR) performed a set of measurement campaigns in order to investigate the effects with influence to the mobile-to-satellite link. The programme comprises narrowband and wideband propagation measurements as well as interference measurements for the satellite up- and down-link.

5.1 Measurement Setup

The measurement set-up basically consists of a transmitter (TX) and a receiver (RX) part. The transmitter is located in an aircraft. The receiver part is fitted in a van. The basic schemes of the transmitter and receiver parts are shown in Fig. 5.1 and 5.2, respectively .

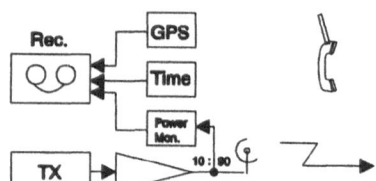

Fig. 5.1. Measurement set-up of the transmitter part.

The operators in the aeroplane and the van can communicate via a VHF speech link.

Wideband Equipment. For the wideband experiment we use the channel sounder CSPE 1800 with a swept time-delay cross correlator (Ref. [17]). The carrier frequency is adjustable in the 1.8 GHz band. The transmit part of the channel sounder transmits a carrier which is modulated by an m-sequence with a chip rate $f_{chip} = 10$ MHz or 30 MHz, respectively. The swept time-delay cross correlator in the receiver correlates the incoming signal with an m-sequence identical to the transmitted sequence, but clocked at a slightly

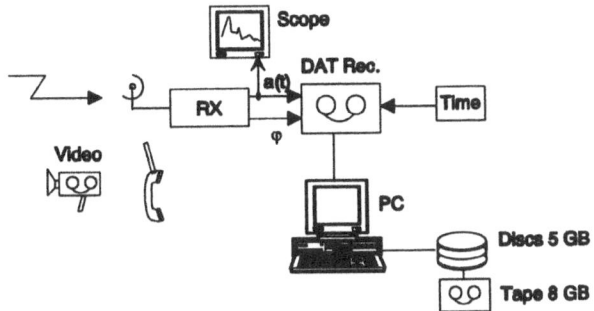

Fig. 5.2. Measurement set-up of the receiver part.

lower rate. The difference of the clock rates is $f_{slide} = 8$ kHz, which is equivalent to the bandwidth of the baseband signal at the output of the correlator. This signal consists of a continuous series of samples of the channel impulse response which are extended in time by a factor f_{chip}/f_{slide} compared to the actual imulse response. For possible combinations of m and f_{chip}, Table 5.1 shows the maximum excess delay d_{max} and the number of profiles per second f_{prof} that can be measured.

The present wideband equipment is able to measure the amplitude $|h(\tau, t)|$ of the impulse response, from which the most important properties of the channel can be derived:

− Receiver Input Power
− Mean Delay Time
− Delay Spread
− Power Delay Profile
− Echo-Paths statistics
− Carrier-to-Multipath Power Ratio.

If phase is available, additional characteristics can be derived:

− Doppler spectra
− Coherence Bandwidth
− Delay-Doppler Function
− Channel Scattering Function.

Narrowband Equipment. For the narrowband experiment, the radiated signal is generated by an oscillator at 1.8 GHz. The receiver part is able to give the baseband I+Q components of **two** antennas simultaneously with a bandwidth of 8 kHz. The dynamic range is better than 40 dB.

Transmitter Setup. The signals are amplified up to +44 dBm. The transmitted power is monitored via a 10:90-coupler. Information of the power level, the time, and the GPS position is recorded. The GPS system displays deviations of the predefined course of the aircraft. The achieved absolute position

m	f_{chip}	d_{max}	f_{prof}
511	10 MHz	51 μs	15.6 1/sec
	30 MHz	17 μs	15.6 1/sec
255	10 MHz	26 μs	31 1/sec
	30 MHz	8.6 μs	31 1/sec
127	10 MHz	13 μs	62 1/sec

Table 5.1. Parameters of the CSPE 1800 and their impact on the performance. $f_{slide} = 8$ kHz

accuracy during flight is better than 100 m. Also the pitch and roll angle are recorded during the flight.

Receiver Setup. Video recording of the environment is provided in order to identify the environment during the evaluation. Time and date are inserted in the video record. The received complex signal is recorded with time information on a digital audio tape (DAT) recorder. Also the GPS position is continuously recorded.

Online monitoring of the recorded channel impulse responses is possible on an oscilloscope. After recording, the stored data are logged (in the laboratory) for further evaluation on a PC with mass storage facilities and network interfaces.

Antennas. The transmit antenna has an omnidirectional characteristic. Mounting on aircraft without taking precautions would severely disturb the antenna pattern. Once mounted on the aircraft, the channel sounder would not be able to identify secondary reflections from the plane since its minimal spatial resolution corresponds to approximately 10 m. Therefore, the antenna is placed in a glass fiber radome with absorber material.

Interference Equipment. The measurement setup basically consists of an antenna, a low-noise preamplifier and a spectrum analyzer. A laptop records the sampled spectra via GPIB-bus. The equipment is carried either by a vehicle or an aircraft. Three frequency bands have been investigated:

1. 1.98 - 2.01 GHz for the satellite uplink,
2. 2.17-2.2 GHz for the satellite downlink, and
3. 2.48 - 2.5 GHz for the satellite downlink.

For the land mobile interference campaign a dipole antenna was used mounted on a van. For the airborne measurements a high gain YAGI antenna ($>$ 17 dBi) was used. The antenna was mounted in a radome also being capable to rotate in a horizontal plane with a speed of approximately 360 degree/min.

GPS Reception. The GPS reception is stored mainly for the evaluation of the channel data. In order to achieve a high accuracy for the received power level, the distance between aircraft and receiver has to be known with high accuracy. A third GPS receiver is working as a reference station at a well known position. With the help of the reference station, errors of the GPS signals can be corrected. The resulting position accuracy is then in the range of 1...10 cm.

5.2 Environments and scenarios

Measurements have been performed in Germany and in England. Different environments have been investigated in each region:

1. rural, wide open area
2. rural, tree-shadowed
3. urban (two sides of a road)
4. suburban
5. in-building, residential
6. in-building, office

TX Scenarios. Two strategies for the flight paths have been adopted to characterize the time-variant behaviour of the satellite's elevation and azimuth angle:

1. circles centered at the RX in order to get a database for all elevation and azimuth angles. In a constant flight altitude all azimuth angles belonging to one elevation angle are characterized. Eight elevation ranges have been defined from 10...90 degrees.
2. a flight path to simulate the real MEO time behaviour of azimuth and elevation angle. A time scaling is necessary due to the very low satellite angular speed. An orbit with a period of 6 hours (corresponding to an altitude of ≈10.000 km) was considered.

RX Scenarios. Three scenarios have been investigated for the handheld receiver:

1. The handheld antenna is fixed in normal head height on a tripod.
2. A *random* user simulating a person during a phone call.
3. Antenna diversity (vertically separated) fixed on a tripod.

Interference Scenarios. For the interference measurements, the scenarios were different:

– for the land mobile campaign, the measurement vehicle was driving in several environments at different times of day.
– airborne measurements characterized the influence of FS towers by flying circles around the tower in different elevation angles.
– a set of airborne trials scanned the horizon in wide parts of Germany with the YAGI antenna rotating in a horizontal plane.

5.3 Results

Narrowband Measurements. Figs. 5.3 and 5.4 show the time behaviour of the received power level. An antenna diversity experiment was performed with vertical separation of the antennas. The antennas have been mounted on a tripod with a vertical distance of $\lambda/4$. The aircraft was simulating an MEO orbit with a maximum elevation of 70 degree. The figures contain three curves: the envelopes of maximum and minimum values and the short term mean value (dotted) of the received fast fading power. The power level is heavily affected by fading due to specular reflections. The fade depth is up to 30 dB. The differences between the two figures are interesting: one can easily see that deep fades caused by the specular reflections are slightly shifted in time due to the geometry of the antenna arrangement. An intelligent receiver could take advantage of the diversity. Thus, an improvement of the service quality can be achieved. From Fig. 5.4 it can be seen that the signal received by antenna 2 is slightly worse compared to the signal delivered by antenna 1. The reason for this is probably the shadowing of antenna 2 by antenna 1.

Wideband Measurements. As an example for the wideband experiments, Fig. 5.5 shows impulse responses measured in an urban environment simulating a satellite having an elevation of 25 degree. The delay resolution corresponds to a bandwidth of 30 MHz. The receiver antenna was handheld, and the user was acting with a typical behaviour of a speaking person. It can be seen that the delay spread is quite small. Only a few echoes are received over this channel. In general, the echoes arrive within 1 μsec.

Interference. An example is presented for the land mobile interference measurements. The influence of microwave ovens is illustrated in Fig. 5.6. Microwave ovens radiate in a spectral range between 2.35-2.5 GHz. In Fig. 5.6 the measured interference power over time is shown with the mobile user standing on a street in a residential area. The power was averaged in the lower picture over the spectral band 2.483-2.5 GHz whereas in the upper picture a single frequency is shown. High interference peaks with up to 20 dB above the background radiation will disturb the handheld reception. A switching process predominates the interference: when micrwave ovens nearby the receiver are turned on or off the interference situation will change dramatically and remain for periods which may be longer than a telephone call.

Acknowledgement. The DLR authors would like to acknowledge INMARSAT and ESA for their technical support to the measurement programme and for partial funding. The authors of University of Roma Tor Vergata wish to acknowledge the Progetto Finalizzato "Telecomunicazioni" of Italian CNR, under which the MPS channel model in Section 2 was developed.

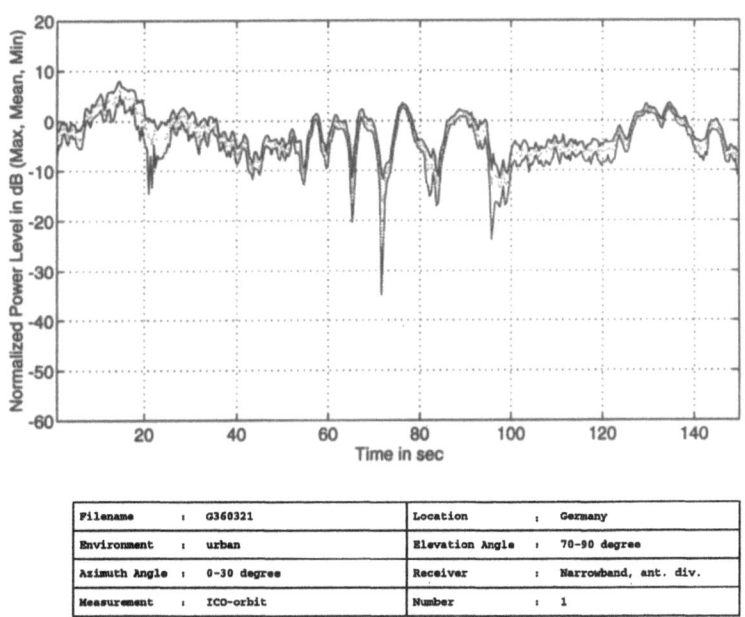

Filename	:	G360321	Location	:	Germany
Environment	:	urban	Elevation Angle	:	70-90 degree
Azimuth Angle	:	0-30 degree	Receiver	:	Narrowband, ant. div.
Measurement	:	ICO-orbit	Number	:	1

Fig. 5.3. Received power level of narrowband experiment. Antenna 1, mounted on a tripod.

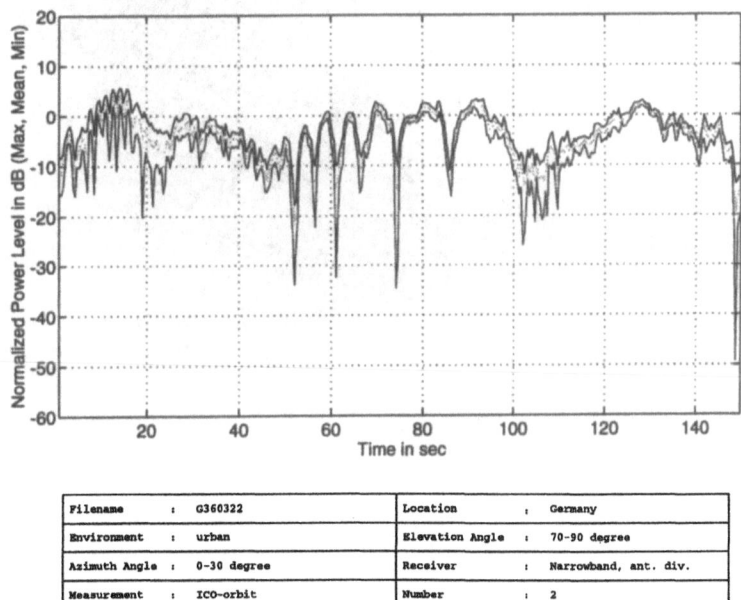

Filename	:	G360322	Location	:	Germany
Environment	:	urban	Elevation Angle	:	70-90 degree
Azimuth Angle	:	0-30 degree	Receiver	:	Narrowband, ant. div.
Measurement	:	ICO-orbit	Number	:	2

Fig. 5.4. Received power level of narrowband experiment. Antenna 2, mounted ca. λ/4 below antenna 1.

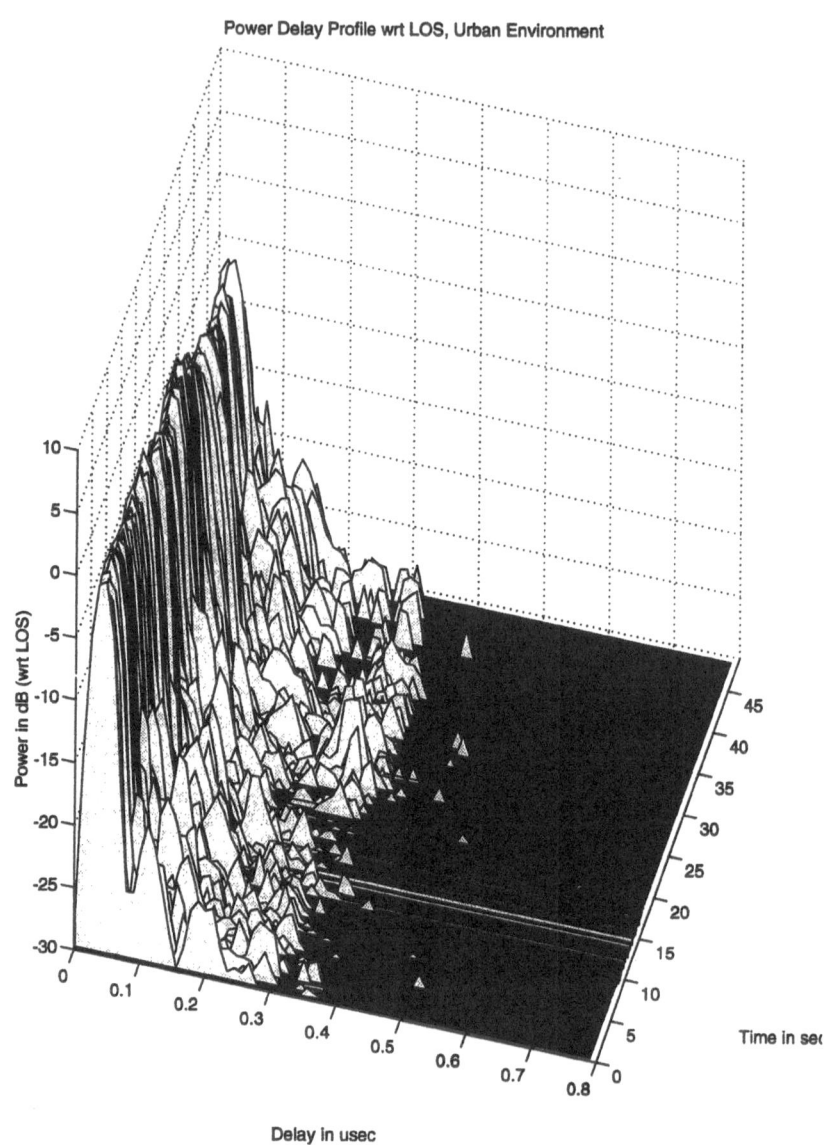

Fig. 5.5. Wideband experiment. Power delay profile with respect to line-of-sight. Urban environment, random user behaviour. Satellite elevation: 45 deg.

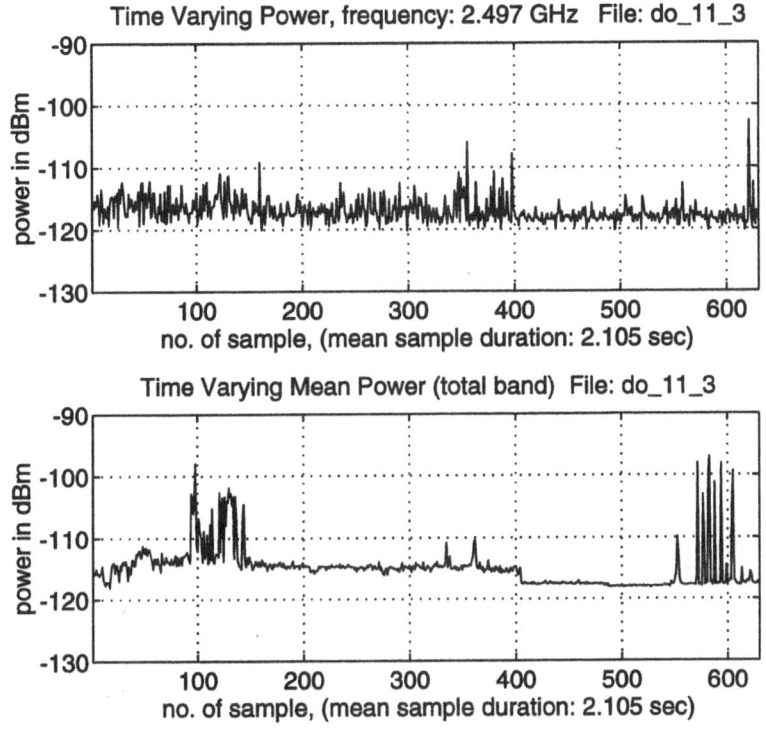

Fig. 5.6. Interference power received by a handheld antenna. Position: standing on a street in a residential area. Dinner time.

References

1. J.G. Proakis, *Digital Communications*, McGraw Hill, 1983.
2. L.F. Fenton, "The Sum of Log-Normal Probability Distributions in Scatter Transmission Systems," *IRE Trans. on Comm. Systems,* March 1960, pp. 57-67.
3. M. Abramowitz, I.A. Stegun, *Handbook of Mathematical Functions,* Dover, 1970.
4. H. Suzuki, "A Statistical Model for Urban Radio Propagation," *IEEE Trans Commun.,* Vol. COM-25, N.7, pp. 673-680, July, 1977.
5. W.C. Jakes Jr. (Ed), *Microwave Mobile Communications,* Wiley, New York, 1974.
6. S.O. Rice, "Statistical Properties of a Sine Wave plus Random Noise," *BSTJ,* vol.27, pp. 109-117, Jan. 1948.
7. W.C.Y. Lee, *Mobile Communications Engineering,* McGraw Hill, New York, 1982.
8. G.E. Corazza, F. Vatalaro, "A Statistical Model for Land Mobile Satellite Channels and Its Application to Non-Geostationary Orbit Systems," *IEEE Trans. on Veh. Technol.,* vol. 43, pp. 738-742, 1994.
9. A. Böttcher, M. Werner, "Personal Satellite Communications: Traffic and Capacity Considerations for the Mobile User Link," *Proc. GLOBECOM,* to be published, 1994.
10. C. Loo, "A Statistical Model for Land Mobile Satellite Link," *IEEE Trans. on Veh. Technol.,* vol. VT-34, pp. 122-127, August 1985.
11. M. Sforza, S. Buonomo, "Characterisation of the Propagation Channel for non-Geostationary LMS systems at L- and S-bands: Narrow Band Experimental Data and Channel Modelling," *Proc. XVII NAPEX Conf.,* June 14-15, 1993, Pasadena.
12. J.E. Hatleid, L. Casey, "The Iridium System: Personal Communications Anytime, Anyplace," *Int. Mobile Sat. Conf.,* IMSC 93, Pasadena, June 16-18, 1993, pp. 285-290.
13. B. Vucetic, J. Du, "Channel Modeling and Simulation in Satellite Mobile Communication Systems," *IEEE J. on Selected Areas in Commun.,* vol. 10, pp. 1209-1218, 1992.
14. E. Lutz, D. Cygan, M. Dippold, F. Dolainsky, and W. Papke, "The land mobile satellite communication channel — recording, statistics and channel model," *IEEE Trans. Vehicular Technology,* vol. 40, pp. 375–386, 1991.
15. J. Goldhirsch and W. J. Vogel, "Propagation effects by roadside trees measured at UHF and L-band for land-mobile satellite systems," in *Mobile Satellite Conference, JPL Pub. 88-9,* pp. 87–94, 1988.
16. H.-C. Haugli, N. Hart, and P. Poskett, "Inmarsat's future personal communicator system," in *Intern. Mobile Sat. Conf.,* IMSC 93, Pasadena, June 16-18 1993, pp. 303–304.
17. J. Parsons, "Sounding techniques for wideband mobile radio channels: A review," *Proceedings of the IEE,* vol. 138, pp. 437–446, 1991.
18. U. Dersch, J.J. Delgado, "CODIT Model for Propagation," *RACE Mobile Telecommunications Workshop,* Metz, June 16-18, pp. 373–377, 1993.
19. R.H. Clarke, "A statistical theory of mobile radio reception," *Bell System Technical J.,* vol. 47, pp. 957-1000, 1968.

Channel Models for Mobile/Personal Satellite Communication Systems

G.Butt , *Prof.B.G.Evans, M.A.N.Parks*
Centre for Satellite Engineering Research
University of Surrey, Guildford, Surrey, GU2 5XH, UK
Tel: (+44) 483 259131
Fax: (+44) 483 259504

1. Introduction

The knowledge of channel behaviour is of paramount importance in order to engineer the various sub-systems of a wireless communication system. The requirement to understand the signal propagation mechanisms, and the ability to model them, is even greater for the frequencies at which satellite-based mobile and personal communication networks are being planned. Satellite-mobile link quality and the availability of such services, to a great extent, will be determined by the propagation channel characteristics. The propagation channel characteristics are believed to be most significantly dependent upon the mobile surrounding environment. Signal amplitude variations of the order of several dBs may be experienced mainly due to the obstruction of the line-of-sight (LOS) path between a satellite and a mobile over a relatively short duration of time. This effect is known as 'shadowing'. Due to generally small dimensions of a mobile terminal, and therefore less or no antenna directivity, signal reception over multiple paths is another propagation degradation called 'multipath' reception. The dynamic nature of the satellite-mobile link results due to the constantly changing phase relationships of various incoming component signals at the terminal. Multipath is essentially cause of faster channel variations. In general, the depth of channel fading is dominated by the shadowing whilst the rate of the fading process (also called fading bandwidth) depends upon multipath.

It is obvious that in the overall design of a mobile communication network, propagation channel characteristics must be taken into account in order to develop robust and efficient modulation, multiple access, speech and forward error correction (FEC) coding techniques. Also for any power control system to operate reliably on a mobile-satellite link, typical channel information is essential for its design. Adequate technical solutions to these sub-systems are significant in terms of cost, capacity and therefore viability of a satellite-based mobile communication system. In general the channel information is grouped into *narrowband* and *wideband* characterisation. The two classes in fact refer to the different types of signalling formats involved in propagation measurements, and hence the information extracted from them. The basic information, related to the mobile-

---

* *e-mail: G.Butt@ee.surrey.ac.uk*

satellite channel, and required by the system design is a quantity called the 'propagation link margin'. The propagation link margin is provided in the link budget in order to support adequate link quality under fading conditions and which is estimated to be sufficient to achieve the desired network service availability. Provision of a specific link margin has direct consequences on the space and ground segment design and cost. In the following sections a review of various channel measurements and models is presented from a systems engineering viewpoint.

2. Narrowband Channel Models

Narrowband channel characterisation is primarily aimed at establishing amplitude (and phase) variations of the signal transmitted through the channel. It is implicit in such characterisation that all frequencies contained in a typical signal would experience similar fading, a phenomenon known as 'flat fading'. A number of channel models have been proposed as a result of different channel investigation campaigns. Both empirical and statistical/analytical type models exist. In general remarkably consistent propagation results have been reported in the literature. These come from various sources and from measurements undertaken in as diverse geographical locations as North America, Australia, and Europe. It has been proven that propagation conditions improve as the elevation angle between a mobile and satellite increases, indicating statistically reduced intervening obstruction in the LOS path.

Models based on statistical distributions are important because they prove useful in understanding the underlying mechanism involved in signal propagation. For the mobile-satellite channel these types of model consist of a combination of *Rician*, *Rayleigh* and *Log-Normal* distributions [1] [2]. Various distribution parameters are fitted from the actual measurement data, and in many cases show excellent agreement with the measured data. Statistical channel models are very useful in the software (and hardware) simulations and comparative analysis of different modulation, access and coding schemes.

Empirical channel models are based on the regression analysis of channel measurement data. Such models provide easy-to-use expressions to directly determine 'link margin' as a function of system level parameters of interest such as percentage availability, elevation angle, frequency etc. Second order statistics such as fade-duration distributions and level crossing rates derived from channel data also provide useful information for system signalling formats and decoder design. Goldhrish and Vogel [3] have published results of their extensive channel investigations which contain a comprehensive account of narrowband channel characterisation. A brief overview of various available empirical channel models is presented in the following sections. It includes description of the models proposed by the CSER (Centre for Satellite Engineering Research at the University of Surrey, UK) based on its successful 'simultaneous multi-frequency narrowband' propagation campaign. Some additional work has now been done on the statistical modelling of the data which will be useful for software system simulations.

1- Empirical Roadside Shadowing (ERS) Model

This model was developed by Goldhirsh and Vogel [3] based on the field trials carried out in the United States. The empirical expression was obtained by applying best fit formulations to the measured fade distributions at 1.5 GHz, and is given by

$$A(P,\theta) = -M(\theta) \cdot \ln P + N(\theta) \qquad \text{dB} \qquad (2.1)$$

where $\qquad M(\theta) = a + b\theta + c\theta^2$

and $\qquad N(\theta) = d\theta + e$

A is the required link margin, θ is the elevation angle (degrees) and P is the percentage of the distance travelled over which the fade is exceeded. The expression is applicable for P in the range of 1-20% and θ in the range of 20^0-60^0. The fit coefficients are :

$$a=3.44, b= 0.0975, c= -0.002, d= -0.0443, e= 34.76$$

2- Modified ERS (MERS) Model

This model has been developed by ESA and is based on the propagation data owned by ESA. The basic form of the model is similar to ERS model, however, the range of elevation angles applicable for the model is increased. The MERS model is as follows [4]:

$$F(\text{Pr},\theta) = -A(\theta) \cdot \ln(\text{Pr}) + B(\theta) \qquad \text{dB} \qquad (2.2)$$

A and B are defined by the following equations

$$A(\theta) = a_1\theta^2 + a_2\theta + a_3 , \qquad B(\theta) = b_1\theta^2 + b_2\theta + b_3$$

The regression coefficients are given as:

$$a_1= 1.117\text{e-}4,\ a_2= -0.0701,\ a_3= 6.1304,\ b_1= 0.0032,\ b_2= -0.6612,\ b_3= 37.8581$$

The applicable range of P_r is up to 30%, which is percentage exceedance of fade as determined by $F()$ in equation 2.2. Also the expression is applicable for elevation angle, θ, between 20^0-80^0. Although the model is for L-band frequencies only, it can be used for S-band frequencies after applying the scaling factor, as shown below:

$$F(f_s) \approx \sqrt{\frac{f_s}{f_L}} \cdot F(f_L)$$

where $F(f_S)$ and $F(f_L)$ are S-band and L-band fade margins, respectively.

3- Empirical Fading Model (EFM)

This model is based on the propagation campaign undertaken at the CSER. The campaign was carried out at three frequencies (L, S, K_u bands) simultaneously and only at high elevation angles i.e. from 60° to 80°. The campaign results have been published in detail and can be found in references [5] and [6]. Again the basic form of the model is similar to that of ERS, but it incorporates the frequency factor which is not the case with either ERS or MERS models. EFM is given by:

$$M(P,\theta,f) = a \cdot \ln(p) + C \qquad\qquad \text{dB} \qquad\qquad \textbf{(2.3)}$$

where a and C are given by

$$a(\theta,f) = 0.029\theta - 0.182f - 6.315$$
$$C(\theta,f) = -0.129\theta + 1.483f + 21.374$$

The model predicts link margin required in a suburban environment for link outages (P) in the range of 1-20% as a function of frequency, f (GHz), in the range from 1.5-10.5 GHz and elevation angle, θ (degrees), between 60°-80°.

This model has been extended by combining the ERS and EFM models to give an overall *Combined EFM (CEFM)* model covering a wider range of elevation angles (20°-80°) [7]. The regression coefficients for the CEFM are as follows:

$$a(\theta,f) = 0.002\theta^2 - 0.15\theta - 0.2f - 0.7$$
$$C(\theta,f) = -0.33\theta + 1.5f + 27.2$$

Figures 1 and 2 compare CEFM and MERS model link margin predictions, for L and S bands respectively, as a function of elevation angle at 1% and 10 % link outage probabilities. Figure 1(a) illustrates, at 1% fade exceedance probability, a comparison between margin predictions by the CEFM and MERS models as a function of elevation angle. The agreement is excellent over most of the elevation angle range, deviations being at the extreme angles. Again similar trends in link margin predictions by the two models can be identified at 10% exceedance level, as shown in figure 1 (b). Figure 2 essentially presents similar comparisons at S-band between CEFM and MERS. However, the only difference is that MERS model calculations at S-band are based on the S/L scaling factor, as mentioned before.

The above mentioned models comparisons show good agreement on a cumulative statistical basis even though the databases from which they have been produced are different. Also the measurement startegies, antenna types and routes for the measurement varied, however routes in general represented channel conditions of rural/suburban type environments where roadside tree shadowing contributes most to the signal fading. Thus we have a good representation of L-band fade margin for the mobile-satellite link, with less amount of data in the S-band but nevertheless some confidence in the prediction of margins in this band as well. Very little data exists for urban areas but it may be argued that mobile/personal satellite communication systems will not be designed for these areas as prime markets. All of the data collected to date has been for mobile (relatively directional) antennas as against personal (omni-directional) terminal antennas. The degree of multipath influence on the overall signal fluctuations in each case can be considerably different.

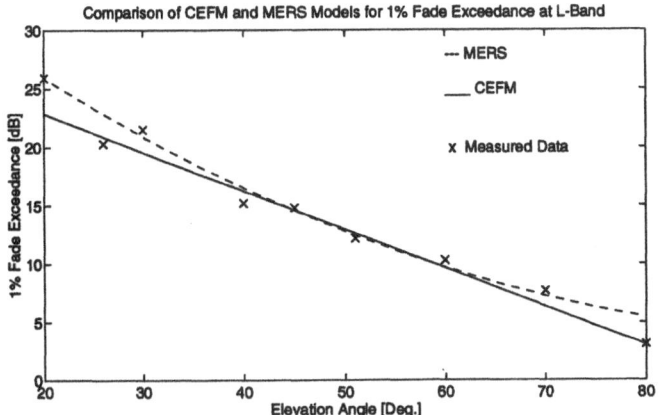

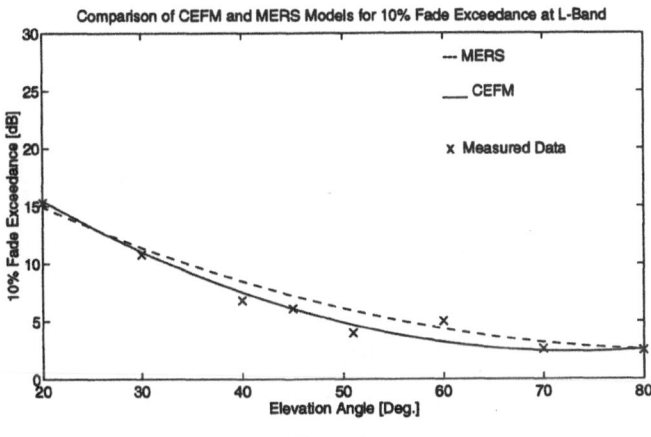

Figure 1

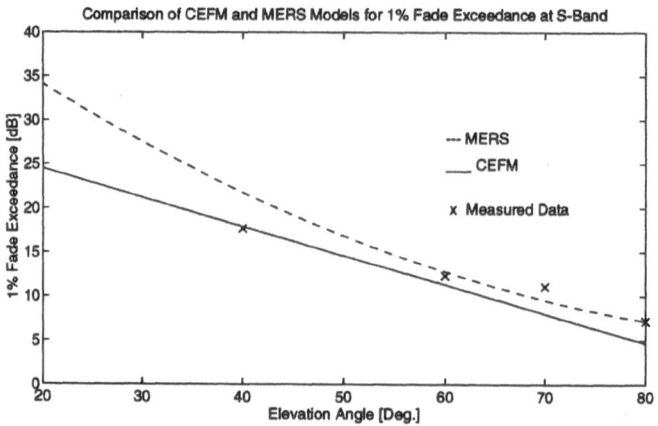

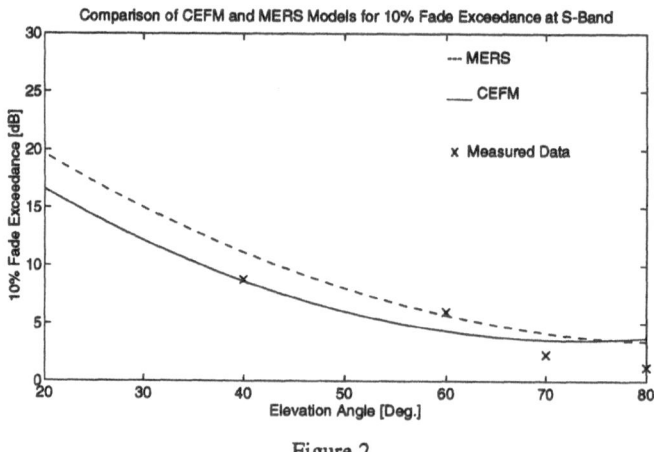

Figure 2

2.1 Statistical Channel Modelling

As mentioned earlier statistical channel models have been developed for mobile satellite channel. Such models are particularly useful for communication system simulations and provide a more 'comprehensive' understanding of the nature of a communication link. Previously reported statistical models, as in [1] [2], mainly describe the channel at lower elevation angles. CSER undertook its simultaneous multi-band narrowband campaign with a view to characterise the mobile satellite channel at high elevation angles, and in particular to cater for a systems such as ARCHIMEDES. Unlike models suggested by Loo [1] and Lutz [2] which consider shadowing as a major problem, it has been found that at high elevation angles the channel (amplitude) variations are Rician distributed with reduced problems from road-side obstacle shadowing. Figure 3 illustrates model fits at the L-band frequency for a measurement run during phase-II of the campaign (Spring '92) in the 'suburban' environment at 60° (fig. 3a) and 80° (fig. 3b). For the same stretch of the measurement route, the simultaneously measured channel responses at S-band are shown in figure 4.

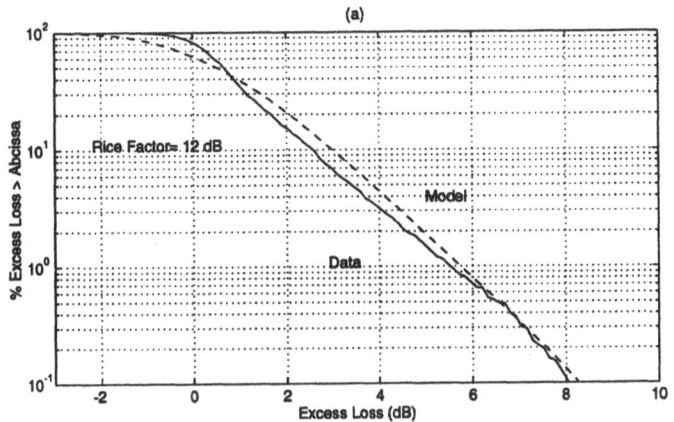

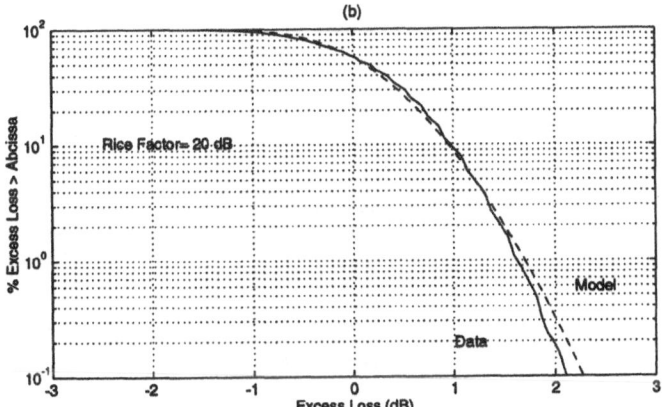

Figure 3: Model Fits for L-band channel data - Suburban - Spring'92

(a) 60° (b) 80°

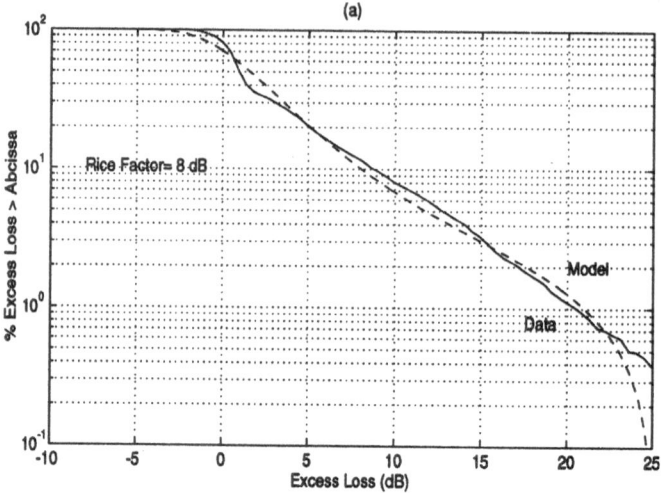

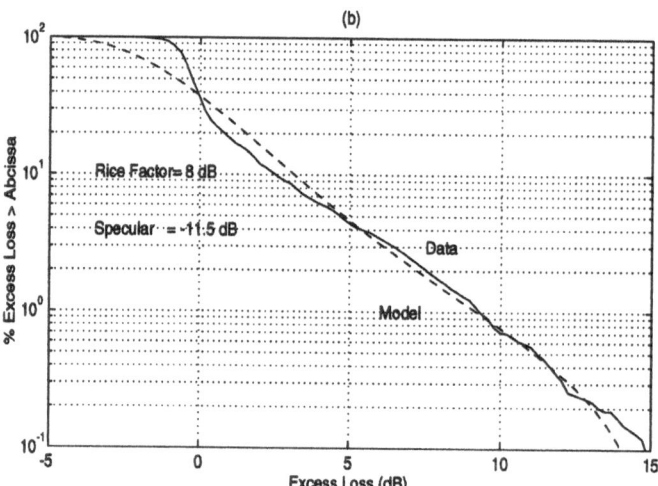

Figure 4: Model Fits for S-band channel data - Suburban - Spring'92

(a) 60° (b) 80°

It is important to note two obvious features in the L-band model fits. Due to the reduced shadowing problem the channel is purely Rician as opposed to Rayleigh and/or Log-normal. Also it clearly improves when the elevation angle is increased; Rice factor increases from 12 dB at 60° to 20 dB at 80°. Such a behaviour is consistent with the common understanding of the mobile satellite channel. Also for the S-band, the channel variations are still Rician distributed but understandably more dominated by the multipath effect. Figure 5 shows the L-band fade duration distributions at 60°, for thresholds at 3 and 7 dBs. The curves indicate 50% of the fades shorter than 5ms. Similarly S-band fade duration distributions are shown in figure 6.

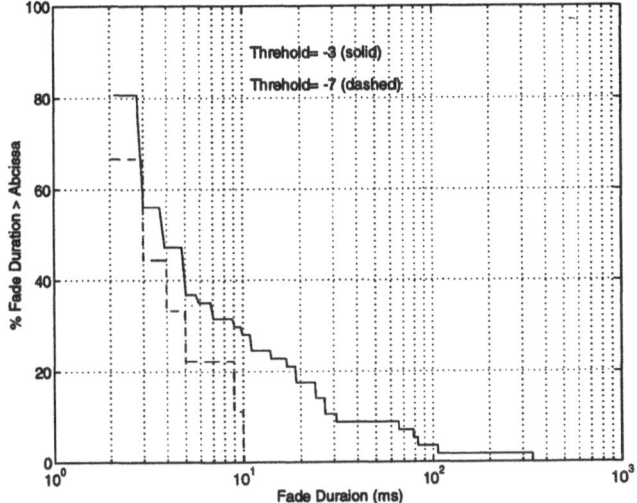

Figure 5: Fade Duration Distributions for L-band - Suburban - Spring '92

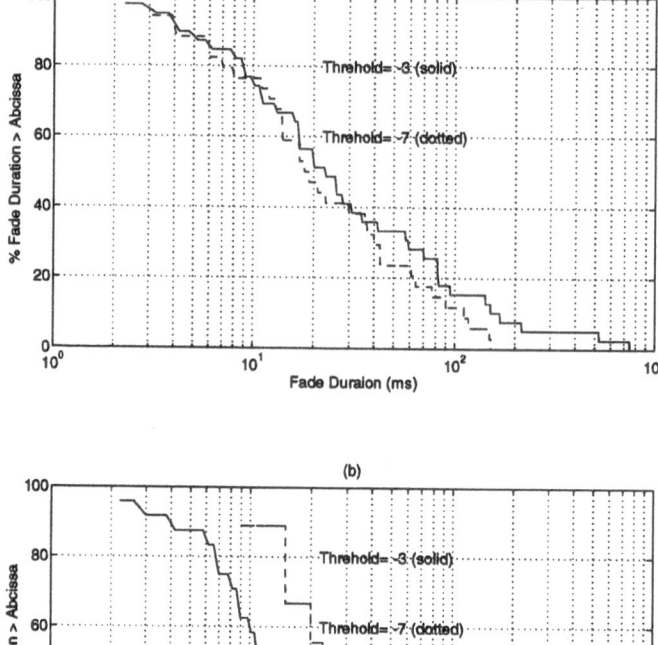

Figure 6: Fade Duration Distributions for S-band - Suburban - Spring '92

(a) 60° (b) 80°

2.2 Mobile vs. Personal Channel : A Comment

It is important to mention that most of the above discussion in terms of models and measurements is applicable to what is now more commonly termed as 'vehicular-mounted' mobile-satellite channel. Presently, satellite-based personal communication networks (S-PCNs) are actively being planned. The nature and operation of a personal communication handset is different from a vehicular (or transportable) terminal and therefore necessitates channel characterisation accordingly. The fading rates for both vehicular-mounted mobile and hand-held personal communication channels are expected to be of the same order. The LOS signal component varies slowly, typically less than 10 Hz. The diffused multipath signal components contribute most to the fast fading and have been found to cause variations in the range 150-200 Hz. Additionally the close proximity of the user's body may attenuate the signal even further under certain link conditions.

At present very little information is available in the public domain regarding the satellite-hand-held channel. Iridium Inc. and Inmarsat, the two major players in the future S-PCN market, are believed to be engaged in extensive channel characterisation for this type of channel [8], [9]. Reference [8] contains some details of propagation measurements carried out for the Iridium system. It is reported that large fade margins are required even in the LOS conditions to support adequate link quality. This may be due to a combination of reasons such as moving satellite and closeness of human body. A 15.7 dB propagation link margin will be incorporated into the Iridium system design. Another satellite PCN competitor system is Globalstar. The system design figures for Globalstar provide 10 dB for the propagation link margin when shadowing occurs on all diversity paths. It is understood that a power control system would be available in each of these systems to adaptively allocate suitable link margin on a link-by-link basis.

3. Wideband Channel Characterisation

Wideband measurements primarily provide information on the excess time delays of the echoes received at the terminal due to multipath propagation of the signal through the channel. The information can be processed to extract the coherence bandwidth of the channel. Knowledge of such parameters is important in designing signal formats to avoid distortion. Excess time delays can cause serious inter-symbol interference in digital systems which can either be handled by choosing appropriate transmission rates or appropriate receiver architectures. Very little data is available on wideband characterisation of either vehicular based or hand-held channels. Some wideband trials carried out in connection with the Iridium program [8] suggest that the delay spread for echoes arriving at the receiver are in order of less than 1 µs, thus indicating no serious problems associated with excess delays in a personal communication channel. This may attributed to the higher path elevation angles involved in such links. Similar conclusions can be made for vehicular based mobile channels as well.

At CSER planning for a wideband propagation campaign is well underway, under an ESTEC contract, to investigate excess delay and other wideband parameters at L and S bands. The campaign is expected to address both vehicular based mobile and hand-held channels. It is intended that the data will be used to construct what is termed as a 'Tapped Delay Line' wideband channel model. This type of model is believed to be most suitable for system simulation purposes such as CDMA performance evaluations etc.

4. Conclusions

An up-to-date picture on the information available on the propagation and channel models for mobile satellite applications has been presented. Much of the available information in this area is more applicable to vehicle-mounted channels and may not necessarily reflect the severity of hand-held (personal) type channels. Both the empirical and statistical models have been discussed. Empirical models have been improved over the recent years to include full range of elevation angles, which is an applicable scenario given that almost all satellite PCN operators to-date have chosen, although different but non-geo-stationary (moving) satellites constellation options as optimum space segment. But sadly the link margin predictions from these models may not be directly applicable to design such systems. On the statistical modelling side, Lutz model has been around for some time and may be used but applicability may only be representative up to $40°$ elevation. At present CSER statistical model is applicable only in the range of $60°-80°$. Attempts will be made to obtain a more general database in order to develop a comprehensive statistical model covering the full elevation angle range. CSER's most recent activities in the area of wideband channel sounding are expected to produce the additional data needed to complete a wideband 'Tapped Delay Line' model to facilitate full analysis of mobile and personal communication systems.

References

1. C.Loo et.al., "Measurements and Modelling of Land Mobile Satellite Signal Statistics", IEEE 36th Vehicular Technology Conference, May 1986

2. E.Lutz et.al., "The Land Mobile Satellite Communication Channel - Recording, Statistics, and Channel Models", IEEE Transactions on Vehicular Technology, vol.40, no.2, May 199, pp.375-385

3. J.Goldhrish, W.J.Vogel, "Propagation Effects for Land Mobile Satellite Systems : Overview of Experimental and Modeling Results", NASA Reference Publication 1274, February 1992

4. M.Sforza et.al.,"Global Coverage Mobile Satellite Systems : System Availability versus Channel Propagation Impairments", IMSC '93, June 1993

5. G.Butt et.al.,"Narrowband Channel Statistics from Multiband Propagation Measurements Applicable to High Elevation Angle Land Mobile Satellite Systems", IEEE J-SAC, SAC-10(8), vol.10, no.8, October 1992, pp.1219-1226

6. G.Butt et.al.," Results of Multiband (L, S, Ku band) Propagation Measurements for High Elevation Angle Land Mobile Satellite Channels", Proceedings of International Conference on Antennas and Propagation ICAP '93, Edinburgh, March 1993

7. M.A.N.Parks et.al.,"Empirical Models Applicable to Land Mobile Satellite System Propagation Channel Modelling", IEE Colloquium on Communications Simulation and Modelling Techniques, 28 September 1993, Digest no.1993/139

8. N.Kleiner, W.J.Vogel,"Imapct of Propagation Impairments on Optimal Personal Mobile Satellite Communication System Design", November 1992, Australia

9. H.C.Haugli et.al.," Inamrsat's Future Personal Communicator System", IMSC '93, June 1993

Propagation models for the Land Mobile Satellite Channel: validation aspects

E. Damosso (*), G. Di Bernardo (**),
A.L. Rallo (**), M. Sforza (***), L. Stola (*)

(*)	CSELT S.p.A.	Turin (Italy)
(**)	Space Engineering	Rome (Italy)
(***)	ESTEC	Noordwijk (The Netherlands)

Abstract

In recent years there has been significant evidence that personal communications will evolve as hybrid fixed/mobile/satellite networks: system concepts have been developed to cater to specific markets but conventional satellite designs appear not to be able to provide economical service to hand-held terminals, thus different concepts have emerged. Specifically, geosynchronous orbit satellites (GSO) operating at the Ka-band (20-40 GHz) or low earth orbit (LEO) and highly elliptical orbit (HEO) are contenders to provide the order of magnitude increase in communication capacity required for operation of extremely small terminals used in mobile communication networks. Congestion in currently used frequency bands and the demand for small earth stations is bringing EHF satellite communications to the point of economical viability, even if signal deteriorations due to hydrometeors and vegetation/building blocking phenomena are major problems. The future Universal Mobile Telecommunication Systems (UMTS), to be deployed at the turn of the century, is therefore designed to have a terrestrial and a satellite component: furthermore, work towards defining 4th generation Mobile Broadband Systems (MBS), for use beyond 2005, is already in progress. It is indeed of strategic importance to meet the requirements for wireless communication with very high throughput and to develop the industrial capability to produce the necessary system components: with better reason, in order to achieve the required high bit rates (up to 154 Mbit/s), it is necessary to consider high frequency spectrum resources. In this context, research to determine the optimum spectrum allocation (either in the K- or Ka-band, or in the 60 GHz region) is needed.

Accordingly, the scope of this paper is to provide information about the channel characterisation of a land mobile satellite channel, emphasizing the typical propagation blocking impairments encountered at the high microwave and quasi millimeter wave bands, addressing potential solutions and outlining a software tool (ARAMIS©), which can be used for a simulation of the transmission channel in a variety of environments. An overview will then be given of two measurement campaigns, to be run in the near future in Turin and Rome, using a K-band Land Mobile Beacon Terminal, designed and built in the framework of an ESA contract, that will be made available for propagation measurements and model validation purposes.

Introduction

Current mobile communications market analyses indicate that Europe is mobile aware and entering the consumer market. The most successful market supports a penetration of about 10%. To achieve the predicted higher penetration levels will require broadening the general user community and at the same time offering more advanced and different services.

Recent initiatives taking place in Japan, North America and within the European Community indicate that to meet the challenges of globalisation and international competition, very significant effort must be devoted to the development of personal telecommunications networks, products and services.

These new generations of mobile communication systems must be considered as systems that aim to support over a seamless radio infrastructure not only the different service offerings of the second generation ones, but also a much wider range of broadband services (data, voice, video, multimedia), commensurate with the technological developments taking place in fixed telecommunication networks. The progressive migration from second to third generation systems must hence be undertaken while ensuring that the current user markets will perceive such a service evolution as relatively seamless, attractive and natural.

In Europe, in particular, the objective of accommodating the foreseeable demand for personal communications beyond the year 2000 and to permit the European industry to retain its leadership position in this area is a milestone of two ambitious projects: UMTS and MBS. In both cases there has been significant evidence that personal communication networks will evolve as hybrid fixed/mobile/satellite systems. Certain international organisations and large multinational corporations need fixed, temporary fixed and mobile communications and often such organisations operate in developing countries with no telecommunications infrastructure. Another category of user is the personal roamer, operating in developing as well as in developed countries: to provide telecommunication services to these categories of users will require the internetworking and service support of fixed and mobile satellite or terrestrial systems that support narrow and broadband services. Individual requirements may demand handover between terrestrial and satellite systems with terminal equipment capable of accessing both. Relevant to the work in this area are the developments taking place regarding the satellite components of future mobile systems, e.g. IRIDIUM, GLOBALSTAR, INMARSAT P21, ESA-ARTES, etc. In this framework, the European RACE SAINT project, aiming at evaluating and identifying the requirements for integrating satellites into future personal mobile telecommunications is of paramount importance.

In this perspective, the satellite component of MBS, as with UMTS, has to be carefully investigated. In addition to the task of determining the optimum spectrum allocation (which is still an open question), other critical issues need addressing, such as channel characterisation, radio access, digital processing, dual mode terminals and impairment mitigation techniques. The scope of this paper is to address the main issues concerning the propagation medium identification, with particular reference to the natural and man-made blocking effects at K-band; two specific topics (the effects of buildings and trees) will be examined, from a modelling standpoint, and a software simulation tool (ARAMIS©), capable to account for the relevant propagation phenomena from a deterministic standpoint, will be presented. Moreover, two experimental campaigns will be described, designed to provide results to support the development of such models and to validate the software tool in the field: an ad-hoc section will be also devoted to illustrate the technical characteristics and performance of the mobile beacon terminal, developed by RESCOM under ESA contract, which will be used in the measurement validation campaign.

Scenario Considerations

A typical scenario for the propagation channel in a land mobile satellite system (LMS) is depicted in Fig. 1. The total field incoming on a mobile station is the vector sum of three major components. The direct Line-Of-Sight (LOS) coherent component between satellite and mobile antenna is paired by a ray (or cluster of rays) which accounts for the specularly reflected component due to the ground reflection in the mobile antenna direction. Finally, there is an additional diffuse multipath component, which represents the power scattered into the antenna main beam and side lobes from surrounding objects (terrain irregularities, buildings, roadside trees, ecc.). Then the total multipath contribution is the result of both the geometrical characteristics of the environment and the antenna directivity.

Since Faraday rotation considerations, as well as signal deterioration due to tropospheric effects (rain/wet snow attenuation, rain/ice depolarisation) are out of the scope of this paper, blocking, specular and diffuse reflection only (which occur in the terrestrial environment) will be considered in the following, with particular reference to their contribution to shadowing and multipath phenomena.

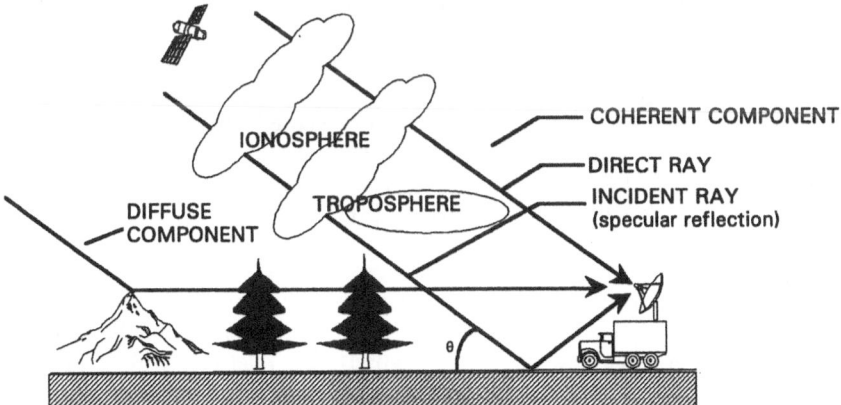

Fig. 1 - Typical propagation scenario in a land mobile satellite environment

Shadowing (or long-term fading), generated by natural or man-made obstructions on the Line-Of-Sight propagation path, is typically flat within a given bandwidth and shows a strong dependence on the satellite elevation angle. Multipath (or short-term) fading originates from interference of various signals reflected and/or scattered by the objects surrounding the mobile terminal, which usually add in a destructive way to the (eventually blocked) direct path, resulting in deep frequency selective fades.

The behaviour of the multipath fading is correlated to the shadowing: if the direct signal is obstructed (bad channel state), the multipath effects are severe; otherwise, if the direct signal is dominant (good channel state) the effects are much less significant. These are typically environment-dependent situations: in urban areas, for example, more obstructions and reflections are expected than in rural conditions; moreover, the antenna pattern could play a significant role in the whole process.

A number of propagation measurements have been performed to investigate the two phenomena, particularly near or at the L-band (for example at 860 and 1550 MHz [1, 2]). Analysis of the data confirms the similarity with the most relevant propagation characteristics exhibited by terrestrial mobile links; namely, over small distances (some tens of wavelengths in urban areas and several hundreds of wavelengths in open or rural areas), the statistical variations of the received power (short term fading, Rice distributed) exhibits a rather constant local mean value (subject to long term fading, following a Log-Normal distribution), mainly due to the shadowing effects caused by the surrounding obstacles.

The above quoted measurements do not allow to separate the effects of LOS and diffuse scattering attenuation contributions, in particular in built-up areas, where the most relevant propagation mode is via diffuse scattering, and the LOS component can experience high attenuations (at least for GSO satellites, at elevation angles below about 40 degrees). Furthermore, a lack of suitable measurement information prevents a satisfactory formulation of models able to cope with the effects of the mobile antenna characteristics on the received signal. In fact, in a high dense urban environment the multipath components play a double role: first of all they are a significant portion of the received energy; furthermore, they are the main potential interference source to the wanted signal. Then, if a directive antenna reduces the multipath interference effects, it also reduces the average level of the useful received power. Finally, the limited sample size of the analyzed experimental results data base suggests that caution should be exercised in extending their application to other situations [3].

In any case, the majority of the models proposed in the literature so far, being of empirical/statistical nature, cannot be suitably used for a local field strength point prediction, in that they are only able to provide information about the statistical parameters (mean value and standard deviation) of a relatively wide area signal distribution. This is a very remarkable drawback, particularly at high frequency bands, where a more deterministic approach (based on ray-optical techniques) is unavoidable: the ARAMIS© software, as described in the following, belongs to this kind of tools and allows to account for local field strength values (taking into account actual scattering and diffraction phenomena), as well as the influence of the antenna radiation pattern.

In the previous scenario, the potential advantages of the K-band for future LMS (from a system/subsystem standpoint) can be summarized as follows:

- large spectrum resources would dramatically relax the frequency re-use requirements and efficient on-board channel routing and processing, currently envisaged for L-band systems

- mobile terminal and spacecraft antenna sizes could be considerably reduced, hence both increasing market potential and lowering payload mass

- share of the on-board resources of Fixed Satellite Services (FSS), namely EUTELSAT with its EUTELTRACKS system, could be of remarkable benefit for satellite operators

As a counterpart, from a transmission channel standpoint, additional attenuation effects due to tropospheric phenomena (rain, clouds, gaseous absorption) and increased blockage due to man-made (or natural, i.e. vegetated matter) structures have to be taken into proper account: all these phenomena (multipath, absorption and blockage) have not been yet thoroughly studied, at least for typical mobile environments, in this frequency range. As a practical consequence, higher link margins (with respect to L- or S-band) should be budgeted.

In order to get adequate information on the K-band mobile channel, RESCOM developed under ESA contract a Land Mobile Beacon Terminal (LMBT), originally designed to be used with Olympus satellite, which can be adopted to carry out passive experimental campaigns in different European locations: the Turin and Rome measurement sets outlined in this paper are two examples of these experimental facilities.

The LMS channel simulator ARAMIS©

ARAMIS© is a SW package implementing an innovative concept for the study of the land mobile radio channel in urban environments [4, 5]. Getting over all the previous packages based on a statistical approach, a deterministic modeling of the propagation, based on accurate definition of the urban structure, on ray tracing techniques and on the GTD (Geometric Theory of Diffraction) has been implemented, aimed at the managing of frequencies above ~ 1 GHz. The first experimental validation of the tool has been performed at ESTEC Compact Antenna Test Range (CATR) on a scaled hardware model, made up of four metallic boxes, placed on a supporting frame to simulate a simple urban site. Several wide band measurements in the frequency range 28.5-40 GHz (so as to maintain the ratio between wavelength and typical building dimensions) were carried out, for various source elevation angles and mobile terminal positions within the model.

Results have been very encouraging with respect both to the selected number of interactions and type of scattering mechanisms considered, and to the amplitude and phase calculated for each contribution to the total received local e.m. field [6]. An additional outdoor validation campaign of ARAMIS© at L-band and X-band will probably be funded by ESA in the near future.

Increasing frequency causes two opposite effects on the accuracy of the deterministic approach used in ARAMIS©; on the one hand, as the ratio between the wavelength and the typical urban building dimensions decreases, GTD approach efficiency increases. On the other hand, the higher the frequency, the greater are the resolution and the accuracy required in the knowledge of the urban structure under consideration; as a consequence, the suitability of the rather simple boxes currently used by ARAMIS© for this purpose has to be assessed better. In order to get information about such topics, which have not been yet adequately investigated, two ad-hoc measurement campaigns are foreseen in the near future, in Turin and Rome, respectively, using the above quoted RESCOM mobile terminal working in the K-band, as outlined in the following. Such experiments should allow to point out the level of accuracy, needed in the considered frequency range, as far as the required knowledge of both location and orientation of buildings transferred from the real world into the urban model, and of their relevant geometrical (length, width, height, etc.) and radioelectrical (reflectivity and transmittivity) characteristics.

The K-band RESCOM Land Mobile Beacon Terminal

The K-band Land Mobile Beacon Terminal (LMBT) is a narrowband receiver installed in a van, designed to perform propagation researches on a mobile satellite channel. It is currently equipped with a K-band antenna, but has been designed as a multi-frequency receiver, so that data can be collected and stored at up to three different frequency bands, simultaneously; such a feature could allow for the possibility of identifying frequency scaling empirical laws in a rather rigorous way .

As said before, the LMBT was originally developed to operate with the B1 Olympus polarization-switched beacon at 19.8 GHz; due to the premature demise of the satellite, the RF front-end was modified in order to allow for operation with the ITALSAT F1 18.7 GHz beacon, even if such a beacon has a reduced European coverage in comparison with Olympus, and provides a slightly less dynamic measurement range [7].

Fast signal acquisition and tracking, combined with stringent operational constraints, (van speed and relevant Doppler shifts, medium to long fade durations and consequent re-acquisition performance) were the driving factors in the design of the LMBT. Particular attention was also given to Data Acquisition System (DAS) capability and video subsystems, in order to cope with real-time data handling needs. The video subsystem, in particular, was conceived to provide the experimenter with real-time and synchronous optical information on the environment and the landscape, using two video cameras, the former slaved to the antenna (by means of a servo loop, that allows to point at the satellite direction with an accuracy of ± 0.5°), the latter forward-looking (wide-angle, 107°): the video signals from both cameras are combined in a mixer and time synchronized with the DAS (see Fig. 2 for the antenna radiation patterns).

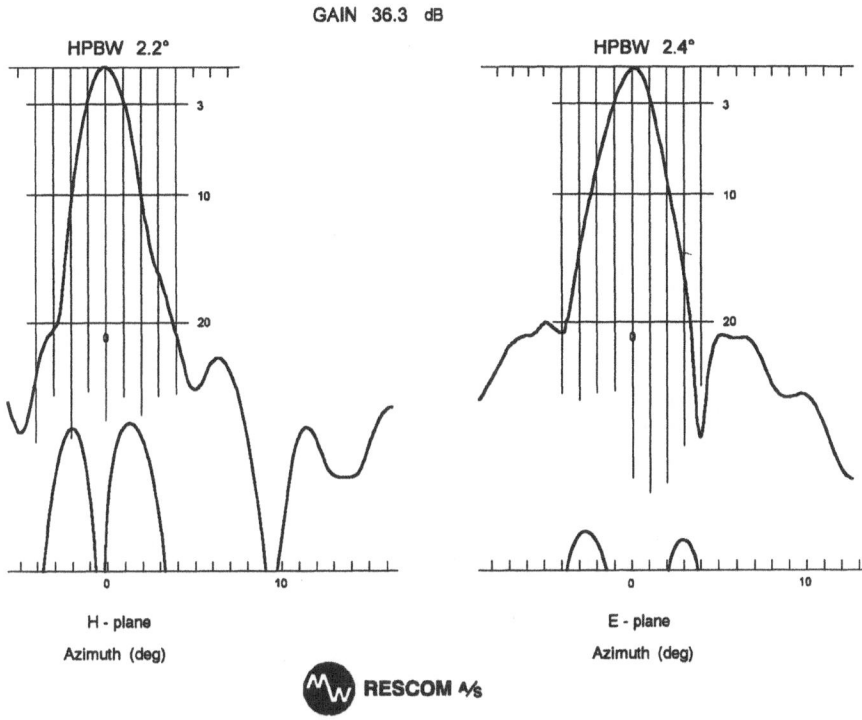

Fig. 2 - Land Mobile Beacon Terminal (antenna radiation patterns)

The antenna was designed to provide a gain able to ensure a minimum dynamic range of 30 dB (with a 50 Hz PLL bandwidth, locked mode) in all operational scenarios and to achieve a good pointing accuracy. As it can be seen in Fig. 2, the half power beamwidths for the selected Cassegrain-type antenna with a 50 cm main reflector diameter and a gain (at 18.7 GHz) of about 36 dB, are approximately 2.2° and 2.4° in the H-plane and E-plane, respectively. In order to meet the very stringent requirements at Ka-band with such a narrow antenna beamwidth, a Phase Lock Loop (PLL) was adopted, with four bandwidths (50, 100, 200 and 400 Hz).

A Central Control Subsystem (CCS) is responsible for all the interfacing LMBT network and allows the operator to monitor and control all the functions. Furthermore, the whole receiver needed to be built and integrated in the van, to ensure the required robustness in a mobile operational scenario, as reported in the following Fig. 3, where an overview of the on-board installed equipment is given.

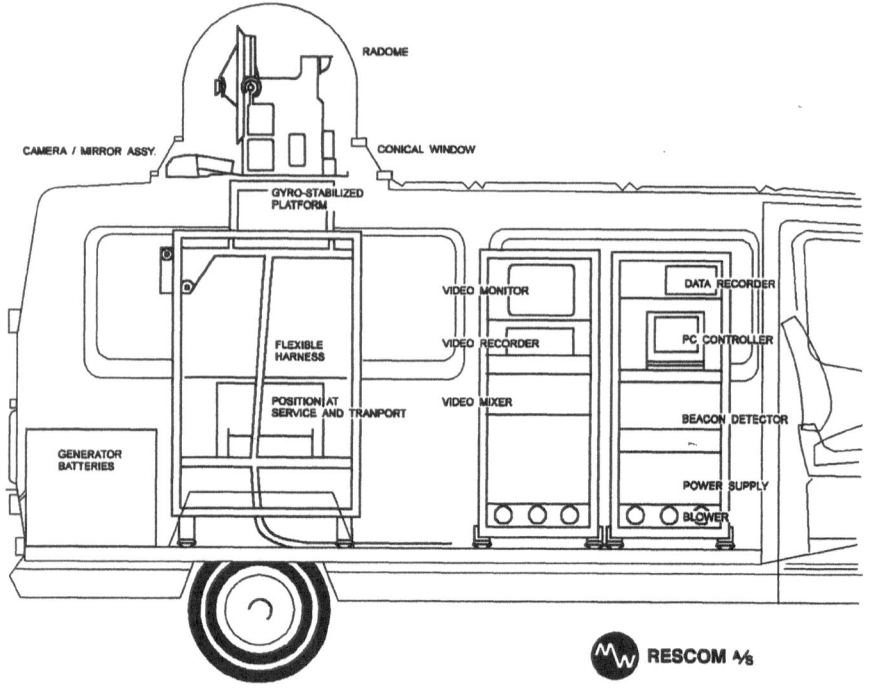

Fig. 3 - Land Mobile Beacon Terminal (van installation)

The final LMBT configuration presently consists of the following subsystems [7]:

- Antenna: Cassegrain-type, with rotatable feed and radome

- Pointing: gyro-stabilized elevation-over-azimuth platform (~ 0.6 dB maximum pointing losses)

- Receiver: RF units mounted on the platform and a rack-mounted detector unit (LNA gain of 28.0 dB, bandwidth of 1 GHz, NF of 1.4 dB)

- DAS: high capacity tape streamer (5 Gbytes) for data recording and computer-assisted acquisition and measurement data handling (sampling rate selectable from 1 to 15 kHz)

- CCS: operator PC (Intel 386 CPU, 33 MHz clock, 2 Mbytes RAM)

- Video System: two colour CCD cameras and a video mixer

- Power System: motor/generator unit (2 kW), plus a 24 V battery pack

- Van: Mercedes-Benz 208/33D passenger, with air conditioning unit

Measurements of the propagation channel characteristics

This first series of measurements will be carried out in the metropolitan area of Turin, with the purpose to assess the propagation characteristics (essentially, shadowing effects) of a mobile satellite transmission channel at 18 GHz in various topographical environments. Moreover, since a considerable interest has been generated in finding electromagnetic deterministic solutions to field strenght coverage, multipath signals (and related echo delays) in microcellular environments, the above measurements could also contribute to the in-field validation of ARAMIS©, which is the main scope of the companion campaign, scheduled in Rome.

Due to the enormous range of possible topological configurations, the measurement environments have to be chosen in a systematic way with respect to a ranking into some representative classes; although there are not clear criteria that allow a well assessed classification, the following tentative categories could be used:

- LOS wide routes (suburbs, villages), without buildings;

- LOS routes, with buildings on both sides;

- NLOS (Non Line-Of-Sight) routes, where one or more buildings can affect the propagation.

- POLOS (Partially Obstructed Line-Of-Sight) measurements in irregular topological situations, such as squares, parks, etc. or routes with scattered buildings or trees: in this case the effect of vegetation will be in particular investigated.

In every case an accurate description of the environment will be given: width of the streets, height of buildings, azimuth and elevation angles, presence and type of vegetation, etc. The effects of foliage, in particular, will be investigated against a roadside trees model [8], schematically reported in Fig. 4. Both the tree trunks and branches are assumed as represented by structures of rectangular cross-section (thickness s_0 and height h) on both sides of the road, while the foliage is represented by a circular crown section structure, bounded by inner and outer circles of radii r_1 and r_2, respectively. Furthermore, two additional lateral rectangular structures (of thickness s_{01}) are also considered, to account for possible effects due to second order rows of trees. For example, according to the situation of Fig. 4, the attenuation experienced at 19.77 GHz by a mobile terminal running along the right side of the road (at 3 m from the first row of trees) has been computed (see Fig. 5 for the values of the parameters), for various source elevation angles, using specific attenuation values as suggested in [9].

It can be pointed out that, for elevation angles from 30 to 40 degrees (source on the right side of the road), the attenuation values are in the range 14 to 18 dB, whereas if the source is on the left side (elevation angles from 140 to 150 degrees), the experienced attenuation is considerably less, in the range 8 to 10 dB.

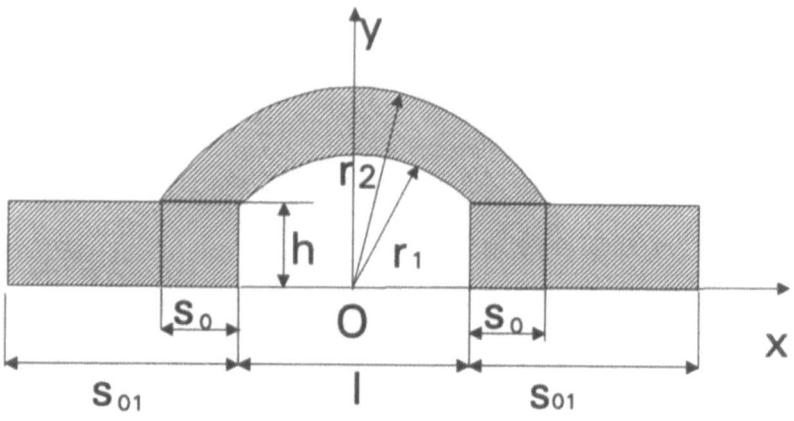

Fig. 4 - Modelling of a fully vegetated road
(Trunks, branches and foliage are represented by the shaded areas)

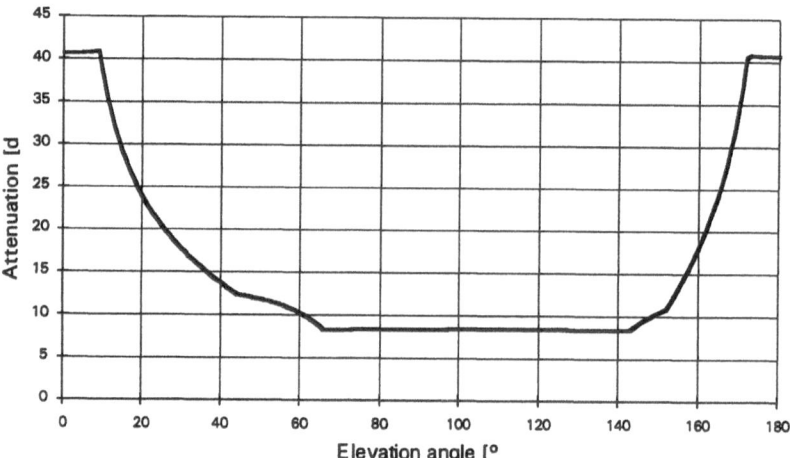

Fig. 5 - Attenuation due to vegetation (mobile terminal on the right side of the road)

Mobile antenna height: 1.5 m; Frequency: 19.77 GHz; Road width: 15 m; Trees height: 10 m; Distance of the terminal from the right side of the road: 3 m

Moreover, the potential effect of the vehicular traffic will examined too; at least in two typical situations (LOS and NLOS, respectively) measurements will be repeated in two different day hours, 9 a.m. and 10 p.m., when traffic conditions are

272

expected to be quite different, in order to point out the effects due to different reflection conditions by the street surface.

For a more specific characterization of the potential environments to be tested, reference will be made to the following working assumptions, recently identified in the framework of a study performed for DCS@1800 terrestrial microcellular systems [10, 11]. In that case, in order to get a rather complete classification of urban/suburban situations, four sample italian cities were examined (Milan, Turin, Novara, Gattinara), characterized by sizes, structures, population and traffic density quite different from one another, and supposed as reasonably representative of actual Italian urban environments. The first phase of the work was devoted to choose some topographical reference typologies (as listed in Table 1), namely areas which have different geometrical structures (different kinds of streets and squares, building types and densities, etc.); this allows to represent an actual rather complicated town as a composite jigsaw of such reference typologies [11]. Table 1 reports the adopted reference geometries; the high density term denotes an area with quite narrow streets (15m wide on average), whereas low density zones include avenues and boulevards (45m wide on average) with large distances between adjacent streets.

Table 1 - Reference typologies

1: Low density areas with perpendicular cross-roads.

2: High density areas with perpendicular cross-roads.

3: Low density areas with oblique cross-roads.

4: High density areas with oblique cross-roads.

5: Squares as enlargement of streets.

6: Squares as central node structures.

7: Low density areas with curved streets.

8: High density areas with curved streets.

9: Green and/or low built-up areas.

10: Specific cases (bridges, over passes, subways, etc.)

11: Industrial areas, sheds, barracks.

12: Hilly zones and undulating terrain.

13: Rural (not built-up) vegetated areas.

The four sample towns have been analysed with the aid of land register maps (scale 1:5000, or better), which provide the required resolution. Such maps have been divided by means of a square grid with a pixel size of 200m; for each element of the grid, a reference typology was extracted.

The histogram of Fig. 6 shows the relative occurrence frequencies of the typologies, evaluated for the whole urban area of the four towns: it is interesting to note that if typologies 9, 12, 13 are removed, the final result (see Fig. 7) is that all the four cities show a very similar mix of structures.

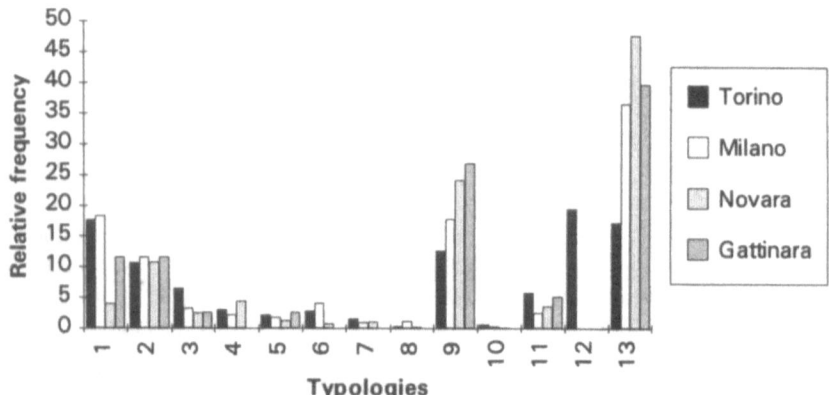

Fig. 6 - Relative occurence frequency of urban/suburban typologies

Therefore, measurements performed, for example, in the metropolitan area of Turin can be easily considered as representative of most situations typically encountered in urban/suburban environments of Italian towns.

Statistics

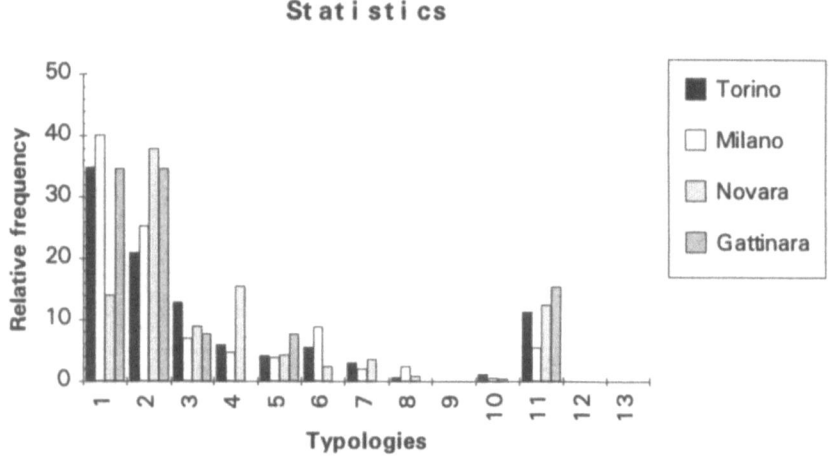

Fig. 7 - As in Fig. 6, but with typologies 9, 12, 13 removed

The ARAMIS© validation campaign

The second phase of the measurement campaign will be held in Rome. At this stage of the activity, the SW tool ARAMIS© will intensively be used as a support for predicting the Directions of Arrival (DOA) from which the e.m. power is expected to reach the RESCOM terminal, so as to be able to point the antenna towards the satellite or the investigated scattering source.

A recently built quarter has been selected for measurements, being characterized by large green areas, by building blocks made up of concrete, having different number of floors, and by offices having metallic structures and large glass windows. The presence of two underpasses (with concrete vault) will allow the assessment of the signal transmission also in such conditions. The selected area has already been inserted in ARAMIS© Urban Modeler, entering the building coordinates and the streets layout, taken from land register maps and aerial photography, by means of a digitizer. It has still to be assessed if this information will allow the identification of DOAs with the accuracy needed using the 2.4° beamwidth RESCOM antenna.
Otherwise, a more accurate geometric description of the zone will be obtained by on site inspection; moreover, a preliminary test on the dynamic range of the receiving equipment has to be carried out.

Different azimuth angles will be available during RESCOM envisaged routes, whereas ITALSAT elevation as seen from Rome is about 41.6 degrees. The main purpose of the campaign will be the checking at K-band the e.m. model used in ARAMIS© for the elevation of the signal received at a mobile terminal, when the Line-Of-Sight path from satellite is obstructed by the buildings. The collected set of experimental results will be used to create a data base of transmittivity coefficients for buildings of different geometrical shape and made up of different materials. As soon pointing problems related to the small antenna beamwidth will be solved, building reflectivity and other rays contributions in various conditions will be measured. Experimental data will possibly be available for investigating the correctness at K-band of the specular wall reflection assumptions. After validation campaign completion, the set of measured results will be compared with responses provided by ARAMIS©, this phase being expected to stimulate a refinement of the parameters characterizing the e.m. behaviour of the materials used in the simulator. As a matter of fact, ARAMIS© can model walls either as dielectric slabs or transmitting/reflecting media, but there is still a lack of information about the parameters to be used at K-band (complex dielectric permittivity in a former version, complex reflectivity and transmittivity in the latest).

Finally, a full Narrow Band Analysis will be performed using ARAMIS© both on measured and computed signal time series, includind 1st (PDF & CDF) and 2nd order statistics (Average Fade Duration, Level Crossing rate, Time Share and Distribution of Fades and of Connections).

Conclusions

An overview has been given about the software tool ARAMIS©, devised to assist a land mobile satellite system planner, and capable to provide the necessary information (coverage, echo delays, bandwidth distortions) using a deterministic approach (integrated by a GTD solver) to evaluate local field strength values.

Two companion measurement campaigns at K-band frequencies were also outlined, to be carried out in Turin and Rome respectively, in order to assess the propagation characteristics of the high frequency/quasi mm-wave mobile satellite transmission channel in different terrestrial environments, and to get an in-field validation of the software tool. The two measuremet sets will made use of an ad hoc Land Mobile Beacon Terminal, developed by RESCOM under ESA contract, whose main technical characteristics are described in some detail.

Acknowledgements

The authors wish to thank Ms. L. D'Amato (CSELT) for providing the roadside tree attenuation model and relevant computations, Mr. C. Mattiello (CSELT) for his invaluable contribution to drawing and illustrations, and Mr. E. Saggese (Space Eng.) for the helpful discussions; a sincere appreciation to ESTEC people of the Electromagnetic Division for their availability to supply the Mobile Terminal for the measurement campaigns.

References

1. Hess GC. Land mobile satellite path loss determination. Final report (3rd year ATS-6 Experiment), Motorola Inc. 1978
2. Anderson RE. et al. Satellite aided mobile communication: Experiments, applications and prospects. IEEE Transactions 1981; VT- 30, 2:54-64
3. CCIR. Factors affecting the choice of antennas for mobile stations of the land mobile-satellite service. Rept. 925-1, XVth Plenary Assembly, Düsseldorf, 1990
4. Propagation model for the Land Mobile Satellite radio channel in urban environment. ESA Contract 9788/92/NL/(SC), Final Report, vol. 1, 1993
5. Di Bernardo G. et al. Propagation model for the Land Mobile Satellite radio channel in urban environment. In: Proceedings of 3rd ESA Workshop on EMC&CEM, 1993
6. Propagation model for the Land Mobile Satellite radio channel in urban environment. ESA Contract 9788/92/NL/(SC), Final Report, vol. 2, 1993
7. Sforza M, Buonomo S, Arbesser-Rastburg B. Channel characterisation for future Ka-band Mobile Satellite Systems and preliminary results. In: Proceedings of XVIII NAPEX Conf., 1994
8. D'Amato L Private communication, June 1994
9. CCIR. Propagation over irregular terrain with and without vegetation. Rept. 1145, XVth Plenary Assembly, Düsseldorf, 1990
10. Daniele P et al. Propagation models for microcellular environment. In: Atti del VI Convegno Sistemi Radiomobili, Pontecchio Marconi, Bologna, (I), 1993
11. Lo Gatto E, Perucca M. Authority maps utilization for urban microcellular coverage. CSELT Techn. Repts. 1993; vol. XXI, n. 5:983-994

Advances in Modulation and Coding for the Satellite Mobile Channel

E. Biglieri[*]

Dipartimento di Elettronica • Politecnico di Torino • Torino (Italy) Phone:
+39 11 5644030 • Fax: +39 11 5644099
e-mail: biglieri@polito.it

Abstract

This review paper summarizes some of the recent advances in coding and modulation that may find application to the satellite mobile channel. They include unequal error protection, block detection, turbo-codes, and joint source-and-channel coding.

1 Introduction

In the perennial trade-off between performance and complexity, the recent advances of digital circuitry keep on moving up the threshold of an acceptable complexity. As a consequence, better and better system performance is made available to communication systems. We may categorize the achievable improvements as follows:

- Better overall performance through more sophisticated coding schemes for which optimum (or marginally suboptimum) decoding schemes exist with a reasonable complexity. Among these schemes we may mention turbo-codes, product codes, and generalized concatenated codes.

- Better overall performance by making use of all the available information that can be extracted from the transmitted signal (joint source and channel coding).

- More robust performance by allocating the error-control resources only where they are specially needed (unequal error protection).

- Better receiver performance for the same system choices, by making use of more sophisticated detection algorithms.

In this paper we discuss some of these recent solutions.

---

[*]This work was supported under contract by the Italian Space Agency (ASI).

2 Unequal error protection

The need for unequal error protection (UEP) stems from the recognition that in the transmission of digital speech or digital TV only a fraction of data is extremely sensitive to channel errors. Consequently, it is a waste to provide uniform error protection to all the bits transmitted, and optimum utilization of transmission resources requires unequal error protection. Moreover, for transmission over channels that might be affected by fading and shadowing, and consequently provide a transmission quality that changes with time, it is important that at least the most important data be reiceived error-free even in the worst channel conditions, so that even when the quality of the signal received is poor, the decoded signal will be degraded, but not catastrophically.

While early work on unequal error protection has focused on use of error-control codes (both block and convolutional: see [18, 26] for recent advances in the field), more recent activity in this area was aimed at designing a (possibly coded) modulation scheme matched to the need for UEP (see, e.g., [27, 29]).

Here we describe two different solutions to the problem of achieving UEP.

2.1 UEP for coded DS-CDMA mobile radio

Hoeher [20] has recently described a family of techniques that achieve UEP in coded and uncoded direct-sequence code-division multiple access (DS-CDMA) mobile radio systems. These techniques, that may also be combined, include different channel code rates, different number of chips/ symbol or different symbol durations, and different symbol energies. It is assumed that the source bits are sorted, according to their significance, in J blocks of n_j bits each, $1 \leq j \leq J$. The source data are first encoded by a rate-R convolutional encoder, then symbol-by-symbol interleaved, differentially encoded, and further spread by a pseudo-noise code with N chips/ symbol. The processing gain is the bandwidth expansion N/R.

The first technique consists of varying the rates R_j used to encode the different classes of bits. Rate-compatible punctured convolutional codes [15] can be used to increase the code rate, while symbol repetition decreases it. UEP may be achieved by selecting R_j and adjusting N_j so as to obtain a constant ratio N_j/R_j, or by selecting R_j and using the same N for all j. With the second technique UEP is achieved by varying N_j ("UEP by different symbols durations"). The third technique consists of varying the symbol energies. A comparison among these techniques and their performance can be found in [20].

2.2 UEP by multilevel coded modulation

Wei [29] advocates use of two-level coded modulation [7], where the two classes of "important" and "less important" data are separately encoded by two convolutional encoders. The signal constellation is non-uniformly spaced, obtained as the direct sum of two two-dimensional constellations (e.g., 4-PSK and 8-QAM), whose relative energies are selected so as to achieve the desired degree of unequal error protection. Decoding is done in two stages in order to guarantee a limited complexity. The result is a range of coding gains "that can hardly be achieved using conventional coded modulation with equal error protection" (up to 10 dB at BER=10^{-3} for the important data).

3 Block detection

The idea of block detection stems from the fact that in the presence of disturbances that do not act independently from symbol to symbol, operation of a symbol-by-symbol demodulator is not optimum, and a maximum-likelihood sequence detector (MLSD) should be used instead. When the latter is too complex for implementation, suboptimum solutions should be sought that have a reduced performance penalty. One of these is based on the concept of block detection. This is performed by detecting independently blocks of more than one symbol, so that the detection process will be improved by a longer observation time that smooths out the disturbances.

A number of applications of the block detection concept can be found in the recent literature, with results that show the extent to which block detection can improve upon standard detection at the price of a moderate increase in receiver complexity.

The first reference known to the author in which the concept of block demodulation is applied to detection of M-ary PSK is [22]. Subsequently, Leib and Pasupathy [21] observe that a reason for the performance loss of differentially coherent relative to coherent detection of PSK is that its phase reference is impaired by channel noise in the same way as the information phase. Thus, a receiver is proposed where the phase reference is obtained by averaging the past phase references. These are obtained by removing the information phases from past symbols, and hence requires past decisions to be fed back. The performance of this new receiver for M-PSK and for MSK is found to bridge the gap between coherent and differentially coherent detection.

In [11], analysis of block demodulation is carried out for uncoded M-PSK over the additive white Gaussian noise (AWGN) channel. If N denotes the number of symbols in the observation interval, it is proved that as $N \to \infty$ the bit-error-rate (BER) performance of block-detected M-PSK approaches that of ideal coherent-detection with differential encoding, and that this limiting performance is approached with small values of N. In particular, for binary PSK, extending N from 2 to 3 recovers more than half the power loss of differential versus coherent detection with differential encoding. For 4-PSK, the power gain achieved by increasing N from 2 to 3 is more than 1 dB.

In [12], the analysis of block demodulation over the AWGN channel is extended to trellis-coded M-PSK. Use of multiple trellis-coded modulation (TCM) [7, Chap. 7] with multiplicity (i.e., trellis-coded symbols per input symbol) equal to $N-1$ in combination with block detection yields significant performance improvement over conventional detection even for small N.

Ho and Fung [19] study uncoded QPSK with a correlated Rayleigh fading channel model. Here the signal s_k is multiplied by a complex random variable u_k, a sample of a zero-mean correlated Gaussian process. They find that block detection is a very effective strategy for eliminating the irreducible error floor associated with a conventional differential detector. In all the cases investigated in [19] a detector with N as small as 3 is sufficient for this purpose. Moreover, it is discovered that block detection is not sensitive to the mismatch between the decoding metric and the channel fading statistics.

Divsalar and Simon [14] consider a wide variety of situations where block detection may be succesfully used to improve performance over conventional differential detection. In particular, they examine uncoded and trellis-coded M-PSK and QAM transmitted over Rayleigh and Rice fading channels, with unknown slowly and fast-varying phase, and with or without channel-state information. The channel is endowed with a perfect (i.e., infinite-depth) interleaver.

Block detection may also be applied to modulation schemes other than PSK. In [13], uncoded full-response continuous-phase modulation (CPM) is examined. Again, it is shown that in the limit as $N \to \infty$ the BER of the block detector approaches that of the coherent receiver. The special case of GMSK is dealt with in [1]. Here it is shown that for $N \geq 5$ the BER performance of block detection comes close to ideal (i.e., maximum-likelihood sequence) detection by a few tenths of one dB.

A generalization of block demodulation is presented in [24]. The receiver here is a sliding-block detector. It processes a block of N consecutive received samples and makes a decision about the symbol in the middle of the block. Once a decision has been made, the block is moved ahead by one symbol interval. The complexity of this receiver may be reduced if the previous K decisions are fed back to the decoder. Analysis of this structure for differentially encoded PSK over the 2-ray frequency-selective Rayleigh fading channel shows that with $N = 5$ its performance is superior to conventional differential detection, with no error floor in nearly all the cases examined.

In [6], the concept of block detection is applied to Double-Differential PSK. This modulation scheme was first introduced in the early 1970's, and recently resurrected because it provides insensitivity to frequency offsets, due for example to Doppler effect. While with DPSK the transmitted phases are associated with differences of information phases

$$\theta_i - \theta_{i-1},$$

with standard DDPSK they are associated with differences of differences, i.e.,

$$\theta_i - 2\theta_{i-1} + \theta_{i-2}.$$

A problem arising with symbol-by-symbol double differential detection is that in the presence of noise it causes a high noise correlation in the receiver output, which results in a considerable performance degradation (on the order of 3–4 dB) with respect to conventional differential detection. A significant portion of the performance degradation (about a couple of dB) can be bought back by modifying the encoding scheme so that the transmitted phases are associated with the modified differences

$$\theta_i - \theta_{i-1} - \theta_{i-2} + \theta_{i-3}.$$

Simulation results obtained over an AWGN channel and a Rician-fading channel have shown that block detection has the capability of reducing the gap between DDPSK and PSK.

4 Joint source and channel coding

In contrast to the separation of channel and source encoding and decoding, there may be some advantage in doing joint source and channel coding [16]. In fact, often nonstationary sources cause the source-encoded symbol stream to contain residual redundancy, which causes them to be highly correlated. In these conditions it might be advantageous to pass "source significance information" from the source coder to the channel coder. On the receiving side, "source-controlled channel decoding" uses a-priori and a-posteriori information (obtained from the source coder and the source decoder, respectively) about the source bits. These techniques share one fundamental principle, of "telling the receiver as much as you know about the source" (and possibly about the channel).

5 Turbo codes and staged decoding

It has long been known that coding schemes can be designed which can approach the theoretical performance limit usually referred to as "Shannon bound." However, from a practical point of view the search for these codes has been discouraged by the fact that "good codes are messy," i.e., the more powerful a code is, the more difficult it is to decode it.

Thus, while the effectiveness of powerful error-control schemes in digital communications is undisputed, a major obstacle to their use is the complexity of the decoder. The complexity barrier has motivated many suboptimum detection algorithms with the usual price of performance degradation, or the choice of coding schemes that do not exploit in full the potential of coding theory.

Recently, a leap forward in the search for codes that are both good and practical has been made through the discovery of error-control coding schemes based on the concept of *generalized concatenated coding* [9, 30]. These schemes are especially powerful (many among them improve upon previously known codes) and are endowed with a special structure that allows practical decoding. The complexity, which may prevent real-time operation of the decoder (particularly if optimum, i.e., soft-decoding, algorithms are used) is reduced by using parallel algorithms and structures that are suitable for VLSI implementation [5, 10]. These codes may provide a robust transmission scheme suitable for the most modern digital satellite environments like (but not restricted to) mobile radio communication or personal-access systems. Moreover, they may provide unequal error protection at no additional cost in complexity.

More recently, the concept of *iterated decoding* has been developed. This improves upon plain staged decoding, by leaving the receiver structure invariant but feeding back signals from a decoder stage to the previous one [8]. Successful applications of this concept include the recently discovered *turbo-codes*, whose performance is surprisingly good (but whose behavior has not been fully understood yet), and the soft-decoding of the well-known *product codes* (see [2, 3, 4, 23, 25, 28]).

References

[1] A. Abrardo, G. Benelli, and G. Cau, "Multiple-symbols differential detection of GMSK," *Electronics Letters*, Vol. 29, No. 25, pp. 2167–2168, 9th December 1993.

[2] G. Battail, C. Berrou, and A. Glavieux, "Pseudo-random recursive convolutional coding for near-capacity performance," *Proc. Communication Theory Miniconference, IEEE Globecom'93*, Houston, TX, pp. 23–27, Nov. 29–Dec. 2, 1993.

[3] C. Berrou and A. Glavieux, "Turbo-codes: General principles and applications," *6th International Tirrenia Workshop on Digital Communications*, Tirrenia, Italy, September 1993.

[4] C. Berrou, A. Glavieux, and P. Thitimajshima, "Near Shannon limit error-correcting coding and decoding: Turbo-codes," *Proceedings of ICC'93*, pp. 1064–1070, Geneva (Switzerland), May 1993.

[5] E. Biglieri, "Parallel demodulation of multidimensional signals," *IEEE Trans. Commun.*, Vol. 40, No. 10, pp. 1581–1587, October 1992.

[6] E. Biglieri, E. Bogani, and M. Visintin, "Block demodulation — An overview," *ISSSTA'94*, Oulu (Finland), July 4–6, 1994.

[7] E. Biglieri, D. Divsalar, P. J. McLane, and M. K. Simon, *Introduction to Trellis-Coded Modulation with Applications*. New York: Macmillan, 1991.

[8] E. Biglieri and A. Spalvieri, "Generalized concatenation: A tutorial," in: E. Biglieri and M. Luise, (eds.) *Coded Modulation and Bandwidth-Efficient Transmission*. Amsterdam: Elsevier, 1992.

[9] É. Blokh and V. V. Zyablov, "Coding of generalized concatenated codes, *Problemy Peredachi Informatsii*, Vol. 10, No. 3, pp. 45-50, July-September 1974.

[10] G. Caire, J. Ventura-Traveset, and E. Biglieri, "High-speed staged decoding for BCM: Systolic array solutions," *to be published, 1994*.

[11] D. Divsalar and M. K. Simon, "Multiple-symbol differential detection of MPSK," *IEEE Trans. Commun.*, Vol. 38, No. 3, pp. 300–308, March 1990.

[12] D. Divsalar, M. K. Simon, and M. Shahshahani, "The performance of trellis-coded MDPSK with multiple symbol detection," *IEEE Trans. Commun.*, Vol. 38, No. 9, pp. 1391–1403, September 1990.

[13] M. K. Simon and D. Divsalar, "Maximum-likelihood block detection of noncoherent continuous phase modulation," *IEEE Trans. Commun.*, Vol. 41, No. 1, pp. 90–98, January 1993.

[14] D. Divsalar and M. K. Simon, "Maximum-likelihood differential detection of uncoded and trellis coded amplitude-phase modulation over AWGN and fading channels – Metrics and performance," *IEEE Trans. Commun.*, Vol. 42, No. 1, January 1994.

[15] J. Hagenauer, "Rate-compatible punctured convolutional codes (RCPC codes) and their applications," *IEEE Trans. Commun.*, Vol. COM-36, pp. 389-400, April 1988.

[16] J. Hagenauer, "Source-controlled channel decoding," *submitted for publication*, May 1993.

[17] J. Hagenauer and P. Hoeher, "A Viterbi algorithm with soft-decision outputs and its applications," *GLOBECOM'89*, Dallas, TX, pp. 47.1.1–47.1.7, November 1989.

[18] J. Hagenauer, N. Seshadri, and C.-E. W. Sundberg, "The performance of rate-compatible punctured convolutional codes for digital mobile radio," *IEEE Trans. Commun.*, Vol. COM-38, pp. 966–980, July 1990.

[19] P. Ho and D. Fung, "Error performance of multiple-symbol differential detection of PSK signals transmitted over correlated Rayleigh fading channel," *IEEE Trans. Commun.*, Vol. 40, No. 10, pp. 1566–1569, October 1992.

[20] P. Hoeher, "Unequal error protection for digital mobile DS-CDMA radio systems," *Proc. IEEE ICC'94*, New Orleans, LA, May 1–5, 1994.

[21] H. Leib and S. Pasupathy, "The phase of a vector perturbed by Gaussian noise and differentially coherent receivers," *IEEE Trans. Inform. Theory*, Vol. 34, No. 6, pp. 1491–1501, November 1988.

[22] P. Y. Kam, "Maximum-likelihood digital sequence estimation over the Gaussian channel with unknown carrier phase," *IEEE Trans. Commun.*, Vol. 35, No. 7, pp. 764–767, July 1987.

[23] Y. Li, B. Vucetic, and Y. Sato, "Optimum soft-output detection for channels with intersymbol interference," *IEEE Information Theory Workshop*, Susono, Japan, June 1993.

[24] W. Liu and P. Ho, "On multiple-symbol differential detection of PSK signals in frequency selective fading channels," *Unpublished manuscript*, 1993.

[25] J. Lodge, R. Young, P. Hoeher, and J. Hagenauer, "Separable MAP 'filters' for the decoding of product and concatenated codes," *Proceedings of ICC'93*, pp. 1740–1745, Geneva (Switzerland), May 1993.

[26] D. G. Mills and D. J. Costello, Jr., "Using a modified transfer function to determine the unequal error protection capabilities of convolutional codes," *IEEE International Symposium on Information Theory*, San Antonio, TX, January 1993.

[27] N. Seshadri and C.-E. W. Sundberg, "Multi-level block coded modulation with unequal error protection for the Rayleigh fading channel," *European Transactions on Telecommunications*, Vol. 4, No. 3, pp. 325–334, May-June 1993.

[28] P. Thitimajshima, *Les Codes Convolutifs Récursifs Systématiques et Leur Application à la Concatenation Parallèle*, Ph.D. Thesis, Université de Bretagne Occidentale (France), December 21, 1993.

[29] L.-F. Wei, "Coded modulation with unequal error protection," *IEEE Trans. Commun.*, Vol. 41, No. 10, pp. 1439–1449, October 1993.

[30] V. A. Zinov'ev, "Generalized concatenated codes," *Problemy Peredachi Informacii*, Vol. 12, pp. 5–15, January–March 1976.

Part V

Panel papers

Mobile/Personal Satcoms System Alternatives - Satellite and Network Aspects

Josef F. Huber - Siemens AG, Mobile Radio Networks Division
Munich, Germany

1 Introduction

Present satellite networks are built up as extension to public networks. They provide subscriber access via satellite for those areas where telecommunications do not exist. The switching functions are relatively simple, the radio interfaces have high signal delay and require high power. Such systems will not be acceptable in a future in which personal and mobile communications dominate the wireless market in many countries of the world. Subscribers will compare with mobile standards like GSM expecting a higher level of service. If such a level can be achieved with satellites the demand for satellite based Personal Communications Networks (PCNs) will increase. The targets should be: Global coverage, personal numbering/roaming, voice and data, pager services, handhelds and lower tariffs.

Such networks should be autonomous public mobile networks with global coverage and personal mobility around the world. The enormous progress in digital cellular technology makes it possible to think about bringing a cellular network up into the orbit and to build global personal networks there. However, to put the entire network into the orbit with one technological step is by far too risky and a complicated task. For lowering the risks it seems feasible making smaller steps by assigning certain network elements either to the orbit or to the ground. The present satellite-PCN programs Iridium, Globalstar and others show that this will already happen before the year 2000 and they go different ways in their implementations. There are system alternatives right from the beginning: If we look to the satellite part there are GEO, MEO and LEO systems with their advantages and disadvantages. These solutions will lead to specific requirements to the radio part in the orbit and on the ground. As known TDMA and CDMA principles will be used in the different satellite PCNs. These principles have influence on the network side, on the switching part and on network management.

The issue of subscriber terminals and integration of satellite PCNs into a future Universal Mobile Telephone System (UMTS) is another challenge influencing system alternatives in a positive or negative way. There are technical areas in which existing standards can be adopted and put into a global satellite PCN concept.

2 Satellite Constellations

The constellation of satellites determines the main characteristics of the satellite PCN. Figure 1 shows geostationary (GEO), medium orbit (MEO) and low orbit (LEO) solutions, the table in figure 2 provides some characteristic data.

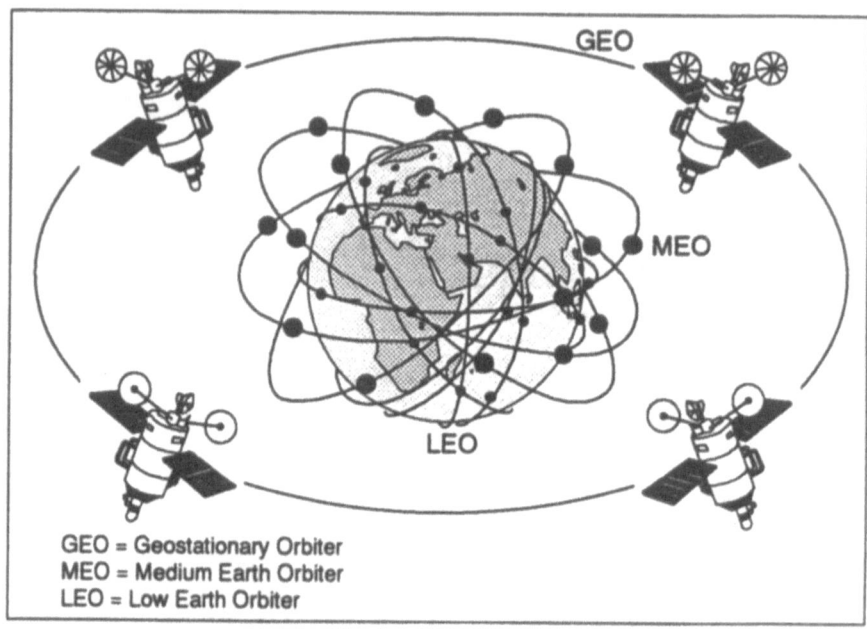

GEO = Geostationary Orbiter
MEO = Medium Earth Orbiter
LEO = Low Earth Orbiter

Figure 1: Alternative system solutions satellite technology

	GEO	MEO	LEO
Altitude	36,000 km	10,000 - 15,000 km	700 - 1,500 km
Number of satellites	3 - 4	10 - 15	> 40
Number of cells Signal delay	≤ 800 > 300 ms	~ 800 150 ms	> 3000 < 50 ms
Costs/Coverage	low	medium	high
Subscriber capacity versus bandwidth	low	medium	high

Figure 2: Characteristics of alternative satellite constellations

It can be seen that the most attractive points about GEO systems are cost savings and about LEO systems the attractive short signal delay and high system capacity. MEO systems seem to be more a compromise between both than a better solution.

The important question is:

Are LEO systems with their technical advantages economically feasible?

3 Radio and Communications Transport System

According to the decision in the radio conference WARC '92 the assigned frequencies in the 1.6 and 2.1 GHz range allow for satellite PCNs a maximum bandwidth of 16.5 MHz uplink and 16.5 MHz downlink. The upcoming project Teledesic belongs to the broadband radio networks and lies in the 20/30 GHz range. It certainly is a real alternative to the satellite PCNs presently under development. It focusses another timeframe beyond the year 2005 and uses the ATM switching and transport principles with radio ATDMA. I avoid to compare this contrast solution and rather concentrate on satellite PCNs planned for 1998 - 2000.

These networks already show alternatives with enormous impacts on the economical, the technical and the regulatory side. The table in figure 3 shows the networks presently under development.

As we can see from the table there is one network 'Iridium' that has on-board processing. This means that the base transceiver station known from terrestrial mobile networks is actually operating up in the orbit. All the intelligence needed for a base transceiver station has to be controlled via the ground station; also functions like software download and radio management have to be handled via the K-Band transmission links with the ground station.

Regarding the frequency bandwidth there presently are discussions in the FCC Groups. The different radio interfaces TDMA and CDMA result in the question of subscriber capacity. As known there exists a frequency assignment regarding TDMA and CDMA operators. CDMA operators can share a common bandwidth of 11.4 MHz: Fullband sharing. TDMA operators get segments of frequency band. At present the FCC discusses 5 MHz to be assigned to Iridium.

There is another topic which highlights Iridium compared with other systems. It is that of intersatellite links - the first global transmission system in the orbit.

	Globalstar	Odyssey	Inmarsat P	Iridium
Satellite Constellation	LEO	MEO	MEO	LEO
Radio Interface	CDMA	CDMA	TDMA/FDMA	TDMA/FDMA
Power/ Sat (Watt) Terminal (Watt)	1000	1800	3780	1200 < 0.39
Frequency band (MHz)	1610 - 1626.5 2483.5 - 2500	1610 - 1626.5 2483.5 - 2500	1980 - 2010 2170 - 2200	1610 - 1626.4
Bandwidth	11.4 MHz Shared with other CDMA operators		5 - 10 MHz	5 MHz
Number of cells	768	~ 800	800	3168
On-board processing	no	no	no	yes
Switch interconnections	terrestrial	terrestrial	terrestrial open	intersatellite global
Coverage	1) Northern hemisphere 2) Global	1) Northern hemisphere 2) Global		

Figure 3: Radio part of some satellite PCNs

The questions for the radio part are:

- Voice and data quality and transmit power for the handheld?

- Are Intersatellite links important for global PCN or not?

4 Network Services and Switching

The main target for a satellite PCN is to provide personal communications services worldwide for voice, fax and data. Some operators also intend to offer a worldwide pager service.

Personal communications means a number of service features which are defined today either in GSM or in IN standards. It is recommendable to compare system alternatives on that basis.

Taking GSM features brings the satellite PCNs closer to the cellular mobile networks on the ground. Taking Intelligent Network features means a closer relation to the wireline networks. The present problem to discuss alternatives here is that there is not enough information available on the planned satellite PCN features. There is, however, one basic difference that can be discussed: the way how subscriber connections are switched through the network. Figures 4 and 5 show two examples.

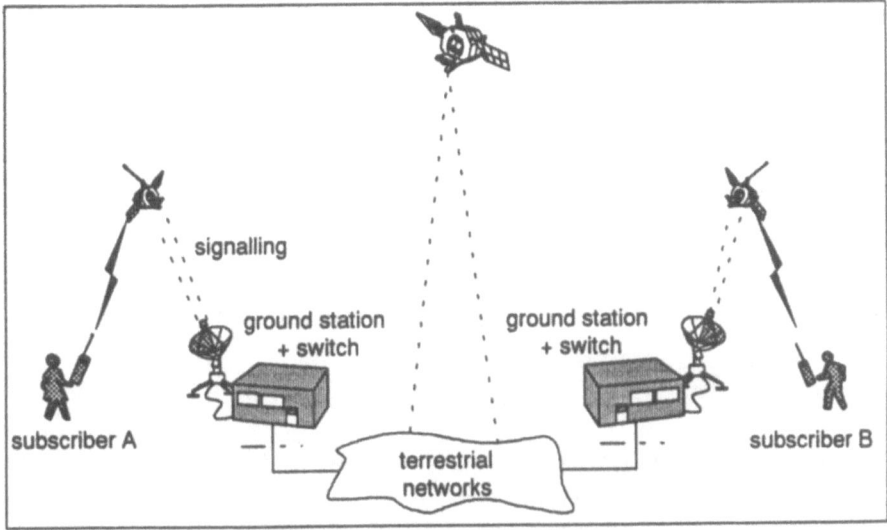

Figure 4: Network connections - example Global Star

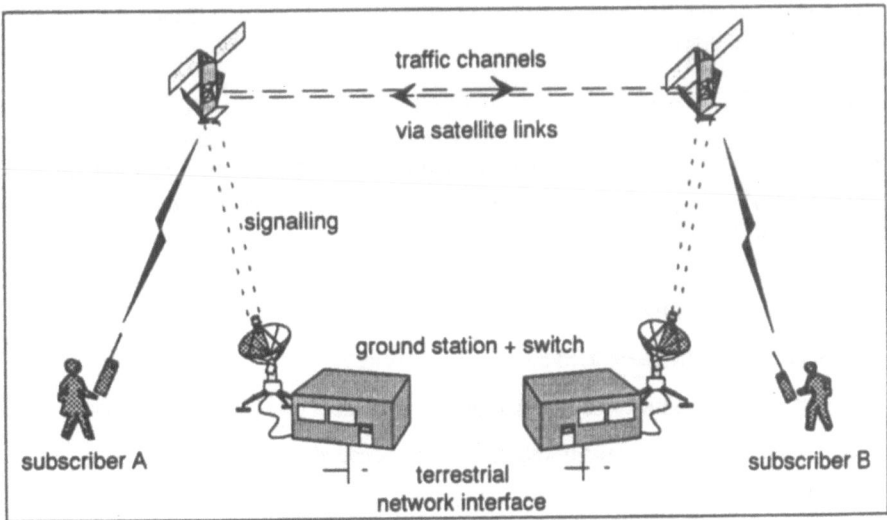

Figure 5: Network connections - example Iridium

The first one has considerable unpredictable signal delays because the traffic channel has to be switched down to the ground and put through wireline networks. If a satellite path is in between these switches there would be a considerable impact on the signal quality. This is of relevance especially for voice and fax services.

The second example provides a guaranteed signal quality between the subscribers A and B which is in the order of satellite interconnections. The analysis work done by Siemens shows this solution as the best of all alternatives known. Siemens as a leading supplier in the switching field contributes to the Iridium program with its worldwide expertise gained from a number of network deployments.

The questions arise:

- Which switching network alternative leads to good signal quality for network control, voice and data ?

- Which impact is there on the costs if terrestrial links are used?

- What solution optimizes interworking with terrestrial networks?

5 Terminals

All the present satellite PCN programs consider the dual-mode terminal. GSM as the most successful digital mobile standard is presently considered as the basis for such development. This means that 900 MHz handhelds (2 Watt) should be combined with 1600/2400 MHz high frequency parts and different digital processing functions. The same development could be done for USDC (United States Digital Cellular) and PDC (Personal Digital Cellular = Japanese standard) standards. In addition multimode solutions could be considered if the customer base would justify this development.

Iridium is the first satellite PCN introducing a sample device for a GSM-Iridium dual-mode application. Due to the fact that Iridium is using only the 1600 MHz frequencies for subscriber up and downlink with TDMA the combination with GSM TDMA in the 900 MHz range seems to be very economical. The CDMA alternative appears more complicated.

The questions are:

- Is it economically feasible to develop this device?

- What are the quantities of such handhelds to be sold on the market, the cost?

- Are weights and sizes of satellite handhelds acceptable by the user?

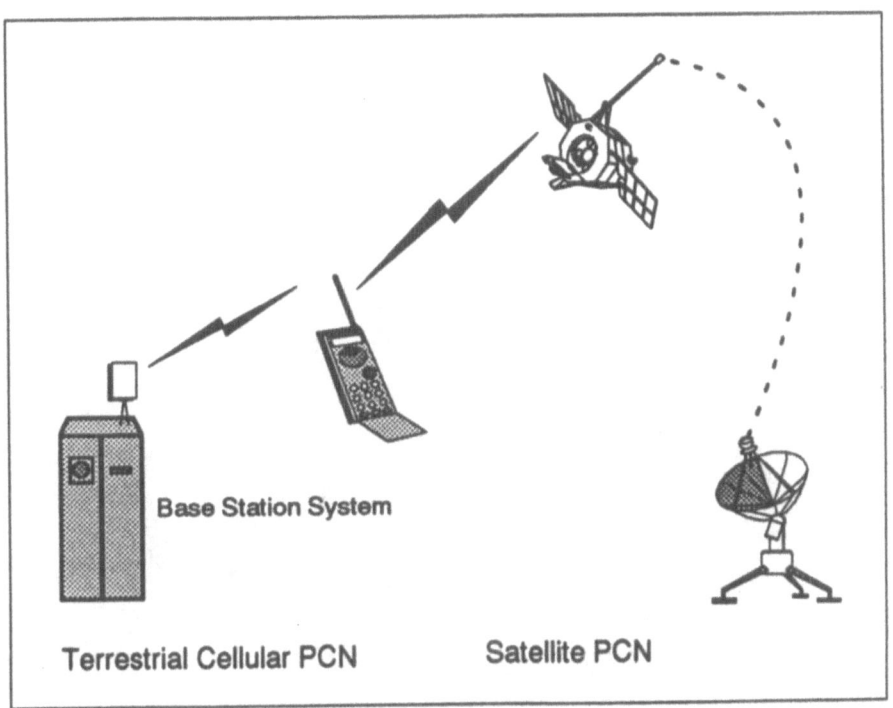

Figure 6: Main objective: Dual-Mode Terminal

6 Investment Costs

Figure 7 shows the presently known investment costs of the various satellite programs. Iridium appears with the highest investment. On the other hand, if compared with terrestrial cellular networks, Iridium seems to be more realistic than the other estimates. On the right side the investment figure estimated for the E1 Network in Germany for approximately the same number of subscribers (about 1 - 2 million subscribers and 3000 - 4000 cell areas) is shown. If we take the lower numbers of cells in the Globalstar and Odyssey networks into account we come to plausible results: e.g. Globalstar's 768 cells correspond to 1/3 Iridium which would correspond to 1.2 billion US dollars, actually Globalstar is planned with 1.7 billion US dollars.

This results in the question:

If we can build a worldwide network in the orbit with similar investment costs as for a national cellular network on the ground, why are we not forcing to do it?

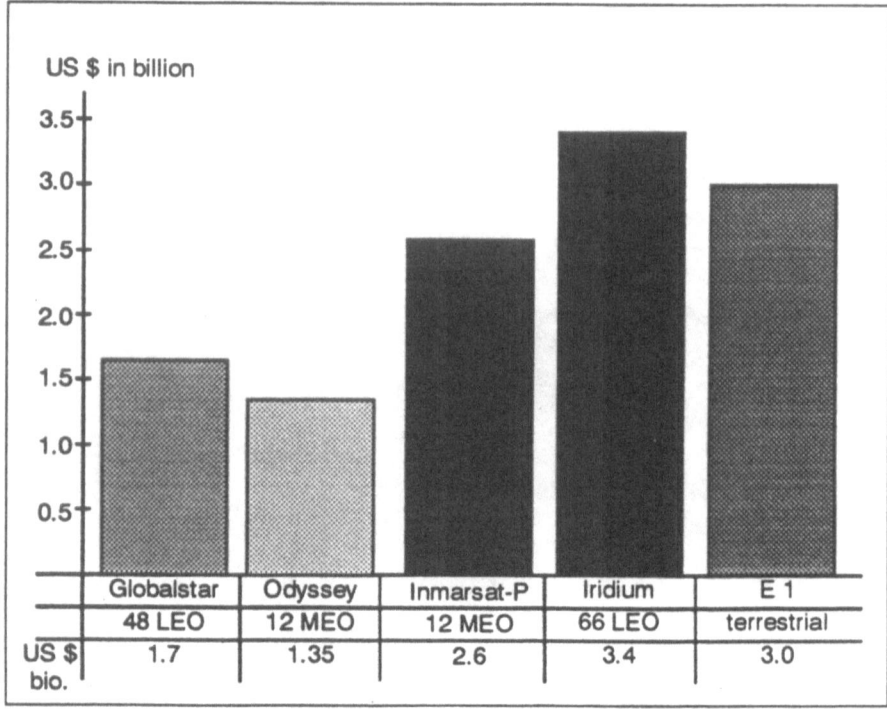

US $ in billion	Globalstar	Odyssey	Inmarsat-P	Iridium	E 1
	48 LEO	12 MEO	12 MEO	66 LEO	terrestrial
US $ bio.	1.7	1.35	2.6	3.4	3.0

Figure 7: Investment costs / examples

7 Summary of the Main Questions

1. Are LEO systems with their technical advantages economically feasible?

2. Are intersatellite links important for global PCNs or not?

3. Which switching network alternative is the best?

4. Is it desirable to put more satellite PCNs into the orbit, because the investment costs are comparable to national terrestrial cellular networks?

Market Prospective: Satellite Mobile Multi-media Services and Systems

Hanspeter Kuhlen
Deutsche Aerospace AG (DASA)
Munich, Federal Republic of Germany

1. The global Market for Mobile Telecommunication

US President Clinton recently declared the implementation of an American National Information Infrastructure (NII) as a key component of his administrations economic agenda. He said "The existence of an NII will not only demonstrate the capabilities of American technology on the verge to the 21st century but it will furthermore support the development in many areas of the American economy. Its importance is compared to the programme of building national highways and international airports in the United States forty years ago"[1].

The network should interconnect business areas, offices, homes, hospitals, education and academic sites. It shall provide information and communication to all groups of the society providing a non-discriminatory access to all areas of information. Due to the expected high demand on capacity the infrastructure mainly comprises glassfibre broadband networks. These networks will have radio link extensions for mobile services into the terrestrial arena (cellular) and into the global arena (Globalstar, Iridium, and others). If the other economic regions in Europe, Asia etc. decide to complement the effort the resulting network could eventually evolve into to a real global information infrastructure (GII). As a consequence, such an international network would inspire a large global market, create many new business and job opportunities since all products will be based on common air interfaces and standards.

Again, it is expected that beside having universal broadband networks for integrated services ranging from low to highest bitrates, there is a growing demand to provide even broadband services to mobile subscriber. Mobility and wide area coverage are the two factors a glassfibre network can never provide. In the mobile communications market, I would like to differentiate between two-way low bitrate services (GSM, Globalstar, etc.) and high bitrate unidirectional services.

The common denominator for both is the demand for small and easy to use media terminals, either as handhelds or lap-top/palm-top personal computers. In this panel paper as well as in the companion session paper to this conference, I will discuss the extension of the terrestrial networks by satellite with special emphasis on the latter category, a new contribution to the mobile scenario, a transmission system for Digital AudioBroadcasting (DAB®) to small user terminals named Archimedes[2]. The word "audio" in DAB is somewhat misleading, since it does not highlight the vast potential this standard offers for the provision of high-rate data services.

The DAB standard, defined by European experts in the last couple of years, has reached market maturity. Therefore, the market entry in 1995 as well as the degree and rate of penetration is definitely one of the most interesting issues of the next years.

Information and knowledge are growing fast. One of the tasks of future communications systems would be to make this information available to all who are interested. Either as raw data or processed; but in any case making it available. It is a matter of fact, that 90% of all scientists who ever lived are living today. The next fifteen years will have the same amount of scientific work as we had in the 2500 years since Aristotle's. It is of great importance that the results of all this work and the work of all other creative people gets a chance to be recognised. In addition, that information transfer is promoted among all people anywhere on the globe.

When talking about a future market, it is impossible to extrapolate from the past since private investments more and more substitute activities of former state monopolies. It would be somewhat the equivalent of comparing the mobile communications market before GSM and after. Over decades, mobile voice services were used only by a very small group of users who, for several reasons, were prepared to pay highest prices for a comparatively modest service. User terminals were not only expensive to buy but required also "appropriate" cars to provide the needed environment in terms of power consumption and installation of bulky hardware.

The total capacity of mobile networks were fairly limited and hence, the subscriber community remained small. Due to lack of competition in the terrestrial networks, highest user fees also kept the demand from subscribers low. This was also true for the satellite based mobile communications market. Today, there are about 30.000 users which are served in the present Inmarsat networks with voice services.

With the advent of sophisticated micro miniaturisation, technology became available which already today, but even more in the near future, allows smallest terminals with smartest performance characteristics. Now, with the key technologies at hand combined with an increasing trend towards liberalisation of markets (GATT) and the termination of telecommunication monopolies in many European countries by 1998, we are probably at the beginning of a new and vastly growing global telecommunications market.

The mobile market for narrow band voice and data services is predicted to serve between three and ten million subscribers, ranging from very pessimistic to very optimistic forecasters respectively, maybe both are wrong and the markets behave totally unexpected either way.

However, even in the case of pessimistic prediction we are facing a global subscriber population which is supposed to be two to three orders of magnitude higher than anybody has experience with in the mobile satcom community. The new markets can be characterised as mass markets with lowest consumer prices for terminal equipment but with highest sophistication. This, what looks like a contradiction at the first glance, can be the key to success. Mass production requires greatest efforts in the non-recurring and lowest repetition costs in the recurring phase of the user terminal value chain. Most of the mass products we use today, PCs, calculators, even watches and cameras are good examples for this. Who would have expected to buy a Ku-band terminal with 70K noise temperature and a 100 channel satellite receiver for less than 100ECU's in the shop around the corner only ten years ago ?

At a terminal price of about 150 ECU and a subscriber community of about 5 million the revenues are 750 MECU. Terminals in the mobile broadcast market are predicted with more than 30 million after six years of operation. Again, if we assume a price of about 350 ECU for a car terminal (radio+data) the revenues just from selling the hardware would be 10 BECU. A hardware market with significance.

The remaining element to create a multi-media terminal in mobile user equipment is the display. Mobile user terminals, especially those installed in vehicles (cars, busses, trains, ships, etc.) will be more and more equipped with displays. These displays will be multi-purpose displays which may also be connected to the information infrastructure of the vehicle, if not even become an integral part of it. Figure 1 shows the trends in sophistication and resolution of future video display.

Great improvement in technology will provide the high resolution true colour video screen (LCD and tube) in various sizes at a quality exceeding by far the picture quality of today's normal TV sets. This trend is further driven by the development in digitally encoded TV broadcast. Once the technology allows the real mass production of these items, prices will collapse. Recent predictions of European Computer Sources anticipate that already from 1995 onwards only high and super high resolution screens will dominate the market (Figure 1).

Thus, the user terminals will very shortly provide all functions to allow new services. Archimedes can be such a new system which will provide the required high data rate broadcast capability to mobile subscribers. These data rates range from a few kbit/s to hundreds of kbit/s. Archimedes will open new markets for low to high rate data and audio broadcast dissemination.

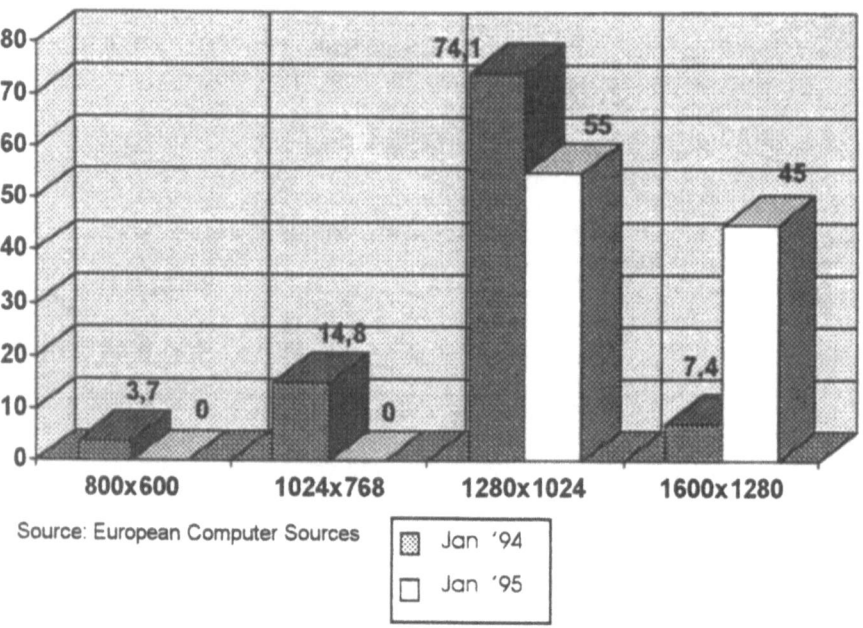

Figure 1: Trend in PC screen formats towards multi-media applications

2. Market Segments

The global mobile market can be segmented as follows

- Hard- and Software supply

- Network Operation

- Service Provision

- Support and Accessories

2.1 Hardware and Software

Transmission and switching equipment: terrestrial, space based; initiated through the standardisation process achieved by GSM standardisation created new generation of digital switches lowering prices by increased competition. The equivalent infrastructure for multiplexers and transmission equipment for DAB is basically available.

User Terminals: Hand-helds, vehicle mounted (car, bus, truck, ship, plane). Figure 2 shows the budget spend per household in Europe for entertainment equipment. Considering that there are about 70 million households in the European Union, the market sizes in the various countries is in the order of 15 BAU.

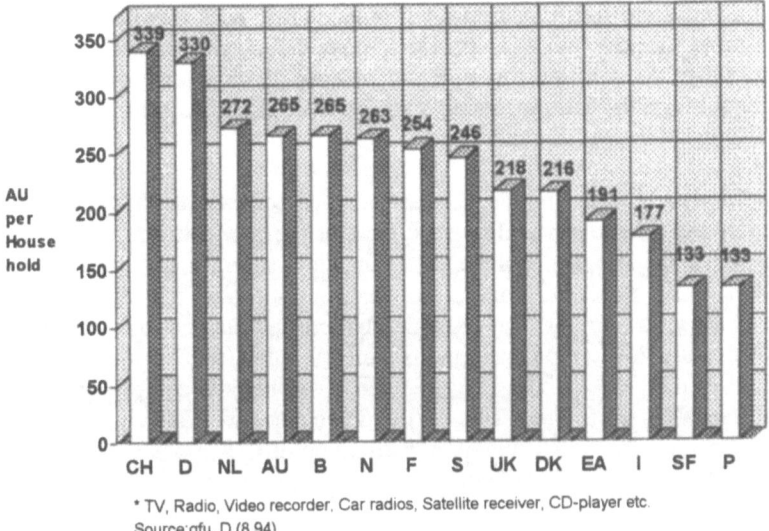

* TV, Radio, Video recorder. Car radios, Satellite receiver, CD-player etc.
Source:gfu, D (8.94)

Figure 2: Expenditures for entertainment electronics[*] in Europe

Software

Software is without doubt the segment with the fastest growing market prospective of all. This includes a wide field of applications ranging from network management software for mobile networks (Mobile Switching Centres) to firmware making the media terminals more user friendly (EEPROMs).

Again, one of the fastest growing segment in the software market is the mass market created by Personal Computer in the home, business and entertainment areas. High sophistication and multiple reproduction are the typical characteristics of the software market, which makes it very similar to the mass markets of components and equipment. The production of software is very time consuming, thus very expensive in the non-recurring phase. But, once it is available, the reproduction and multiplication can be done at lowest costs.

The relatively short lifetime of modern software version makes it very difficult for smaller producers to survive the fierce competition. Therefore, software which was originally considered for a business application shortly later also propagates into the consumer market, since it provides professional performance at lowest consumer prices. So, why creating different versions for a consumer market ? This is true for software for text processing, drawing, spreadsheet as well as databanks, mailboxes, and other applications.

Another important factor in the success story of software is the provision of distribution and sales channel. Presently most of the new software, except those from the major supplier, is distributed as a sampler of several software packages on CD-ROMs under shareware conditions. Shareware means, that I can test all features of a software on my own computer, but created files can neither be stored nor printed, a use in a commercial sense is therefore inhibited. Only after paying by credit card or otherwise, a special code number will be provided by the producer which then enables the full use. This distribution channel is relatively fast and cost effective. However, the production and distribution of CD-ROMs in the long run seems to be only the second best solution.

It is expected, that future software will be more and more distributed through data networks. Due to the fact that software is getting very comprehensive, this will create collisions of interest with other users of the public and private data networks. Networks will be very loaded with bulky file transfers, actually congesting on-going real-time communications of others. The transfer of bulky information or data files unavoidably becomes a very time consuming, process due to the overhead introduced by the handshake protocol. If every transaction would be performed individually, the network might be fully loaded by this type of operation, a nuisance to other network users.

Data broadcast systems such as Archimedes, but also geostationary systems, could provide the necessary one-way high-rate file distribution channel, since they can address many terminals in parallel, actually millions in Europe. Special broadcast protocols need to be implemented to ensure an error free reception of the broadcasted material. Even several repetitions of the file dumps are more efficient than individual file transfers.

2.2 Network Operation

Mobile services are in more and more countries provided through networks not operated by the national PTT under monopoly. Liberalisation of PTT's in Europe has created new fields of business, particularly in the area of network operation. Big investor groups, often led by either a military or an energy provider invested and are still investing billions of accounting units per year into building new networks. The success of this fierce competition race shows first results.

For instance, a total of almost 4 BAU was invested in Germany in the two competing networks called D1 (DeTeMobil) and D2 (Mannesmann). After two years of operation almost 2million subscriber have been attracted. It seems that the market is still not saturated since a couple of months ago, a third system called "E+" (Preussen Elektra), a PCN system, entered the market. Network Operation another innovative market segment with growing significance and importance.

2.3 Provision of Services

The following list is not complete. It should provide a vague idea about the potential services for which, in the next couple of years, the transmission infrastructure will become available. These services will extend the present telecommunication services portfolio (Bearer and teleservices) by introducing supplementary services and, on top, value added services.

Supplementary services modifying or enhancing the basic telecommunication and **network services.** These services are well defined in the GSM standard and will become available with the further development of the mobile networks.

However, it is expected that the **Value Added Services** will become the real big business. Value added services utilise the communications infrastructure to sell information or special knowledge. Typical examples of value added services are expected in the area of traffic and tourist information services, where updated information on road conditions, parking, travel guidance, special events, booking, and many other information bits are provided.

Pseudo Interactive Services. These services are transmitted in broadcast mode. They are received by the user terminal, but only the wanted contents is actually decoded and communicated to the user. "Pseudo" means, that there is no on-line user dialogue to the information source. This operation can be compared with a newspaper. Every reader reads only what is of interest to him. Due to the high capacity transmissions, frequent updates and comprehensive media material can be offered. Intelligent on-board equipment can combine the received information into a more complex service. Artificial voice can guide to a destination combined with a city road map retrieved either from a vehicle resident CD-ROM or from a permanently provided and updated road atlas broadcasted via DAB. These services can be of interest to special user groups (taxi, pleasure bus driver) as well as to private users.

Software Distribution Services. Shareware, public domain software, music, information. This distribution channel for shareware and public domain software is faster, more up-to-date, more comprehensive and lower in costs in comparison with CD-ROM's. The potentially addressable number of consumer, i.e. customer, is unprecedented.

Information Services (Cyber Marketplace/Radio Kiosk). Here, information is broadcasted to a large audience at very low costs. A lot of bulky information (voice, data, sound, video) can be communicated without unnecessarily overloading the networks. More important than a pure entertainment information will be the broadcast of business and administrative information. Examples are request for proposals, service offers, real estate, commercial services, movie announcements/photographic pre-views, latest news, special event information (Concerts, Championships, Olympic games), emergency messages (disaster and distress information), organ and blood-donor services, search for people, specialists, offer from department stores, test reports on consumer equipment, publishing information of public interests: new laws, legal decisions, contacts between specialists, hobbyists, etc.

Transport Management Services. Fleet Management for cars, trucks, planes. The broadcast with an encrypted transfer of graphical material (order sheets, photos of delivery terminals) can provide even ad-hoc information into the fleet on the roads. A return channel from a truck cockpit can be via Euteltracs, EMS, GSM, Globalstar.

Audio Broadcast Services (music and voice). Special music channel, news, pan-European programmes. Special programmes for people roaming through Europe, trucker, bus driver, shipper, train traveller, sales representatives, reading of fairy tales, short stories etc. as traveller entertainment programmes.

Navigation and position determination

Remote control (telecommand/telemetry for industrial and in-situ applications). Connected to the feeder uplink via a data network (Internet, CompuServe®, etc.) control commands could be applied simultaneously to instruments, sensors, or other remote systems in the entire coverage area.

2.4 Support

In addition to the services discussed above the market comprises much more which can be categorised under support activities. Herewith are understood all services which contribute, prepare, maintain or are otherwise mandatory to provide services. These are:

Charging systems (pay services): Smart card concept for a low cost billing system.

Encryption strategies and procedures. The development, test and implementation of reliable procedures to ensure a smooth and secure operation of commercially sensitive services.

Fraud prevention. Avoidance, or at least minimisation of abuse or other criminal actions.

Legal issues. global co-ordination and co-operation for common frequency assignments, taking care of IPR, music and copyrights, patents, licences, etc.

Maintenance and repair

Accessories

Accessories for user terminals ranging from auxiliary battery sets (a 100 million DM business in 1993 in Germany !) to more fancy terminal design, leather boxes, VIP outfits, outdoor versions, etc.

3. User terminals for Archimedes Mobile Multi-media Services

By far the largest market segment in the hardware is the user terminal. The population in Europe aged over 16 and under 65 in the year 2000 will be around 214 million. A market study, performed by Touche Ross [3] predicted a penetration of user terminal for the European market as shown in Figure 3. The diagram shows the penetration of Digital Audio Broadcast equipment.

The study predicts quote "If DAB becomes the standard, accepted way of listening to radio almost all of these people will, in time, own a suitable receiver. This may, of course, take many years and some of them, perhaps many will own more than one. About 74% of the population have cars and the vast majority of cars (97%) have radios. There is thus an eventual opportunity to supply both cars and homes with a total of about 350 million receivers of either a fixed, portable or mobile variety." unquote

The prediction may be somewhat optimistic for the start-up phase, but it in the long run, at saturation after six years of operation, DAB has the potential to attract a user population of more than 30 million people only in Europe. The grade of sharing between terrestrial and space borne services is of less relevance. Both transmission systems will have to play a complementary role in the provision of audio and data services. Both profit of each other. The terrestrial network is better for local and regional services while the satellite's domain will be the services with a wider geographical spread of the subscriber population.

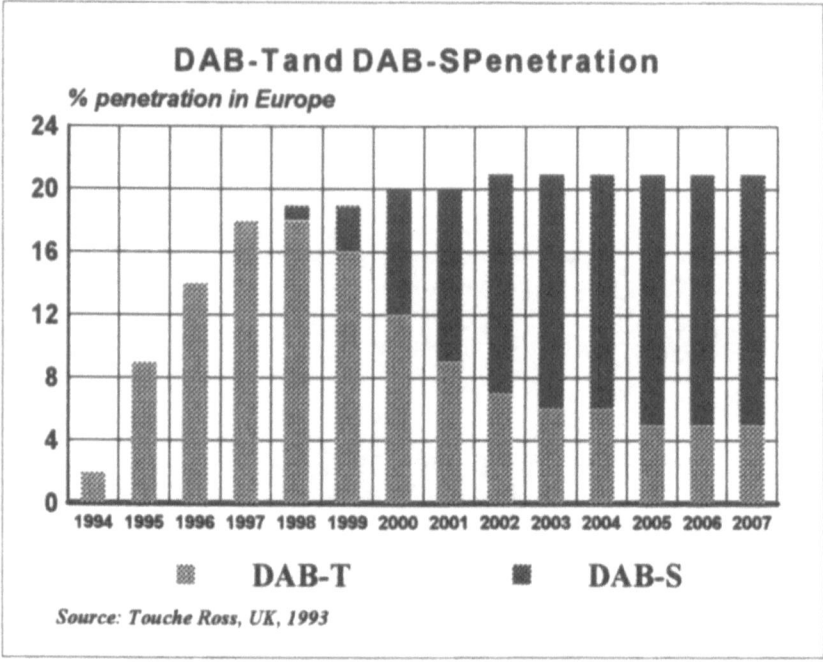

Figure 3: Market prediction for terrestrial and space based DAB implementation

With the introduction of GSM networks the range of potential mobile communication's services has been widened significantly, although some of the defined GSM services are not yet implemented to the full extend. The development of new services in terms of performance (voice, data, Fax), charging to the minute, re-routing, call barring, international roaming and so on is comparable to the developments of applications in the early computer times. The general statement, that "a computer can do everything" is probably correct, but did not sell a single machine.

Only when the application software became available in a user friendly way, the great success of personal computer really began. Today, for many people telecommunications is still limited to the use of the telephone. In a few years from now, when the computer kids will be grown up, computing and telecommunications will have merged in a way that it will be difficult to explain how we ever worked without it.

Presently we see billions of strategic investment dollars flowing into the global infrastructure for mobile and fixed telecommunications. One element of the global multi-media services, which has not yet been mentioned, is the position determination and navigation aids. The present global navigation infrastructure, comprising the US-GPS and the CIS-Glonass, is a heritage of the cold war. Both systems were implemented to enable global military navigation, guidance of long range cruise missiles and position determination of troops and equipment. Although these systems created an additional civil market for terminals there is no legal commitment that both systems will remain commercially available. Without going into detail, it is clear that the provision of signals for position determination is also a segment of the global mobile services market with increasing importance and growing commercial relevance. What remains open for the time being is the question who are the strategic investors in the next generation of the space segment. A fully commercial mission for a concurrent global navigation satellite system (GNSS) is presently under investigation by industrial groups and Agencies.

4. Summary and conclusions

This panel contribution to the prospective of multi-media mobile communications markets has highlighted the various market segments and their interrelations. It is impossible to rank the significance of any of the elements since they are all equally mandatory, and each segment only makes sense in commercial terms if the provider of the other segments also play their roles.

It is sometimes very difficult to start an activity from one segment knowing that those who eventually profit are not necessarily those who kicked the initiative off. This chicken-and-egg paradigm is often the reason that nothing happens at all. Therefore, it is very admirable how individuals in the American industry really did overcome the initial phase of new system. I would go so far as to say, that no matter which of the mobile communications systems, Iridium or Globalstar or any other, eventually arrive at their commercial targets it is of merit that this new market, which is just about to come into existence, has been so well nurtured.

I also believe that the European industry can and will contribute in many ways to the Global Information Infrastructure. One element of Europe's contribution should be an operational Archimedes system.

5. References

[1] "Communications: Policy-makers expected to remove many regulatory barriers" Special Report; The Bureau of National Affairs, Inc., Washington, DC 20037; Ref.: 0148-8155/94. Also "The National Information Infrastructure: Agenda for Action"

[2] "Archimedes" Phase A study. ESA Contract No. 10607/93/NL/US Final presentation December 1994

[3] ARCHIMEDES Market study, January 1993; Touche Ross & Co., London, UK

GLOBALSTAR System : An Overview

JB. Lagarde, D. Rouffet, M. Cohen
Alcatel Espace
Nanterre, France

1.   GLOBALSTAR SYSTEM

GLOBALSTAR is a worldwide telephony and data communication system, using a low earth orbiting satellites constellation.

It can be operated to and from handheld terminals and it is integrated into public switched and mobile terrestrial networks by design : satellites are transparent and all processing functions are performed on the ground.

2.   GLOBALSTAR SYSTEM DEFINITION

GLOBALSTAR system offers a worldwide coverage between 70° latitude north and 70° latitude south and comprises three segments :

- ⇒ **The Space segment**, which is made of 48 LEO satellites (1414 km), 8 spares and two Satellite Operation Control Center (SOCC).
- ⇒ **The Terminals**, which can be hand-held, vehicle mounted or fixed and can either be operated in the GLOBALSTAR system.only, or in the GLOBALSTAR system and a terrestrial cellular like GSM or DCS 1800.
- ⇒ **The Gateway station**, which interconnects the users terminals with PSTN or any PLMN. The gateway station will also allow mobility management by updating the subscribers databases.

This integration into cellular terrestrial system will allow the bi-mode subscribers to lease a unique calling number for terrestrial cellular system as well as for GLOBALSTAR system.

The gateway coverage will take into account geometric coverage and regulatory aspects.

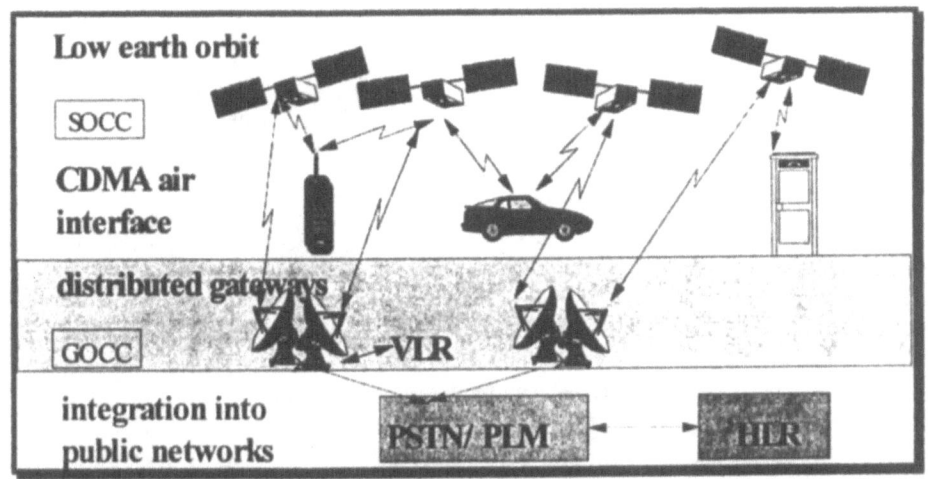

THE GLOBALSTAR SYSTEM

3. SERVICES

Basically, GLOBALSTAR will provide digitized voice services using variable rate vocoders, with a mean rate of 2.2 kbps and instant rate ranging from 0.6 kbps to 9.6 kbps. Frame error rate is 10^{-2} and bit error rate is 10^{-3}.

GLOBALSTAR will also provide data transmission with a higher rate of 9.6 kbps to and from dedicated terminals with a BER of $10^{-6.}$

Mobility service will also be provided by GLOBALSTAR. Worldwide roaming is allowed thanks to the integrated terrestrial switched and mobile networks.

Service areas will be mainly suburban and rural areas, as cellular networks are covering high density urban areas.

Due to its transparency and integration in public networks, the system will offer the same services as the terrestrial systems.

4. AIR INTERFACE

- Propagation conditions in a mobile satellite link are characterized by fading and shadowing due to trees or blocking due to buildings. Above a certain elevation angle to user, multipath levels is very low and satellite channel is Ricean like with a strong line of sight signal. To combat fading, the first strategy is to use high margins. But due to the constellation, two satellites are simultaneously visible, most of the time, quite everywhere in the covered area and to combat fading, it is more efficient to establish a second link through the second satellite (satellite diversity). This is also possible thanks to the chosen multiple access (CDMA) and the Rake Receiver in the demodulator.

- An efficient power control algorithm will improve satellite capacity.
- An other important point in the LEO system is the spot beam handovers (every two minutes) and the satellite handovers (every ten minutes). During those handovers, resources can change (PN code), but for a satellite handover, receiver must acquire new Doppler shift. The Doppler shift is predictable by the gateway station and can be transmitted to the receiver if this one is localised. It is necessary to localise terminals within ten kilometers approximately for transmission purpose.
- Transmission characterisations are as follows :
  ⇒ on the forward link, data are encoded by a Viterbi code (R=1/2, k= 9). The transmitted signal consists of two quadrature carriers, quadriphase modulated by a pair of PN codes, quadriphase modulated by an assigned orthogonal Walsh function (one among 128) and quadriphase modulated by coded, time interleaved and scrambled data,
  ⇒ on the return link, data encoded by concatenation of Viterbi code (R=1/3 ; k= 9) and orthogonal Hadamard code (6/64) before being spread by a pair of PN codes. The pair of carriers are biphase modulated by data and quadriphase modulated by PN codes,
  ⇒ in both links, chip rate is 1.2288 Mcps and signal bandwidth is 1.25 MHz.
- The frequency bands are the following :
  ⇒ 1610-1626.5 MHz for the return mobile link,
  ⇒ 2483.5-2500 MHz for the forward mobile link,
  ⇒ 5025-5250 MHz for the feeder uplink,
  ⇒ 6850-7075 MHz for the feeder downlink.

The choice of feeder links frequency bands will be finalized after WRC 1995.

5. SPACE SEGMENT

Constellation is a Walker constellation of 48 satellites in 8 plans with an inclination of 52°.

This constellation presents the advantage of providing nearly everwhere double satellite visibility with a minimum elevation of 18°. The altitude has been chosen to use standard electronic devices. At higher altitudes, we are nearer of Van Allen belts, which implies highly protected electronics.

Inclined orbits have been chosen to provide transfer orbits. It is possible to launch satellites by groups of 12 or 4.

Control during the station acquisition is performed by gateway stations and two control stations. TTC is in C-band. Satellite mass is 450 kg with an expected lifetime of 7.5 years. Fiability at end of life is 0.85.

Mobile link is using a 16 spot beam antenna (forward and return). There is a central spot beam, six spot beams in the inner ring and nine spotbeams in the outer ring. Global EIRP is 39 dBW and G/T is -13dBW. Power consumption is depending on the supported traffic. Traffic is quite null over the oceans, allowing battery charge.

6. GATEWAY STATIONS

Final architecture of the gateway stations is still under consideration. Anyway, according to the operators point of view, and especially those who are operating GSM networks, it is reasonable to think that final architecture will be very close to the following description.

The gateway will comprise a radio station (transmission and reception), a modem and multiplexing unit (GSC) and a switch to perform mobility management and access to mobile databases and public networks.

This switch is called G-MSC (GLOBALSTAR Mobile Switching Center) and will include a Visitor Location Register (VLR). There is no HLR (Home Location Register) in this station.

This architecture allows to use switches directly derived from the cellular networks switches.

The interface between GSC and MSC should be similar to the GSM « A » interface.

It allows to interconnect subscribers from GSM, DCS1800 or DECT.

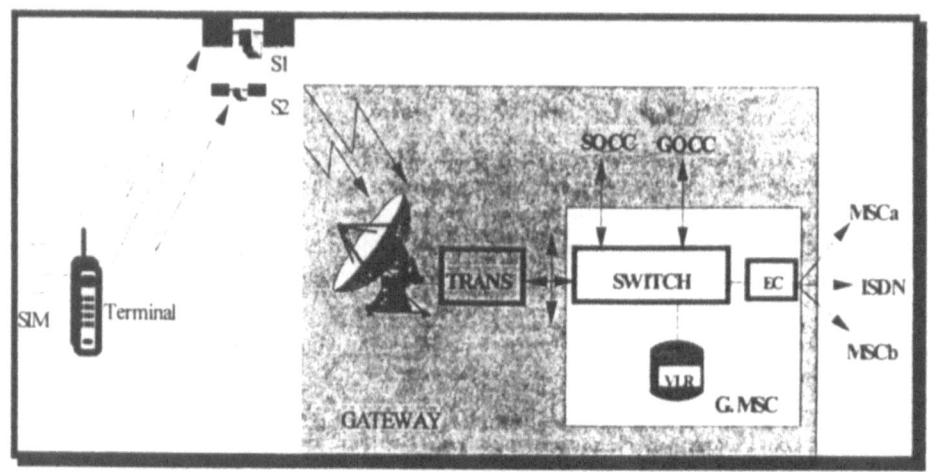

GLOBALSTAR GENERAL ARCHITECTURE

SOCC	Satellite Operation Control Center
GOCC	Ground Operation Control Center
MSC	Mobile Switching Center
EC	Echo Canceller
SIM	Subscriber Identity Module

7. CONCLUSION

GLOBALSTAR system will provide complementary coverage to the existing cellular network, telecommunications infrastructure for developing countries, rural communications for developed and developing countries. Moreover, it will provide communications services to the so-called global roamers.

Strategy of Alenia Spazio in the Mobile Communications

C. Mastracci - D. Santilli
Alenia Spazio Rome (Italy)

1 Introduction

The mobile TLC industry continues the great expansion experienced throughout the last decade, confirming the evolution from a minority, high cost application into a mainstream TLC service
- According to a recent market analysis, the worldwide cellular will continue to grow rapidly through out the present decade:
34 million cellular subscribers at the end 1993 (around 7 million in Europe at the end July 1993) and over 118 million by 1999, when 45 million will be on digital cellular with a major share for GSM (more than 50%)
- In Italy the mobile subscribers have grown from 6.400 in 1985 to about 1.2 million by 1993
- The market for cellular terminals will reach by 1999 a value of 10 billion $ corresponding to 35.2   million units sold in that year

Satellite complements the geographic coverage of terrestrial carriers for personal services (mobile, transportable, fixed) in remote, rural or other difficult to serve areas

The dimension and type of market demand in the different world regions depends on a number of factors: availability of terrestrial services, population, level of economic development, cost and quality of satellite services

Market projections for truly global mobile communications addressable by SATCOMS via hand-held terminals, vary between 10 and 20 million subscribers by the year 2010. Such estimates are based on the assumption of capturing from 2 to 5% of the ground cellular market in both developed and developing countries, plus important shares of rural fixed communications and auxiliary services such as worldwide paging and radiolocalization.
Globally the satellite mobile market might be worth over 1 B $ by the year 2000

2 Presence of Alenia Spazio in the mobile communications

Alenia Spazio perceived from the beginning the great opportunities offered by the mobile SATCOM market and since mid 85's decade promoted exploratory studies to define a commercial strategy an acquire an adequate role.
An important start role was the possibility to offer, on the 2nd flight model of ITALSAT, a spare capacity which is adequate to promote an early experimentation of voice/data mobile services over Europe after the satellite launch in 1995

To-day two main achievements were met by the company in the mobile SATCOM
- The responsibility of the space segment in the ESA LMSS regional program
- The participation to the Globalstar project in a partnership position with the proponent company SSL

3 The ESA regional LMSS program

The mobile communication program developed in the last year by the European Space Agency (ESA) is oriented to the Private Mobile Radio (PMR) applications

Alenia Spazio is responsible of the two first mobile payloads that will be embarked on two spacecrafts developed under company leadership
- EMS (European Mobile System) payload on Italsat 2
- LLM (L-band Land Mobile) payload on Artemis

System architecture is based on the concept of mobile satellite business network (MSBN) where a user at a fixed side has direct access to its "fleet" of mobiles through the satellite.
A main aspect is that mobile roaming in a large geographical area can be supported, with the same standard, owing to the large coverage achievable by satellite in GEO orbit
Potential applications are numerous and main goal is to provide mobile users with on-line access to business, personal and societary informations with the important inclusion of "value added" partners.
Main market areas are fleet management, demand/offer management, travel advisory
Estimates concerning traffic forecasts show a wide range of requirements and about 70.000 users before 1999

3.1 EMS and LLM payloads

In the frame of ESA activities, Alenia Spazio has defined two payloads for mobile communication: EMS with global regional coverage for preoperational use and LLM as back-up of EMS and for promoting the development of technologies for future mobile communication with larger capacity

The general configuration consists of two transponders, one for the forward link (fixed to mobile) from Ku-band to L-band, the other for the   return link (mobile to fixed) from L-band to Ku-band. The design matches the MSBN requirements, needing high EIRP for the mobile link, and of a Ku-band VSAT access on the feeder link.
The coverage area includes West and a major part of a East Europe, Turkey and North Africa

The EMS payloads includes L and Ku-band antennas, RF subsystems and two IF chains, each on providing 3 SS-FDMA channels 4 MHz wide. SAW technology at 140 MHz is utilized for channel filtering. Under the hypothesis of CDMA access with voice coded transmission rate at 4.8 Kbit/s which is compatible with public network interfaces, a total of about 60.000 users can be served

The main characteristics of LLM payload are the use of multiple beams that allow frequency reuse and of innovative techniques for efficient utilization of payload resources i.e.:
- Channel to beam connection by telecommandable switching matrixes
- RF power flexibility via a multiport amplifier architecture

3.2   From EMS to LLM. Achievements and prospect

The development of EMS/LLM payloads implied adoption of innovative system solutions and resolution of challenging technology problems
- Efficent channellization
- Phase noise limitation
- Ku-band linearized solid state amplification
- Immunity to multipaction
- Protection from passive intermodulation products

In parallel Alenia Spazio performed several studies aiming to define more advanced payload architectures for future operative systems, making optimum use of the on going developments in Europe

The on-board technology will see in the near future the introduction of advanced digital signal processing in order to accomplish the tasks of channelling, routing and beamforming, retaining analogue functions only for the front-end and transmit sections.

Other potential advantages for the digital solution are high frequency reuse,

transparency to any access scheme (FDMA, TDMA, CDMA) and capability to address different parts of the service area by means of beam repointing
Other technology developments are required for the implementation of high density CMOS in ASIC for most of the digital processing functions, and of very low noise amplifiers, high efficiency low-mass SSPA, highly integrated receive   and transmit sections using hybrid MMIC.

Antennas with a large number of beams will require the development of:
- Large feed array technologies including the use of high dielectric substrates and dual frequency ele   ments
- Large deployable reflectors (presently flown by USA, Russia and Japan) or active and semi-active   antennas

4 Alenia Spazio in the SATCOM global communications

The advent of the cellualar telephony has started the era of personal communications (PC) and has opened an entirely new chapter for mobile SATCOM. Business globalization and the resulting increased international mobility of people and goods, calls for systems capable of providing service continuity through areas marginally served by cellular telephony

Several alternative LEO SATCOM concepts are presently competing for the same or contiguous market segments including cellular services and rural telephony.
They are conceived either as overlays to the terrestrial TLC infratsructures or to complement and interoperate with them

In the emergent market segment of LEO mobile systems the true entry barrier is the capital investment necessary for the development of the project, before the expected revenues from the operation of services in competition with opposing systems

This commercial high risk scenario explains the necessity to join forces in a series of alliances between industries and service providers with the objective of gaining the critical mass to compete with other world competitors.

Alenia Spazio has chosen to partecipate to Globalstar project after a careful evaluation of the innovative system characteristics and the strength of business organization.
The role of company is enforced as it is shareholder of SPACE SYSTEM LORAL and participates as member of the D3A european alliance

4.1 Globalstar system overview

Globalstar system consists of:
- A constellation of 48 satellites
- The user terminals
- The ground segment composed of the satellite and ground operation control centers (SOCC and GOCC) and the gateway stations that are the interconnection point between the space segment and the existing terrestrial TLC networks (PSTN, PLMN)

This characteristic of incorporating existing terrestrial systems into its overall configuration is a key aspect of Globalstar architecture that should guarantee a number of advantages with respect to many competitors (e.g. Iridium). The interoperability and cooperation of Globalstar with terrestrial infrastructures
- Enhances the system's reliability and decreases costs to the end users by decreasing the complexity of the space segment
- Incentives terrestrial service providers for its fast implementation by increasing their revenues

5 A Big Opportunity to get a Commercial Market-Driven Role

Mobile SATCOM has a strategic role for European Space Industries to achieve business and profits in our era of derugulation and privatization.
The funding from European governments tend to become stable and probably will be reduced in the medium term.
The market demand shows significant modifications as the opportunities of classic type tend to decrease, while ever more frequently the space system supplier must assure the final service with an active participation to the satellite operation phase.
The profitability in a competitive market requires big changes in the space industry organisation well beyond the technological excellence. A series of competences must be expanded:

- more attention should be given by the commercial structure to market research and to contacts with service operators within and outside the western world, including organizations which do not deal with satellites

- adequate financial and assurance strategies should be set-up to provide capital for construction, launch and operation of the proposed systems.

Furthermore Space Companies have to join forces by alliances for gaining the critical mass to compete in a worldwide market.

The role of satellites need not be restricted to the so-called "niche applications" like special business services or data dissemination. Indeed they can fulfill a variety of functions as part of public infrastructure, but the success of space market will depend, among other important factors, upon industry's policies, initiatives, expertise and competitiveness

Standards for Satellite Personal Communications Networks

Prof. G. Stette
University of Trondheim
Chairman ETSI TC SES[1]

The need for standards

The implementation of systems for mobile/personal satellite communications systems has many aspects, e.g. technical, economical and regulatory. My contribution to this Workshop will concentrate on standardization as a basis for regulations.

WARC-92, the World Administrative Radio Conference for Dealing with Frequency Allocations in Certain Parts of the Spectrum, made new frequency allocations to the new systems of the LEO (Low Earth Orbit) and of the MEO (Medium altitude Earth Orbit) types. There were also expressed concerns at the WARC that the first global Low Earth Orbit (LEO) system to be implemented by one country might result in *de facto* standards being imposed on the rest of the world.

A WARC Resolution (COM5/11) was agreed which calls for the establishments of standards for the operation of low-orbit satellite systems.

" - - to carry out, as a matter of priority technical regulatory and operational studies to permit the establishment of standards governing the operation of low-orbit satellite systems so as to ensure equitable and standard conditions of access for all countries and to guarantee proper world-wide protection for existing services and systems in the telecommunication networks."

Europe is preparing for licensing of satellite personal communications services. CEPT has been asked by the European Commission to look into the spectrum issues and the commission is about to ask them to review any licensing issues that could be agreed Europe wide, whereas ETSI, as I shall describe later, is reviewing the standardization aspect as a basis for type approval for hand-held equipment and other issues. The inherently multilateral efforts of these bodies can obvious be used as a basis for global agreements [1].

---

[1] Representing the Norwegian Telecommunications Regulatory Authority

ETSI (European Telecommunications Standards Institute)

As the Chairman of the ETSI Technical Committee Satellite Earth Stations and Systems (TC-SES), one of the 13 TCs covering the whole field of telecommunications, I would first give a brief description of the organization and function of ETSI, the European Telecommunications Standards Institute.

ETSI is one of the three European recognized standards institutes (together with CEN and CENELEC). It continues the work of CEPT (European Conference of Post and Telecommunications Administrations) in the standardization field, but unlike CEPT, which only had the European Network Operators as members, ETSI has more than 330 full members of different categories, network operators, regulators, satellite system operators, manufacturers, research establishments and users. Its headquarters are in Sophia Antipolis, near Nice, France.

European Telecommunications Standards, ETSs, are produced and approved according to established procedures. A draft ETS is usually produced by a Sub Technical Committee (STC) within a TC, or by a Working Group within an STC. It is also possible to use a PT, Project Team, to produce the initial text. The draft ETS is then approved by a TC and then sent for Public Enquiry (PE). This means that all parties in the participating countries can express their views on the proposed standard.

The PE comments are then dealt with by the TC in a particular Resolution Meeting, before the Draft ETS in its final form is sent to the National Standards Organizations for Voting. This process ensures a certain stability in the standardization process. There are also procedures for the maintenance and revision of standards.

In addition to the ETSs and the TBRs described below, ETSI also produces ETRs, ETSI Technical Reports, which are official publications of the organization.

The functions of standards

Standards are basically publications in the **voluntary** domain. It is in principle up to a manufacturer's decision to make equipment according to the standards, or not.

We can talk about different types of standards. *Envelope standards*, as they sometimes are called, define such parameters as maximum unwanted radiation from telecommunication equipment. *Interface standards* ensure compatibility between system components from different manufacturers. There are also examples of standards which to the necessary details *specify a total system* (but still leaves room for innovation in implementation). The highly successful GSM system is based on an ETSI standard of this type.

Standards also serve some other important functions. They can be used to derive another official publication of ETSI, TBRs, Technical Basis for Regulations, which together with additional elements of regulatory nature are converted to **mandatory** requirements, CTRs, Common Technical Regulations, by the authorities of the EU (European Union) and EES (European Economic Space) countries.

Equipment meeting the requirements of a CTR has a right to free circulation under the terms of the Directive without the application of further national procedures or national requirements within the scope of the CTR.

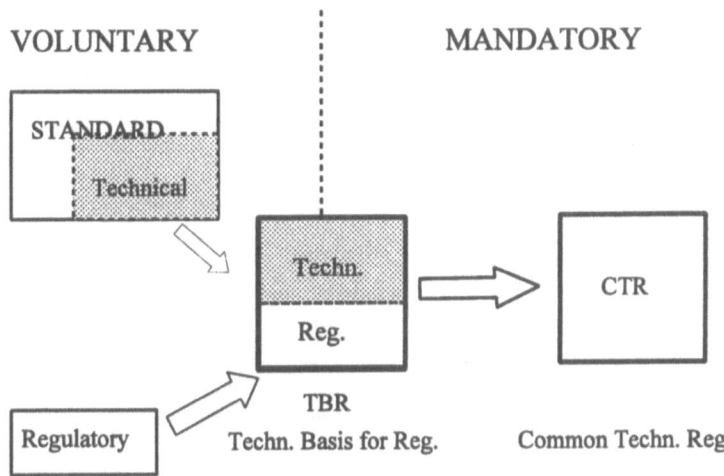

Fig.1   Standards as a basis for CTRs.

Functions of the S-PCNS

S-PCNS are planned for personal communications over a wide range of services, and system design is driven by user requirements. The satellite system can operate as a separate system, independent of the public telecommunications systems, but it is considered of increasing importance to integrate the satellite system with terrestrial mobile systems like GSM.

A role for satellites in the development and deployments of PCNs has been identified both for fixed and mobile services in the general framework of **UPT (Universal Personal Telecommunications)**.

ETSI are planning to integrate a satellite component in the third generation of mobile telecommunications in Europe, **UMTS (Universal Mobile Telecommunications System),** which is the European version of the FPLMTS of the ITU. This system is planned for development over a period up to the year 2000.

ITU, Task Group 8/1 is considering the role of satellites in **FPLMTS (Future Public Land Mobile Telecommunications System)**. The UMTS standard may be compliant with the FPLMTS, which will be world wide, and global roaming is therefore also one of the priorities of the UMTS. The relations between S-PCN and UMTS/FPLMTS need further studies, and it has strong impact on the standardization issues.

Levels of Standards for S-PCNs.

The CEU, having a responsibility for the European role in the implementation and operation of these systems, therefore provided ETSI with a mandate to prepare a Technical Report in which the standardisation aspect of such systems are investigated in some depth.

The task was given to the ETSI Technical Committee SES, and a Project Team (PT) was recruited to prepare the initial text.

There are two main reasons for introducing standards, some of which could be converted to CTRs:

A certain level of standardization is required to comply with the **"essential requirements"** established by Directives of the European Community.

Additional standards may be useful because of particular technical features of the S-PCNs, their use or interworking with the public networks. This standardization is voluntary, and the need must be assessed in a detailed manner to ensure that it is really necessary.

For additional standards, since they are voluntary, they should be required and supported by organizations that are willing to implement them. It is also a requirement that the requirements of a standard must be measurable.

The telecommunications terminal directive of the CEU

The Directive   defines a set of "essential requirements" to be met by all telecommunications terminal equipment. The "essential requirements" set by the Directive address the following aspects:

- user safety in sofar as this is not covered by the Low Voltage Directive;
- safety of employees of public telecommunications networks operators;
- electromagnetic compatibility requirements;
- protection of public telecommunications networks from harm;
- effective use of the radio frequency spectrum;
- interworking with of terminal equipment with public telecommunications network equipment
- interworking of terminal equipment via the public telecommunications network

An extension applicable to satellite earth stations was published 24 November 1993. "Council Directive supplementing Directive 91/263/EEC in respect of satellite earth station equipment (93/97/EEC)" addresses topics such as the effective use of orbital resources and avoidance of harmful interference between space based and terrestrial systems.

Possible additional standardization

A decision to introduce standards beyond those required to ensure conformity with "essential requirements" under the EU Directives should be taken on the basis of a broad consideration of the needs for such standards, and the consequences that might arise from such standards being **applied or not**.

For closed networks there may be no need to standardize quality and availability. Open user group networks are integrated part of the public network, and standards may cover voice quality and delay, standards of interconnect networks, and availability. There may also be two sets of standards, for voice and non-voice systems.

Frequency issues that must be resolved are frequency coordination with other NGSO systems, frequency coordination with GSO systems and sharing methods, interference avoidance and interference tolerance. Type approval may apply to system, terminal, network control and databases.

Network operations and integration

There are three important interfaces, **network interfaces, user interfaces** applying to operation of the terminals and placing of calls, and **air interfaces** comprising

- modulation and access methods
- signalling
- data framing and format
- bit rates and coding

Gateway interfaces would define the process of system integration, guarantee the protection of the PSTN, and allow multiple manufacturers to supply hardware for the gateway and gateway interfaces.

Ongoing activities.

The Phase I report, ETR 093 [2], was published in September 1993, and the Phase II report [3] is about to be completed. It will address the areas in which technical standards will be needed, to meet the essential requirements under the Terminal Directive (TD), the Satellite Earth Station Directive (SESD), or other Directives, and also, where considered appropriate, to go beyond these Directives to consider other voluntary standards that might be regarded as useful. This report will form one of the bases for European policy in this field, which could be of immense importance in the future.

The Draft ETR will be put on the agenda of the Technical Committee SES for approval at its November 1994 meeting.

Acknowledgement

Many ETSI members have taken an active role in the development of these ETRs. The author, and TC Chairman responsible for this project, wishes to thank in particular the rapporteur, Peter Dondl, Bundesamt für Post und Telekommunikation and the members of Project Team PT37V who have put 18 man-months of competent work into the project, Mark Posen as the project leader, Alberto Ciarniello and Rijnder Hagedoorn.

References:

1.   Verhoef, Paul, CEU, DGXIII. Mobile communications. Proceedings of Satel Conceil's Fourth Symposium, Versailles - France, June 6 - 8, 1994, to be published.

2.   ETSI TC-SES. Possible European standardisation of certain aspects of Satellite Personal Communications Networks (S-PCN) Phase I report. ETSI Technical Report, ETR 093, September 1993, ETSI, Sophia Antipolis Cedex, France

3.   ETSI TC-SES. Phase II Report: Objectives and options for Standardisation. draft ETR DTR/SES-00002, ETSI, Sophia Antipolis Cedex, France, to be published.

System Alternatives For Satellite Personal Communications

A Satellite Manufacturer's View

Stuart C. Taylor
Matra Marconi Space
Stevenage, United Kingdom

1. Introduction

Thank you Mr Chairman and thank you to the organisers for inviting me to speak at this excellent event.

Satellite personal communications (SPC) has been on the agenda now for around 5 years. At first many traditional satellite manufacturers thought it was a dream or a joke, but today few would doubt its technical and commercial merits. There are many proposed system options utilising widely varying constellations and satellite technologies and aiming at a variety of markets. Today I will give a personal view of systems with global coverage from the perspective of Matra Marconi Space; Europe's largest and most experienced manufacturer of satellites. I will discuss technical, market, regulatory and commercial issues and summarise with how these issues relate to the most well known system proposals.

2. Technical Issues

The most important technical issues for a satellite manufacturer are the design, manufacture, integration and launch of the satellites. Testing and reliability are key issues and these should not be compromised for quick commercial gain; the long term survival of the business depends 100% on customer satisfaction.

MMS can reliably build between 8 and 12 communications satellites per year, with typically three years between contract award and on-station commissioning. We would aim to build a generation of satellites in 5 years; longer and there are problems with project continuity. In 5 years MMS could build between 24 and 36 satellites, but for commercial reasons any single project should only take half of this capacity : between 12 and 18 satellites.

This capacity, however, assumes that the satellite platform is based on a current design, Eurostar. Any radically new satellite design, such as would be required by a LEO system, would take longer to develop. We might try to improve productivity, but even in the long term we cannot envisage more then double the present rate. In addition we might look to increasing the capacity by building or acquiring new facilities, but the key resource is skilled personnel and the recruitment, training and development needed would not be achieved quickly.

Then there are the communications payloads. Here new, advanced technology is needed and is being developed. Indeed MMS is currently in partnership with ESA, Alcatel, and others in Europe, leading the development of the advanced Digital Signal Processing payload equipments that are likely to be needed. There is much to be developed, however, and we look forward to further European co-operation. Initially a European payload demonstrator should be developed. We believe it will need the collaboration of most of the European payload industry, ESA, possibly the European Commission, and close liaison with the customer to rapidly develop and manufacture the flight payloads to a world-class and world beating standard. Payloads are becoming more important than their host spacecraft and, therefore, this opportunity is critically important for Europe's space industry in its battle for survival against the dominant US industry and the increasingly competitive Japanese.

From a technical perspective, therefore, the first generation of satellites should build upon current spacecraft technology and on-going European payload developments. With the co-operation of ESA and European industrial partners, MMS could build a constellation of 12 to 18 satellites in the next 5 to 7 years (including up to 2 years for payload technology development prior to full system construction). There is no organisation in Europe which could build more and therefore constellations larger than this will either be built in the US or by US-European collaboration.

3.   Regulatory and Market Issues

The market for mobile communications is huge, but due to the existing European regulatory environment, infrastructure and operators the market for satellite personal communications is relatively small here, as it is also in North America. The major markets are, in descending order, Asia, South America and Africa. This may suggest that regional, rather than global systems, should be developed. However, there will always be a demand for global services from a relatively small, but wealthy group of users. The airline industry's huge infrastructure is required by the vast majority of "Economy Class" passengers, but it is with the "Premium Class" passengers that the airlines make their profits and focus much of their advertising. Satellite personal communications will be a similar type of industry, with huge numbers of "economy class" users requiring mostly domestic services in developing nations, but an elite of "premium class" users roaming the globe and providing much of the revenues and profits. With this type of market, regional and global systems may both be developed; with roaming agreements between them. The services required by users will not be homogeneous; in the vision of the Future

Public Land Mobile Telecommunications System they will range beyond basic telephony to include packet data, high speed data, video links, sound broadcasting and navigation. From the regulatory perspective, therefore, it will be necessary to have generic service bands, and indeed these are envisaged for FPLMTS. There are national and global issues and it is important for system developers, including manufacturers, operators and international regulators, to work closely with regulators and local operators in developing nations. Typically these nations will require incentives, some of which may include technology transfer and local manufacture. From a practical perspective this will primarily be terrestrial equipment, but some developing nations with aspirations towards an indigenous space industry - China, Indonesia, South Korea, India and others - will also want to build space equipment.

As a manufacturer of space equipment, therefore, we are fully prepared to work together with developing nations in order to secure markets for our products. We would hope to be supported by the European Commission, ESA and our national governments in these endeavours.

4.   Commercial Issues

Like many major investments, satellite personal communications requires a large amount of capital many years before there is sufficient cash-flow to generate a return. Financiers are very wary of investing in such projects and will only do so if the proposers (1) demonstrate a sound business plan and (2) show their own commitment to the project by risking their own money. A large amount of the initial investment is to finance the manufacture of the space segment and, therefore, it is reasonable to ask the beneficiaries of this investment, the space segment manufacturers, to also take some risk.

Matra Marconi Space understands this principal. Indeed MMS is a key risk sharing partner in the development of the ORION satellite system, which will see its first satellite, built by MMS, launched in just a few weeks time; a testament to several years of co-operation between system developers, manufacturers and financiers. We have recently seen manufacturers making investment commitments to the Iridium and Globalstar systems and Inmarsat have publicly announced that they expect manufacturers to invest in the development of Inmarsat-P.

MMS has taken a careful look at all the investment proposals from the US and awaits with interest a formal investment proposal from Inmarsat. It is clear that pre-requisites for investment include:

- a manageable degree of direct manufacturing by MMS which builds upon current products and developments;
- costs and schedules which are well known and fully justified by detailed planning;

- a business plan which is attractive and fully justified by supporting data.

Conversely, reasons that make investment unattractive include:

- system proposers not showing serious financial commitment;
- a system design which cannot be verified due to "confidentiality" or US Dept of Defense restrictions;
- a constellation which appears to be too large to build in the stated schedule;
- concern over the availability and utilisation of the planned spectrum.

Using these principals, investments in the Iridium and Globalstar systems are unlikely; and there are serious doubts about the Odyssey and Ellipso systems. The Inmarsat-P system appears promising, but more information is needed.

5. Conclusions

MMS is not currently committed to any of the system alternatives, although Inmarsat-P is the most promising. It is clear that for any system to succeed the key players in the development must take financial risks in order that the whole team : system developers, manufacturers, and financiers are working together for success. It is also important that operators, regulators and users are involved in the development decisions to produce a harmony which is beneficial to all. Thank you.

The Developments of Euteltracs, and Some Lessons for Future Systems

L. Vandebrouck
EUTELSAT
Tour Maine-Montparnasse, Paris, France

1. Introduction

You are probably expecting me to present and discuss the market forecasts of the future LEO, HEO and other mobile satcoms systems. However, today I will be examining and analysing the developments of EUTELSAT's existing mobile Satcoms system Euteltracs and the consequences for future systems in development. The Euteltracs system, the first Land Mobile Satellite Service worldwide today, has achieved a different result than initially expected. We can learn a lot from the experience of past developments on Euteltracs for the succesful development of the future systems.

Indeed, when Eutelsat decided in 1990 to develop and launch the equivalent of the very succesful US Omnitracs in Europe, sales of terminals were forecasted at a very high level after 5 years of operation. Before joining Eutelsat, I also worked for Locstar (a company which is perhaps familiar to you), and Locstar's market forecasts were much higher with around 100 000 terminals in operation by 8 years of operation.

In fact, in Europe, Euteltracs faced a slow growth at the beginning of its commercial development in the beginning of '92, for now facing a more rapid and significant growth.

Before revealing the reasons and thus the lessons for the development of the future systems, I would like rapidly to overview how the Euteltracs system works, its architecture, our commercial policy and organisation and the current Euteltracs services. This will also help you to understand the reasons for Euteltracs' current penetration and market segmentation.

2. How does the Euteltracs system work and what is its current market penetration?

As you know, Eutelsat is an inter-governmental organisation with 42 member countries from continental Europe and the Mediterranean Basin. Eutelsat's Signatories are for the most part PTT authorities or the National Public Telecom Operator of each member country. Eutelsat develops and manages fixed and mobile satellite services for its member countries. For most of the current services, Eutelsat provides and manages access to space segment, except for Euteltracs, our mobile service, where Eutelsat manages the overall network from the mobile to the service provider and user.

Eutelsat started developing Euteltracs in 1990 with Qualcomm in the US for whom its equivalent system Omnitracs has seen considerable success with 60 000 terminals in operation.

It is clear that the system has one main and natural application: the improvement in productivity of road transport companies. Indeed, the system ensures and provides

road transport companies, simultaneously and throughout Europe, with the following services:
- location of a fleet vehicle with an accuracy of 80 metres in real time;
- communication between drivers and their base through a message service;
- data collection and transfer from the vehicle to its base.

Furthermore, Eutelsat, through its service providers, ensures the integration of these data into the client's existing Information System.

Through this system, the road transport industry will have an answer to their current constraints such as:
- increasing competition and efficiency requiring a greater flexibility and a better service quality to the clients;
- decreasing profit margins;
- requirements for the shippers for just in time delivery;
- etc.

As Euteltracs ensures simultaneously vehicle location, communication with drivers at low cost and the access to information such as cargo status, the benefits for the user are as considerable:
1. Optimum management of the fleet in real time, giving the opportunity to take new opportunities at any moment for example by re-routing of a vehicle to a new destination.
2. Improved customer service, e.g. customer knows exactly delivery times.
3. Greater efficiency in management of loads and logistic sites, e.g. carrier knows in real time the exact content of a cargo, its location, destination, time of unloading, the shipper to invoice, etc.
4. An improved security through the monitoring in real time and the preventive maintenance increasing the security of hazardous transport.

Transmissions are rapid (less than 30 seconds for one communication), fully guaranteed and confidential. All transmissions go through the Hub in Paris and are then directly routed by PSTN, X25 or a satellite link to the final user PC via a national service provider.

With this information and background in mind, Consulting Companies participating in the elaboration of Euteltracs' and Locstar's business cases, decided that such services and benefits should allow the systems to be generalised to all main vehicle fleet in Europe. Thus, they worked out that the number of vehicles equipped should reach 100 000 terminals after 8 years of operation for each system (Locstar, Euteltracs, and Inmarsat). After 3 years of development Euteltracs has reached 5000 terminals in operation and forecasts a target of 35 000 operational terminals in 2000.

Furthermore, all the previous and original segmentation carried out by consulting companies and ourselves pointed out that such RDSS systems could penetrate the following markets:
- Road transport industry (75%);
- Personal communication for business travellers (12%);
- Remote surveillance and telecontrol for fixed applications (5%);
- Railways applications (3%);
- Marine applications (4%);
- Aviation (1%).

Today, Euteltracs has 99% of its terminals installed on trucks.

Why?

There are several reasons. They are structural, political, economic and to a great extent marketing.

3. The reasons for the current market penetration and the lessons for the global systems in development

The political reasons mainly consist in the monopolistic situation and the non-deregulated environment of the telecommunication market in Europe with scattered and heterogeneous regulations. This situation required to Eutelsat to apply considerable effort, country by country, to be licensed for the free circulation and the free operation of Euteltracs in each country. We have today 32 licences, which means that Eutelsat in two years has negotiated with 32 National Authorities of the Continental Europe for the provision of this licence. Each licence requires a long period of presentation, discussion, and negotiation with the Frequency Management Body and the Ministry of Communication of each country.

The structural reasons consist in the status of Eutelsat: as an inter-governmental organisation which normally merely provides space capacity for the use of its members, Eutelsat is restricted to providing direct access to final users. Every transaction must be handled by the appropriate Signatory, usually the Public Operator of the member country. In conclusion, to enable the Euteltracs service to be operational in one country, an agreement had and has to be signed between Eutelsat, Alcatel-Qualcomm and the Signatory should the latter want to become the National Euteltracs Service Provider, i.e. to deal with the end users. But in most cases, the public TOs are unwilling to commercialise the system alone and give Eutelsat the the responsibility to identifying and selecting one private service provider. When identified, this potential service provider must negotiate an agreement to operate Euteltracs with the Eutelsat Signatory of his country. Not an easy task in certain countries as you can imagine. Nevertheless, we succeeded in establishing 16 Service Providers which cover and operate the service in 18 countries, despite all the drawbacks. Two more will be soon operational.

So today, you have:
- Eutelsat which manages the overall network;
- Alcatel-Qualcomm which provides the system (terminals installed in the vehicles and the software installed at the HQ) to the Service Providers; and
- 16 Service Providers, which are in some cases the public TO or a private company such as ANS in the UK, Alcatel SEL in Germany, etc., and which deal with the final users in the truck industry.

The economic and marketing reasons are due to the dramatic financial situation of the road transport companies during the last five years with an increasing competition, lower and lower margins, bankrupticies, etc.: even if the Euteltracs system improves the productivity by around 10 to 15 % of each client, the client needs the resources to purchase the system. Furthermore, it is clear that the cost of the terminal remains expensive if we consider the financial situation of most clients: 4500 to 5000 ECU per terminal remains expensive for a transport company which manages 1000 trucks.

Due to the terminal cost and a former commercial approach which consisted in presenting the system as a communication tool rather than a management tool, which is in fact the real Euteltracs function, the clients limited the number of terminals to the part of the fleet circulating in remote places where the communication infrastructures are poor such as the ECE, CIS countries, the Middle

East, etc.. In conclusion, even if a road transport company has a fleet of 1000 vehicles or more, in 90% of the cases they have limited the Euteltracs terminals to 20 to 80 vehicles.

This may be a pessimistic view, but in fact it is just to underline that:
- examining client's ways of doing business, their daily constraints and productivity problems;
- developing a turn-key, integrated and competitive solution to meet the exact requirements and needs of each market segment;
- launching and commercialising a pan-European mobile satellite system and service;
- negotiating all the required agreements at the administration and regulatory level;
- etc.;

...require a long time, longer than estimated usually, and I am sure that all the LEO, HEO, ..., systems in development will of course also face these constraints and problems. Furthermore, most of the time we underestimate the time required for the education and the training of the demand and the time required for the technical development to enter a new market segment. These systems of communication will be geographically global, but the offer will have to remain different, market segment by market segment.

4. Future development and growth

Euteltracs however has developed a highly competitive solution dedicated to the trucking industry with new management software systems installed in the HQ of the transport companies. We are also doing a lot of effort on the financial side to become more and more flexible and adapted to our clients' cash constraints.

Thanks to this dedicated offer and this specialisation in one business, with a sound knowledge of the transport industry's business constraints and expectations, we expect to increase our market penetration and forecasts significantly between '95 and '98 with 10 000 terminals in operation by the end of 1995 and 30 000 terminals in 1998, only in the transport industry.

Furthermore, we just finalised and tested a new product range dedicated to marine applications which increasing require similar services to those provided to road transport industry. It is clear that our system is perfectly suited to meet the needs of the fishing vessels and ship-owners as they face problems of competition, management and access to information. We have for instance developed a solution which allow the fishing vessels to receive information from the Fish Market Places such as the fish market rates allowing them to route the cargo to the most profitable market place. Simultaneously, this system allows the vessels to communicate with their base or with another vessel, to be located by its base as well as send distress message to its base or to a control and rescue centre. Thanks to the performance of this new offer, our system has been selected by the UK, Spain and possibly French Government Authorities to control the fishing vessel operation in the EU waters.

We expect a new development in this market segment and expect it to represent 5% of our total market from 1996.

We also finalised a new software package which allows third party access to certain information transmitted by a mobile to its base. This package is currently used within the DRIVE R&D projects of the CEC to control and monitor vehicles transporting hazardous goods in Europe. Through our package, certain Civil Security forces can control and monitor the vehicles entering their zones of responsibility.

In conclusion, from our original business case to the current one, two major conclusions can be drawn:

1. It is difficult to simultaneously target different market segments and business which have different expectations and needs and thus require different hardware and software solutions as well as dedicated commercial approach and arguments.
2. Regulatory and Administrative procedures remain long and difficult for the development of a mobile satcoms system in Europe.

But through an intensive commercial drive to understand our clients' needs and constraints better, and after two years of operation, we are looking forward to significant growth thanks to the elaboration and marketing of competitive solutions, not only at the cost level but above all at the technical concept level of the hardware and software solutions provided.

FIRST EUROPEAN WORKSHOP
ON MOBILE / PERSONAL SATCOMS

(EMPS'94)

VENUE: ESA/ESRIN, Via Galileo Galilei, Frascati, Roma (Italy)
DATE: October 13-14, 1994

PURPOSE

To focus on the state-of-the art and perspectives of mobile/personal satellite commmunications in the framework of hand-held and vehicular land mobile developments. To provide a technical forum to the main European actors in the field where to debate and identify the perspectives, and to improve the interactions.

WORKSHOP CO-CHAIRMEN

Prof. **Fulvio Ananasso**, Telecom Italia/Telespazio S.p.A., Roma, Italy
Prof. **Francesco Vatalaro**, Università di Roma Tor Vergata, Roma, Italy

TECHNICAL PROGRAMME COMMITTEE

Prof. **Fulvio Ananasso**, Telecom Italia/Telespazio S.p.A., Roma, Italy
Prof. **Ezio Biglieri**, Politecnico di Torino, Torino, Italy
Prof. **Enrico Del Re**, Università di Firenze, Firenze, Italy
Prof. **Barry G. Evans**, University of Surrey, Guilford, Surrey, England
Dr.   **Mario Lopriore**, ESA/ESTEC, Noordwijk, The Netherlands
Dr.   **Erich Lutz**, DLR, Oberpfaffenhofen, Germany
Prof. **Gérard Maral**, ENST - TELECOM Paris, Toulouse, France
Dr.   **Roger Rogard**, ESA/ESTEC, Noordwijk, The Netherlands
Prof. **Guido Tartara**, Politecnico di Milano, Milano, Italy
Dr.   **Pietro Porzio Giusto**, CSELT, Torino, Italy
Prof. **Francesco Valdoni**, Università di Roma Tor Vergata and IIC, Genova, Italy
Prof. **Francesco Vatalaro**, Università di Roma Tor Vergata, Roma, Italy
Mr.   **Paul Verhoef**, CEC-EEC, Bruxelles, Belgium

SPONSORS

- Alenia Spazio
- European Space Agency (ESA)
- Istituto Internazionale delle Comunicazioni (IIC)
- Telecom Italia / Telespazio
- Università di Roma Tor Vergata

WORKSHOP FINAL PROGRAMME

1ˢᵗ Day: Thursday, October 13, 1994

Session 1: Services, markets and regulatory issues
(Chairman: *M. Lopriore*, European Space Agency)

F. Ananasso, "Service, market and regulatory aspects for satellite PCN"
P. Lippens De Cerf, P. Verhoef, "European Commission views on satellite personal communication networks"
P. Porzio Giusto, G. Quaglione, "Technical alternatives for satellite mobile networks"
Discussion on Session 1

Session 2: Concepts, systems and key technologies
(Chairman: *B.G. Evans*, University of Surrey)

F.A. Petz, J. Ventura-Traveset, I. Stojkovic, "Key technologies for future mobile and personal satellite communications payloads"
H. Kuhlen, "Multi-media satellite mobile services and systems"
R. De Gaudenzi, T. Garde, F. Giannetti, M. Luise, "An overview of CDMA techniques for mobile and personal satellite communications"
M. Lisi, M. Piccinni, A. Vernucci, "Payload design alternatives for geostationary personal communications satellites"
Discussion on Session 2

Session 3: Processing and network aspects
(Chairman: *G. Maral*, ENST - TELECOM Paris)

E. Del Re, P. Iannucci, B.G. Evans, W. Zhao, R. Tafazolli, "Interworking procedures between cellular networks and satellite systems"
F. Delli Priscoli, "Network aspects on the integration between the UMTS network and satellite systems"
P. Capodieci, R. Del Ricco, A. Vernucci, "Gateway earth stations for future LEO communications satellite systems"
A. Böttcher, G.E. Corazza, E. Lutz, F. Vatalaro, M. Werner, "Aspects of satellite constellation and system connectivity analysis"
C. Cullen, A. Sammut, R. Tafazolli, B.G. Evans, "Networking and signalling aspects of a satellite personal communications network"
Discussion on Session 3

Panel A: Market prospectives
(Chairman: *P. Lippens De Cerf,* European Community Commission)

Participants: *P. Fischer* (Iridium Inc.), *H. Kuhlen* (DASA), *J.Yeomans* (KPMG), *L. Vanderbrouck* (EUTELSAT)
Discussion on Panel A

2nd Day: Friday, October 14, 1994

Session 4: Channel and radiofrequency aspects
(Chairman: *E. Lutz,* DLR)

G.E.Corazza, A.Jahn, E.Lutz, F.Vatalaro, "Channel characterization for mobile satellite communications"
B.G.Evans, G.Butt, M.A.N. Parks, "Channel models for mobile/personal satellite communication systems"
G. Di Bernardo, A. Rallo, E. Damosso, L. Stola, M. Sforza, "Propagation models for the land mobile satellite channel: validation aspects"
E. Biglieri, "Advances in modulation and coding for the satellite mobile channel"
Discussion on Session 4

Panel B: System alternatives
(Chairman: *E. Del Re,* Università di Firenze)

Participants: *M. Cohen* (Alcatel Espace), *J. Huber* (Siemens), *M. Lopriore* (ESA), *C.Mastracci* (Alenia Spazio), *G. Stette* (University of Trondheim and ETSI), *S. C. Taylor* (Matra Marconi Space)
Discussion on Panel B

COOPERATING ORGANIZATIONS

- AEI - Associazione Elettrotecnica ed Elettronica Italiana
- ETSI - European Telecommunications Standards Institute
- IEEE Communications Society, IEEE Region 8, IEEE Satellite and Space Communications Committee